Linear Systems and Operators in Hilbert Space

Paul A. Fuhrmann
Ben Gurion University of the Negev

Dover Publications, Inc., Mineola, New York

Bibliographical Note

This Dover edition, first published in 2014, is an unabridged republication of the work originally published by McGraw-Hill, Inc., New York, in 1981.

Library of Congress Cataloging-in-Publication Data

Fuhrmann, Paul Abraham.
    Linear systems and operators in Hilbert space / Paul A. Fuhrmann, Ben Gurion University of the Negev.
        pages cm — (Dover books on mathematics)
    Summary: "A treatment of system theory within the context of finite dimensional spaces, this text is appropriate for students with no previous experience of operator theory. The three-part approach, with notes and references for each section, covers linear algebra and finite dimensional systems, operators in Hilbert space, and linear systems in Hilbert space. 1981 edition"— Provided by publisher.
    "This Dover edition, first published in 2014, is an unabridged republication of the work originally published by McGraw-Hill, Inc., New York, in 1981."
    Includes bibliographical references and index.
    ISBN-13: 978-0-486-49305-3 (pbk.)
    ISBN-10: 0-486-49305-9
    1. Hilbert space. 2. Linear systems. 3. Linear operators. I. Title.
QA322.4.F83 2014
515'.733—dc23

2013032391

Manufactured in the United States by Courier Corporation
49305901   2014
www.doverpublications.com

TO
THE MEMORY OF MY PARENTS

CONTENTS

PREFACE

Great progress has been made in the last few years in the direction of establishing a system theory in the context of infinite dimensional spaces. Although this direction of research has by no means been exhausted it seems that the available theory has reached a level of maturity where a more systematic description would be in order. This would be of help to other workers in the field.

My aim in this book is to reach different sets of readers—the mathematically oriented researcher in system theory on the one hand and the pure mathematician working in operator theory on the other. I think that the power, beauty, and elegance of that part of operator theory touched upon in this book are such that the interested system scientist who is ready to invest some, maybe even considerable, time and effort in its study will be rewarded with a significantly increased set of methods for tackling multivariable systems and a deeper understanding of the finite dimensional theory. The operator theorist might find that system theory provides a rich ground of interesting problems to the mathematician which might be otherwise overlooked. Mathematics has always benefited from the transplanting of ideas and motivations from other fields. It seems to me that system theory besides being intellectually exciting is today one of the richest sources of ideas for the mathematician as well as a major area of application of mathematical knowledge.

I have tried to present the fairly diverse material of the book in a unified way as far as possible, stressing the various analogies. In this sense the concept of module is fundamental and the key results deal with module homomorphisms, coprimeness, and spectral structure.

The book is divided into three uneven chapters. The first one is devoted to algebraic system theory and serves also as a general introduction to the subject. The various possible descriptions of linear time invariant systems are described. Thus transfer functions, polynomial system matrices, state space equations, and modules are all touched upon.

In the second chapter the necessary operator and function theoretic background is established. The material includes a short survey of Hilbert space theory through the spectral theorem. We use here the classical approach based on integral representations of certain classes of analytic functions. This approach is taken to stress the close connection between representation theory and realization theory. We continue with a sketch of multiplicity theory for normal operators. Next we study contractions, their unitary dilations, and contractive semigroups. The Cayley transform is extensively used to facilitate the translation of results from the discrete to the continuous case. A special section is devoted to an outline of the theory of the Hardy spaces in the disc and in a half plane. Shift and translation invariant subspaces are characterized. Next we describe the main results concerning shift operators as models including the functional calculus, spectral analysis, and the theory of Jordan models.

In the last chapter we study the mathematical theory of linear systems with a state space that is a Hilbert space. Emphasis is on modeling with shift operators and translation semigroups. The operator theoretic results developed in the second chapter are brought to bear on questions of reachability, observability, spectral minimality, and realization theory all in discrete and continuous time. Isomorphism results are derived and the limitations of the state space isomorphism theorem are delineated. A special section is devoted to symmetric systems.

Many of the ideas and results as well as the general structure of the book have been conceived during my two years' stay with Roger Brockett at Harvard. Without his help, influence, and encouragement this book would not have been written. It is a pleasure to acknowledge here my deep gratitude to him. I would also like to recall the many stimulating exchanges over the past few years with my colleagues J. S. Baras, P. Dewilde, A. Feintuch, J. W. Helton, R. Hermann, S. K. Mitter, and J. C. Willems. For her excellent typing of the entire manuscript I want to thank Mrs Y. Ahuvia. I gratefully acknowledge the support of the Israel Academy of Sciences—the Israel Commission for Basic Research for its support throughout the writing of this book.

Most of all I want to thank my wife Nilly for her love, moral support and encouragement.

LINEAR ALGEBRA AND
FINITE DIMENSIONAL LINEAR SYSTEMS

INTRODUCTION

Finite dimensional linear systems discussed in this chapter are given by the dynamical equations

$$x_{n+1} = Ax_n + Bu$$

$$y_n = Cx_n$$

where A, B, and C are appropriate linear maps in linear spaces over an arbitrary field F. Thus the study of linear systems amounts to the study of triples of linear maps (A, B, C).

Historically the study of linear transformations was based upon the study of matrix representations and reduction to canonical forms through a proper choice of basis. The more modern approach studies a linear transformation in a vector space X through the naturally induced polynomial module structure on X. This reduces the problem to that of a description of finitely generated torsion modules over $F[\lambda]$ which is done through a cyclic decomposition. The use of polynomials greatly simplifies the operations involved and compactifies the notation. A case in point is the reduction of the problem of similarity of matrices to that of equivalence of corresponding polynomial matrices for which there exists a simple arithmetic algorithm.

In this chapter this point of view is adopted, but we go one step further by replacing the usual matrix representations by polynomial and rational models. These functional models present a natural setting for the study of linear systems and provide a common ground both for Kalman's emphasis on studying linear systems as polynomial modules as well as for Rosenbrock's extensive use of polynomial system matrices. Moreover this approach provides the natural link to the study of infinite dimensional systems.

The chapter begins with the introduction of the necessary algebraic concepts, focuses on polynomial modules, both free and torsion modules, represents the latter ones as quotient modules and describes all the module homomorphisms. This in turn is used to study the structure of linear transformations as well as linear systems. Coprime factorizations of transfer functions are used in realization theory and related isomorphism results. Associating a polynomial model with a factorization of a transfer function leads to a natural introduction of polynomial system matrices. We conclude with the study of feedback by the use of polynomial models.

1. RINGS AND MODULES

We review in this section the algebraic concepts needed for the understanding of the structure of linear transformations in finite dimensional vector spaces.

A *ring R* is a set with two associative laws of composition called addition and multiplication such that:

(*a*) With respect to addition R is a commutative (abelian) group.
(*b*) The two distributive laws hold.
(*c*) R has a multiplicative unit denoted by 1, that is, $1x = x$ for all $x \in R$.

A ring R is called *commutative* if $xy = yx$ for all $x, y \in R$. If x and y are nonzero elements in R for which $xy = 0$ we say that x and y are zero divisors. A commutative ring with no zero divisors is called an *entire ring*.

Given two rings R and R_1 a *ring homomorphism* is a map $\varphi: R \to R_1$ that satisfies

$$\varphi(x + y) = \varphi(x) + \varphi(y), \qquad \varphi(xy) = \varphi(x)\,\varphi(y)$$

$$\varphi(1) = 1 \quad \text{and} \quad \varphi(0) = 0$$

An invertible element in a ring R is called a *unit*. A *field* is a commutative ring in which every nonzero element is a unit.

In a ring R we have a natural division relation. We say that b is a *left divisor* of a, and denote it by $b|_l a$, if there exists an element c in R such that $a = bc$. In this case we say that a is a *left multiple* of c and a *right multiple* of b. A *greatest common left divisor* (g.c.l.d.) of elements a_α in a ring R is a common left divisor c of the a_α such that every other common left divisor c' of the a_α is a left divisor of c. The units of R are left divisors which are called trivial. Ring elements a and b are called *left coprime*, denoted by $(a, b)_l = 1$, if they do not have a nontrivial common left divisor or equivalently if every g.c.l.d. of a and b is a unit. A *least common left multiple* (l.c.l.m.) of elements a_α is a common left multiple which is a right divisor of any other common left multiple. We say that two elements a and b in R are *left associates*, a left equivalent, if each one is a left divisor of the other. Thus, in an entire ring, two left associates differ by a unit factor on the right. The relation of left associateness is an equivalence relation in R. In an analogous way we define right divisors and all related notions. If R is a commutative ring we drop the adjectives left and right and write simply divisor, g.c.d., etc.

A subset J of R is called a *left ideal* if it is an additive subgroup of R and $RJ \subset J$, and hence $RJ = J$ as R contains an identity. A *right ideal* is defined analogously and thus satisfies $JR = J$. A (two sided) *ideal* is a subset of R which is simultaneously a left and a right ideal.

A left ideal J in R is *principal* if $J = Ra$ for some a in R. A commutative ring in which each ideal is principal is called a *principal ideal ring*. An entire ring which is also a principal ring is called a *principal ideal domain*.

There is a close connection between ring homomorphisms and two sided ideals. In fact if $\varphi : R \to R_1$ is a ring homomorphism then $\operatorname{Ker} \varphi = \{x \in R \mid \varphi(x) = 0\}$ is an ideal in R. Conversely given an ideal in R we can construct the factor ring R/J consisting of all cosets $a + J$ with addition and multiplication defined by

$$(a + J) + (b + J) = (a + b) + J$$

and

$$(a + J)(b + J) = ab + J$$

These definitions make R/J a ring and the map $\varphi : R \to R/J$ given by $\varphi(a) = a + J$ is a ring homomorphism whose kernel equals J.

For the development of structure theory for linear transformations and linear systems it will be important to study rings of polynomials and modules over such rings.

Given a ring R we define $R[\lambda]$ the ring of polynomials in the indeterminate λ with coefficients in R as the set of all expressions of the form $p(\lambda) = \sum_{i=0}^{n} a_i \lambda^i$, $n \geq 0$. The operations of addition and multiplication are defined as usual, that is, if $p(\lambda) = \sum a_i \lambda^i$ and $q(\lambda) = \sum b_i \lambda^i$ then

$$(p + q)(\lambda) = \sum (a_i + b_i) \lambda^i \tag{1-1}$$

and

$$(pq)(\lambda) = \sum c_n \lambda^n \tag{1-2}$$

where

$$c_n = \sum_{i+j=n} a_i b_j \tag{1-3}$$

Given a ring R let $p \in R[\lambda]$. If $p(\lambda) = \sum_{i=0}^{n} a_i \lambda^i$ and $a_n \neq 0$ we say a_n is the leading coefficient of p and call n the degree of p, denoted by $\deg p$. If R is entire so is $R[\lambda]$ and $\deg(pq) = \deg p + \deg q$.

The most important property of polynomial rings is the existence of a division process in $R[\lambda]$. Let $q, p \in R[\lambda]$ and p such that its leading coefficient is a unit then there exist unique h and r in $R[\lambda]$ such that $q = ph + r$ and $\deg r < \deg p$. A similar result holds for right division in $R[\lambda]$. We have also two evaluation maps from $R[\lambda]$ into R given by $p \to p_L(c)$ and $p \to p_R(c)$, respectively, given by $p_L(c) = \sum c^i a_i$ and $p_R(c) = \sum a_i c^i$ where $p(\lambda) = \sum a_i \lambda^i$ and $c \in R$.

As a result of the division process in $R[\lambda]$ it is easily established 9] that $\lambda - c|_L q$ if and only if $q_L(c) = 0$ and similarly for right division. If F is a field then every nonzero constant is a unit and it is an easy consequence of the division rule that the polynomial ring $F[\lambda]$ is a principal ideal domain.

We proceed with the introduction of modules. Let R be a ring. A *left module* M over R (or *left R-module*) in a commutative group together with an operation of R on M which satisfies

$$r(x + y) = rx + ry, \qquad (r + s)x = rx + sx$$

$$r(sx) = (rs)x \qquad \text{and} \qquad 1x = x$$

for all $r, s \in R$ and $x, y \in M$. Right modules are defined similarly.

Let M be a left R-module. A subset N of M is a submodule of M if it is an additive subgroup of M which satisfies $RN \subset N$. Given a submodule N of a left R-module we can define a module structure on the factor group M/N by letting

$$r(a + N) = ra + N$$

This makes M/N into a left R-module called the quotient module of M by N. Let M, M_1 be two left R-modules. A map $\varphi: M \to M_1$ is an R-module homomorphism if for all $x, y \in M$ and $r \in R$ we have

$$\varphi(x + y) = \varphi(x) + \varphi(y) \qquad \text{and} \qquad \varphi(rx) = r\varphi(x)$$

Given two R-modules M and M_1 we denote by $(M, M_1)_R$ the set of all R-module homomorphisms from M to M_1. As in the case of rings we have the canonical R-module homomorphism $\varphi: M \to M/N$ given by $x \to x + N$, with $\text{Ker}\,\varphi = N$. Also given an R-module homomorphism $\varphi: M \to M_1$ $\text{Ker}\,\varphi$ and $\text{Im}\,\varphi$ are submodules of M and M_1, respectively.

Given R-modules $M_0, \ldots, M_n$ a sequence of R-module homomorphisms

$$M_0 \xrightarrow{\varphi_1} M_1 \xrightarrow{\varphi_2} \cdots \xrightarrow{\varphi_n} M_n$$

is called an *exact sequence* if $\text{Im}\,\varphi_i = \text{Ker}\,\varphi_{i+1}$. An exact sequence of the form

$$0 \to M_1 \to M_2 \to M_3 \to 0$$

is called a *short exact sequence*.

If N is a submodule of a left R-module M then the sequence

$$0 \to N \xrightarrow{j} M \xrightarrow{\pi} M/N \to 0$$

is a short exact sequence. Here j is the injection of N into M and π is the canonical projection of M onto M/N.

One way to get submodules of a given left R-module M is to consider for a given set A of elements in M the set $N = \{\sum r_i a_i \,|\, a_i \in A, r_i \in R\}$. This is clearly a left submodule called the submodule generated by A. The set A is a set of *generators* for N. If M has a finite set of generators then we say that M is *finitely generated*. A subset $\{b_1, \ldots, b_k\}$ of an R-module M is called R-linearly independent, or just *linearly independent*, if $\sum r_i b_i = 0$ implies $r_i = 0$ for all i. A subset B of an R-module M is called a *basis* if it is linearly independent and generates M. If M has a set of generators consisting of one element we say M is *cyclic*. By a *free* module we mean a module which has a basis or the zero module.

Theorem 1-1 Let R be a principal ideal domain and M a free left R-module with n basis elements. Then every R-submodule N of M is free and has at most n basis elements.

PROOF Let $\{e_1, \ldots, e_n\}$ be a basis for M. Then every element $x \in M$ has a unique representation in the form $x = \sum r_i e_i$. We prove the theorem by induction. For the zero module the theorem is trivial. Let us assume the theorem has been proved for modules with $(n - 1)$ basis elements. Let N be a submodule of M. If N contains only terms of the form $\sum_{i=1}^{n-1} r_i e_i$ then the theorem holds by the induction hypothesis. Thus we may assume N contains an element $\sum_{i=1}^{n} r_i e_i$ with $r_n \neq 0$. Let $I = \{r_n | \sum_{i=1}^{n} r_i e_i \in N\}$. Clearly I is an ideal in R, and R being a principal ideal domain $I = (\rho_n)$ the ideal generated by ρ_n. Thus N contains an element of the form $f = r_1 e_1 + \cdots + r_{n-1} e_{n-1} + \rho_n e_n$. Hence for every element a in N there exists an $r \in R$ for which $a - rf$ belong to a submodule of the free module generated by $\{e_1, \ldots, e_{n-1}\}$. Hence by the induction hypothesis there exists a basis $\{f_1, \ldots, f_{m-1}\}$ of that submodule with $m - 1 \leq n - 1$. Clearly $\{f_1, \ldots, f_{m-1}, f\}$ generate N. To show that $\{f_1, \ldots, f_{m-1}, f\}$ are R-linearly independent assume

$$r_1 f_1 + \cdots + r_{m-1} f_{m-1} + rf = 0$$

If $r = 0$ then, $\{f_1, \ldots, f_{m-1}\}$ being R-linearly independent, it follows $r_1 = r_2 = \cdots = r_{m-1} = 0$. But $r \neq 0$ implies $r\rho_n = 0$ which is impossible.

Assume now R is a principal ideal domain. An element a of a left R-module M is called a *torsion* element if there exists a nonzero $r \in R$ for which $ra = 0$. The set of all torsion elements of M is a submodule called the *torsion submodule* of M. M is a *torsion module* if all its elements are torsion elements. Given any nonzero element a in M then $J_a = \{r \in R | ra = 0\}$ is a left ideal in R which is nontrivial if and only if a is a torsion element. Since R is principal $J_a = R\mu_a = (\mu_a)$ the ideal generated by μ_a. μ_a is called a minimal annihilator of a. Two minimal annihilators differ by a unit factor. In particular if M is a cyclic module over R then M is isomorphic to $R/(\mu_a)$. So if M is cyclic either M is isomorphic to R which means M is free or it is isomorphic to a proper quotient ring of R and in that case M is a torsion module.

Since a finite direct sum of torsion R-modules is also a torsion module it follows that given elements μ_i in R then $R/(\mu_1) \oplus \cdots \oplus R/(\mu_k)$ is a finitely generated torsion module over R. The converse is also true and is summarized in the fundamental structure theorem for finitely generated torsion modules over principal ideal domains.

Theorem 1-2 Let M be a finitely generated torsion module over a principal ideal domain R. Then M is isomorphic to $R/(\mu_1) \oplus \cdots \oplus R/(\mu_k)$ where μ_i are nonzero elements in R and $\mu_{i+1}|\mu_i$. The sequence of ideals (μ_i) is uniquely determined.

We call the elements μ_i the *invariant factors* of the module M. The invariant factors are determined up to unit factors. We will give a proof for the special case that $R = F[\lambda]$ the ring of polynomials over a field F. However, the proof holds for the general case with only minor modifications.

2. POLYNOMIAL MODULES

In the previous section we introduced $R[\lambda]$, the ring of polynomials over a ring R. In this section we study related objects obtained when starting with a module M. Thus let M be an R-module. By $M((\lambda))$ we denote the module of all truncated Laurent series, that is, the set of all formal series of the form $\sum_{i \geq k} m_i \lambda^i$, $k \in Z$ and $m_i \in M$, $M[[\lambda]]$ the submodule of all formal power series, that is, sums of the form $\sum_{i=0}^{\infty} m_i \lambda^i$. Finally $M[\lambda]$ denotes the polynomial module with coefficients in M, that is the submodule of $M[[\lambda]]$ of all formal power series with only a finite number of nonzero coefficients. For a ring R the module $R((\lambda))$ is actually a ring with addition and multiplication defined by

$$\sum p_i \lambda^i + \sum q_i \lambda^i = \sum (p_i + q_i) \lambda^i \tag{2-1}$$

and

$$\left(\sum p_i \lambda^i\right)\left(\sum q_j \lambda^j\right) = \sum r_n \lambda^n \tag{2-2}$$

with

$$r_n = \sum_{i+j=n} p_i q_j \tag{2-3}$$

Multiplication is well defined as in each of the rings appearing in (2-3) there is only a finite number of nonzero terms. Given a left R-module M then $M((\lambda))$ becomes a left $R((\lambda))$-module if the action of $R((\lambda))$ on $M((\lambda))$ is defined by (2-2) with $\sum p_i \lambda^i \in R((\lambda))$ and $\sum q_j \lambda^j \in M((\lambda))$. It will be convenient to consider also the module $M((\lambda^{-1}))$ which then contains $M[\lambda]$ as an $R[\lambda]$-submodule. Let j be the injection of $M[\lambda]$ into $M((\lambda^{-1}))$ and let π_- be the canonical projection of $M((\lambda^{-1}))$ on the quotient module $M((\lambda^{-1}))/M[\lambda]$. Then we have the following short exact sequence of module homomorphisms

$$0 \to M[\lambda] \xrightarrow{j} M((\lambda^{-1})) \xrightarrow{\pi_-} M((\lambda^{-1}))/M[\lambda] \to 0 \tag{2-4}$$

Moreover we can identify $M((\lambda^{-1}))/M[\lambda]$ with $\lambda^{-1}M[[\lambda^{-1}]]$ the set of all formal power series in λ^{-1} having zero constant term. An element f of $M((\lambda^{-1}))$ is called *rational* if there exists a nonzero $p \in R[\lambda]$ such that $pf \in M[\lambda]$ and *proper rational* if f is rational and $f \in M[[\lambda^{-1}]]$. $M((\lambda^{-1}))$ is clearly an $F[\lambda]$-module and thus the quotient module $\lambda^{-1} M[[\lambda^{-1}]]$ has also an induced $M[\lambda]$-module structure. The action of λ in $\lambda^{-1} M[[\lambda^{-1}]]$ is called the left shift. We denote by π_+ the projection of $M((\lambda^{-1}))$ onto $M[\lambda]$, i.e., $\pi_+ = I - \pi_-$.

For a ring R we will denote by $R^{n \times m}$ the R-module of all $n \times m$ matrices with elements in R. Thus $R[\lambda]^{n \times m}$ is the set of all $n \times m$ matrices with elements in $R[\lambda]$. We call these polynomial matrices. There is a standard isomorphism of $R[\lambda]^{n \times m}$ and $R^{n \times m}[\lambda]$ the set of all polynomials with $n \times m$ matrix coefficients. It will be convenient to have both interpretations at hand.

Let R be a commutative ring. A matrix $U \in R^{n \times n}$ is called *unimodular* if $\det U$ is a unit in R. By Cramer's rule a matrix U in $R^{n \times n}$ is invertible if and only if U is unimodular.

There is a natural equivalence relation in $R^{n \times m}$. We say that A and B in $R^{n \times m}$ are *equivalent*, or *unimodularly equivalent*, if there exist unimodular matrices U and V in $R^{n \times n}$ and $R^{m \times m}$, respectively, such that $B = UAV$. It is trivial to check that this is indeed an equivalence relation.

From now on we assume F is a field and V an n-dimensional vector space over F. By choice of basis in V it is clear that V is isomorphic to F^n and as a consequence $V[\lambda]$ and $(V, V)_F[\lambda]$ are isomorphic to $F^n[\lambda]$ and $F^{n \times n}[\lambda]$ (or $F[\lambda]^{n \times n}$), respectively.

Theorem 2-1 A subset M of $V[\lambda]$ is a submodule of $V[\lambda]$ if and only if $M = DV[\lambda]$ for some $D \in (V, V)_F[\lambda]$.

PROOF That a set of the form $DV[\lambda]$ is an $F[\lambda]$-submodule is clear. Conversely, assume M is a submodule of $V[\lambda]$. By Theorem 1-1 M is free with $m \leq n$ generators. Let $e_1, \ldots, e_n$ be a basis for V as a vector space over F, then $e_1, \ldots, e_n$ is also a set of free generators for $V[\lambda]$ as an $F[\lambda]$-module. Let $d_1, \ldots, d_m$ be a set of generators for M and $D \in (V, V)_F[\lambda]$ be defined by $De_i = d_i$ for $i = 1, \ldots, m$ and $De_i = 0$ for $m < i \leq n$. Obviously $M = DV[\lambda]$.

The partial order, by inclusion, of submodules of $V[\lambda]$ can be related now to division relation of the associated matrix polynomials.

Theorem 2-2 Let $M = DV[\lambda]$ and $N = EV[\lambda]$ be submodules of $V[\lambda]$. Then $M \subset N$ if and only if $E|_L D$.

PROOF If $E|_L D$ then $D = EF$ for some F in $(V, V)_F[\lambda]$, and hence $M \subset N$. Conversely assume $DV[\lambda] \subset EV[\lambda]$. Let $e_1, \ldots, e_n$ be a basis for V and let $d_i = De_i$. By the submodule inclusion there exist f_i in $V[\lambda]$ such that $d_i = Ef_i$, $i = 1, \ldots, n$. Define $F \in (V, V)_F[\lambda]$ by $Fe_i = f_i$ then the factorization $D = EF$ follows.

Corollary 2-3 Let $M = DV[\lambda]$ be a submodule of $V[\lambda]$. If D is nonsingular and $M = EV[\lambda]$ is any other representation of M then $E = DU$ for some unimodular $U \in (V, V)_F[\lambda]$.

PROOF We have $E = DU$ and similarly $D = EW$ hence $D = DUW$. Since D is nonsingular $I = UW$ which implies that U and W are unimodular and E is nonsingular.

The above corollary leads to the following definition.

Definition 2-4 A submodule M of $V[\lambda]$ is called a *full submodule* if it has a representation of the form $M = DV[\lambda]$ for some nonsingular $D \in (V, V)_F[\lambda]$.

Clearly M is a full submodule of $V[\lambda]$ if and only if it has a basis consisting of n elements. As D is nonsingular if and only if $\det D \neq 0$ we have the following trivial corollary.

Corollary 2-5 A submodule $M = DV[\lambda]$ of $V[\lambda]$ is full submodule if and only if $\det D \neq 0$.

It should be noted that $\det D \neq 0$ means that $\det D \in F[\lambda]$ is not the zero polynomial. It might be identically equal to zero as a function on F. Another easy and useful corollary of Theorem 2-1 is the following.

Corollary 2-6 Let D be a nonsingular element of $(V, V)_F[\lambda]$. Then $(\det D) V[\lambda] \subset DV[\lambda]$.

PROOF By a choice of basis in V we may assume we have a matrix representation of D. Then it follows from Cramer's rule that $(\det D) I = D \cdot \text{adj} D$ where $\text{adj} D$ is the classical adjoint of D, that is, the cofactor matrix of D.

Given two submodules M_1 and M_2 of $V[\lambda]$ then $M_1 \cap M_2$ and $M_1 + M_2$ are also submodules of $V[\lambda]$ and $M_1 \cap M_2 \subset M_i \subset M_1 + M_2$.

Theorem 2-7 Let $M_i = E_i V[\lambda]$ and let $M_1 + M_2 = DV[\lambda]$ and $M_1 \cap M_2 = CV[\lambda]$. Then C and D are a l.c.r.m. and a g.c.l.d., respectively, of M_1 and M_2.

PROOF The inclusion $M_1 \subset M_1 + M_2$ implies $E_1 V[\lambda] \subset DV[\lambda]$ and hence, by the previous theorem, $E_1 = DG_1$ or $D|_L E_1$ and similarly $D|_L E_2$. Thus D is a common left divisor of E_1 and E_2. Let D_1 be any common left divisor of E_1 and E_2 which means $E_i = D_1 E_i'$ or equivalently $M_i \subset D_1 V[\lambda]$. Hence $M_i \subset D_1 V[\lambda]$ and therefore also $DV[\lambda] = M_1 + M_2 \subset D_1 V[\lambda]$. Thus we get $D_1|_L D$ and hence D is a g.c.l.d. of E_1 and E_2.
  Next consider $M_1 \cap M_2$. $M_1 \cap M_2 \subset M_i$ implies $CV[\lambda] \subset E_i V[\lambda]$ or $C = E_i G_i$ for some $G_i \in (V, V)_F[\lambda]$. So C is a common right multiple of E_1 and E_2. Let C_1 be any other common right multiple of E_1 and E_2 then $C_1 = E_i G_i'$. So $C_1 V[\lambda] \subset E_i V[\lambda]$ and hence $C_1 V[\lambda] \subset E_1 V[\lambda] \cap E_2 V[\lambda] = CV[\lambda]$. This is equivalent to $C|_L C_1$ or C is a l.c.r.m. of E_1 and E_2.

As a corollary we obtain the following important result.

Theorem 2-8 Every two matrix polynomials E_1 and E_2 in $(V, V)_F [\lambda]$ have a g.c.l.d. D which can be expressed as

$$D = E_1 F_1 + E_2 F_2 \qquad (2\text{-}5)$$

for some F_1 and F_2 in $(V, V)_F [\lambda]$.

PROOF The existence of a g.c.l.d. has been proved in the previous theorem. Let D be a g.c.l.d. of E_1 and E_2. Then $DV [\lambda] = E_1 V [\lambda] + E_2 V [\lambda]$. Let $d_i = De_i$ for a basis $e_1, \ldots, e_n$ of V. Then $d_i = E_1 f_i^{(1)} + E_2 f_i^{(2)}$. Define $F_i \in (V, V)_F [\lambda]$ by $F_i e_j = f_j^{(i)}$ then (2-5) holds.

With only minor modifications we can prove the same result assuming $E_i \in (W_i, V)_F [\lambda]$. Also a completely analogous result holds for a g.c.r.d. A trivial generalization holds for a g.c.l.d. of a finite number of matrix polynomials.

Corollary 2-9 Let $A_i \in (W_i, V)_F [\lambda]$, $i = 1, \ldots, p$. Then $A_1, \ldots, A_p$ are left coprime if and only if there exist $B_i \in (V, W_i)_F [\lambda]$ such that $I = \sum_{i=1}^{p} A_i B_i$.

Corollary 2-10 If two matrix polynomials have a nonsingular g.c.l.d. D then any other g.c.l.d. D_1 is given by $D_1 = DU$ for some unimodular U.

The availability of g.c.l.d.'s allows us to determine the ideal structure in $(V, V)_F [\lambda]$.

Theorem 2-11 A subset $J \subset (V, V)_F [\lambda]$ is a right ideal if and only if $J = D (V, V)_F [\lambda]$ for some $D \in (V, V)_F [\lambda]$.

PROOF The if part is trivial. Suppose now that J is a right ideal in $(V, V)_F [\lambda]$ then J is finitely generated. Let $A_1, \ldots, A_R$ be a set of generators and let D be a g.c.l.d. of $A_1, \ldots, A_R$. Then clearly $J = D (V, V)_F [\lambda]$.

As in the case of submodules we have $D(V, V)_F [\lambda] \subset E(V, V)_F [\lambda]$ if and only if $E|_L D$. If $J = D(V, V)_F [\lambda]$ is a right ideal for which $\det D \neq 0$ then D is determined up to a right unimodular factor. A right ideal $J = D(V, V)_F [\lambda]$ is called a *full right ideal* if D is nonsingular. Obviously this definition is independent of the representation of J. Of course analogous results hold also for left ideals.

Next we pass to the study of quotient modules of $V [\lambda]$ singling out those which are torsion modules.

Theorem 2-12 Let M be a submodule of $V [\lambda]$ then the quotient module $V [\lambda]/M$ is a torsion module if and only if M is a full submodule of $V [\lambda]$.

PROOF Assume M is a full submodule of $V [\lambda]$, then $M = DV [\lambda]$ for a nonsingular D. As, by Corollary 2-6, $(\det D) V [\lambda] \subset DV [\lambda]$ it follows that $\det D$ annihilates the quotient module $V [\lambda]/DV [\lambda]$. Conversely, assume that $V [\lambda]/M$ is a torsion module. Since it is finitely generated there exists a polynomial p annihilating all of $V[\lambda]/M$. This implies $pV[\lambda] \subset DV[\lambda]$ where $DV[\lambda]$ is any representation of M. Thus $pI = DE$ and hence $\det D$ is nontrivial and M is a full submodule.

The next lemma is in the same spirit and its proof is omitted.

Lemma 2-13 Let M be a submodule of $V[\lambda]$. Then $V[\lambda]/M$ considered as a vector space over F is finite dimensional if and only if M is a full submodule of $V[\lambda]$.

In case $n = 1$ $V[\lambda]$ is isomorphic to $F[\lambda]$ and hence has actually a ring structure. Any submodule of $F[\lambda]$ is an ideal which is necessarily principal. The generator m, a nonzero polynomial of least degree p, is uniquely determined if we assume the highest order coefficient to be 1. By the division process in $F[\lambda]$ we may identify $F[\lambda]/M$ with $F_{p-1}[\lambda]$ the set of all polynomials of degree $\leq p - 1$.

Since it is easier to work with representatives rather than with equivalence classes we would like to imitate the scalar construction in some way. The difficulty arises mostly out of the nonuniqueness of such a representation. One way to overcome this difficulty is through the use of canonical matrices as was done by Eckberg in [32]. A related approach is to study finite dimensional vector spaces over the field of rational functions and special choices of bases as was done by Forney [38]. We proceed differently and study the whole set of such representations.

Thus let π_- be a canonical projection of $V((\lambda^{-1}))$ onto $V((\lambda^{-1}))/V[\lambda]$. We identify $V((\lambda^{-1}))/V[\lambda]$ with $\lambda^{-1}V[[\lambda^{-1}]]$. Let now $M = DV[\lambda]$ be a full submodule of $V[\lambda]$. D is therefore nonsingular and has an inverse in $(V, V)_F(\lambda)$. In matrix language D^{-1} would be a matrix over the field $F(\lambda)$ of rational functions. Define now a map $\pi_D: V[\lambda] \to V[\lambda]$ by

$$\pi_D f = D\pi_- D^{-1} f \quad \text{for} \quad f \in V[\lambda] \tag{2-6}$$

Lemma 2-14 Let D be a nonsingular element in $(V, V)_F[\lambda]$. Then π_D defined by (2-6) is a projection map in $V[\lambda]$ and $\operatorname{Ker} \pi_D = DV[\lambda]$.

PROOF Let $f \in V[\lambda]$ then $D^{-1}f \in V((\lambda^{-1}))$. Let $D^{-1}f = g + h$ with $g \in \lambda^{-1}V[[\lambda^{-1}]]$ and $h \in V[\lambda]$. This decomposition is unique. From this we get $\pi_- D^{-1}f = g$ and $\pi_D f = D\pi_- D^{-1}f = Dg = F - Dh$. As $f - Dh \in V[\lambda]$ the range of π_D is in $V[\lambda]$. That π_D is a projection follows from the equality

$$\pi_D^2 f = (D\pi_- D^{-1})(D\pi_- D^{-1}f) = D\pi_-^2 D^{-1} f = D\pi_- D^{-1} f = \pi_D f$$

that holds for all $f \in V[\lambda]$. Next we show that $\operatorname{Ker} \pi_D = DV[\lambda]$. If $f \in DV[\lambda]$ then $f = Dg$ for some $g \in V[\lambda]$. Hence $\pi_D f = D\pi_- D^{-1}Dg = D\pi_- g = 0$ and $DV[\lambda] \subset \operatorname{Ker} \pi_D$. Conversely if $\pi_D f = 0$ then, by the nonsingularity of D, $\pi_- D^{-1}f = 0$, or equivalently $D^{-1}f = g \in V[\lambda]$. So $f = Dg$ and $\operatorname{Ker} \pi_D \subset DV[\lambda]$ which completes the proof.

Define now K_D as the range of the projection π_D, that is

$$K_D = \{\pi_D f \mid f \in V[\lambda]\} \tag{2-7}$$

Clearly K_D is a vector space over F. From the preceding proof it is clear that the following lemma holds.

Lemma 2-15 An element f in $V[\lambda]$ belongs to K_D if and only if $D^{-1}f \in \lambda^{-1}V[[\lambda^{-1}]]$, that is, is strictly proper rational.

The vector space K_D can be given an $F[\lambda]$-module structure by defining the action of polynomials in K_D through

$$p \cdot f = \pi_D(pf) \tag{2-8}$$

With this definition π_D becomes a surjective $F[\lambda]$-module homomorphism of $V[\lambda]$ onto K_D with kernel $DV[\lambda]$. Thus we have the following important result.

Theorem 2-16 Let $M = DV[\lambda]$ be a full submodule of $V[\lambda]$. Then K_D defined by (2-7) with the module structure defined by (2-8) is an $F[\lambda]$-module isomorphic to $V[\lambda]/M$.

We conclude this section with a digression on compound matrices.

Let A be an $n \times n$ matrix over a commutative ring R, that is, an element of $R^{n \times n}$. We denote by $A^{(p)}$ the matrix of all $p \times p$ minors of A ordered lexicographically. The matrices $A^{(p)}$, $p = 1, \ldots, n$, are called the *compound matrices* of A. As there are $\binom{n}{p}$ ways of choosing p rows or columns out of the n available, $A^{(p)}$ is an $\binom{n}{p} \times \binom{n}{p}$ matrix over R. The following theorem summarizes the well-known properties of $A^{(p)}$. We refer to [8] for additional material.

Theorem 2-17 Let A and B be $n \times n$ matrices over the commutative ring R, then

(a) $I_n^{(p)} = I_{\binom{n}{p}}$
(b) $(AB)^{(p)} = A^{(p)}B^{(p)}$
(c) $(\tilde{A})^{(p)} = \tilde{A}^{(p)}$
   and
(d) $\det A^{(p)} = (\det A)^{\binom{n-1}{p-1}}$

PROOF (a) and (c) are trivial, (b) is a consequence of the Binet–Cauchy formula for determinants and (d) follows easily from triangulation of A by elementary row operations.

As extra easy consequences we get that if A is invertible then $(A^{-1})^{(p)} = (A^{(p)})^{-1}$. Hence if U is unimodular so is $U^{(p)}$.

Corollary 2-18 Let R be a commutative ring. Then if A and B in $R^{n \times n}$ are equivalent then so are $R^{(p)}$ and $B^{(p)}$.

PROOF If A and B are equivalent then $A = UBV$ for unimodular matrices U and V. This implies $A^{(p)} = U^{(p)}B^{(p)}V^{(p)}$ and since $U^{(p)}$ and $V^{(p)}$ are unimodular $A^{(p)}$ and $B^{(p)}$ are equivalent.

We introduce now the *determinant divisors* of matrices over a ring and will subsequently relate them to the invariant factors of a matrix.

Definition 2-19 Let R be a commutative ring and $A \in R^{n \times n}$. We define the *determinant divisors* $D_i(A)$ of A by $D_0(A) = 1$ and $D_i(A)$ as the g.c.d. of all $i \times i$ minors of A.

Lemma 2-20 If A is an $n \times n$ matrix over a commutative ring R then

(a) $D_i(A) = D_1(A^{(i)})$
(b) If A and B are equivalent in $R^{n \times n}$ then $D_i(A)$ and $D_i(B)$ are equivalent in R for $i = 1, \ldots, n$.

PROOF (a) is obvious. In the light of (a) and Corollary 2-18 it suffices to prove that $D_1(A)$ and $D_1(B)$ are equivalent in R. Now if A and B are equivalent then each element a_{ij} of A is a linear function of all the b_{kl}. Thus $D_1(B)|a_{ij}$ for all i and j. Hence $D_1(B)|D_1(A)$. The equality follows by symmetry.

As we shall see later, at least in the case of matrices over a principal ideal domain the converse is also true, that is the equivalence of all determinant divisors implies the equivalence of the matrices.

3. THE SMITH CANONICAL FORM

In the previous section an equivalence relation has been introduced in $R^{n \times m}$ where R was a commutative ring. We specialize the ring to $F[\lambda]$ and study the invariants of the relation and the associated canonical forms. This has important implications for the structure theory developed in the next section.

In order to give a satisfactory answer to the question of normal forms for equivalence we introduce the following class of square matrices over $F[\lambda]$. We do not specify their dimensions as it will be n when multiplying on the left and m when multiplying on the right.

$$
T_{ij}(p) = i \begin{array}{c} \\ \end{array}
\begin{bmatrix}
1 & & & & & & \overset{j}{} & & & \\
 & \ddots & & & & & \cdot & & & \\
 & & \cdot & & & & \cdot & & & \\
\cdot & \cdot & \cdot & 1 & \cdot & \cdot & p & & & \\
 & & & & \cdot & & \cdot & & & \\
 & & & & & \cdot & \cdot & & & \\
 & & & & & & 1 & & & \\
 & & & & & & & \cdot & & \\
 & & & & & & & & \cdot & \\
 & & & & & & & & & 1
\end{bmatrix}
$$

$$
D_i(u) = i
\begin{bmatrix}
1 & & & & & \overset{i}{} & & & \\
 & \ddots & & & & \cdot & & & \\
 & & \cdot & & & \cdot & & & \\
 & & & 1 & & \cdot & & & \\
\cdot & \cdot & \cdot & \cdot & \cdot & u & & & \\
 & & & & & 1 & & & \\
 & & & & & & \cdot & & \\
 & & & & & & & \cdot & \\
 & & & & & & & & 1
\end{bmatrix}
$$

where $u \in F[\lambda]$ is a unit, that is, a nonzero constant polynomial, and

$$
P_{ij} =
\begin{array}{c} \\ \\ i \\ \\ \\ \\ j \\ \\ \end{array}
\begin{bmatrix}
1 & & & \overset{i}{} & & & \overset{j}{} & & \\
 & \cdot & & \cdot & & & \cdot & & \\
 & & 1 & \cdot & & & \cdot & & \\
\cdot & \cdot & \cdot & 0 & \cdot & \cdot & \cdot & 1 & \\
 & & & \cdot & 1 & & & \cdot & \\
 & & & \cdot & & \cdot & & \cdot & \\
 & & & \cdot & & & 1 & \cdot & \\
\cdot & \cdot & \cdot & 1 & \cdot & \cdot & \cdot & 0 & \\
 & & & & & & & & 1
\end{bmatrix}
$$

The matrices $T_{ij}(p)$, $D_i(u)$, and P_{ij} are unimodular and are called *elementary matrices*.

It is easily checked that left (right) multiplication by $T_{ij}(p)$ is equivalent to adding the jth row (column) multiplied by p to the ith row (column), left (right) multiplication by $D_i(u)$ is equivalent to multiplying the ith row (column) by u and finally left (right) multiplication by P_{ij} is equivalent to interchanging the ith and jth rows (columns). These operations on rows and columns are called *elementary operations*. From this it is clear that any matrix A' obtained from a given matrix A in $F[\lambda]^{n \times m}$ through a finite sequence of elementary operations is equivalent to A.

We denote by $\mathrm{diag}(\rho_1, \ldots, \rho_k)$ the $n \times m$ matrix whose elements are $\rho_i \delta_{ij}$ and $k = \min(n, m)$.

Theorem 3-1 Let D be a matrix in $F[\lambda]^{n \times m}$. Then there exists a matrix Δ in $F[\lambda]^{n \times m}$ which is equivalent to D and for which

(a) $\Delta = \mathrm{diag}(\delta_1, \ldots, \delta_r, 0, \ldots, 0)$
   where δ_i are nonzero elements in $F[\lambda]$.
(b) $\delta_{i+1} | \delta_i$ for $i = 1, \ldots, r - 1$
   are satisfied.

The elements $\delta_1, \ldots, \delta_r$ are determined uniquely by the above conditions up to unit factors. Equivalently the ideals (δ_i) are uniquely determined.

The matrix Δ is called the *Smith canonical form* of D. The elements $\delta_1, \ldots, \delta_r$ are called the *invariant factors* of D.

PROOF Consider the matrix D. If $D = 0$ there is nothing to prove. Otherwise let d_{ij} be an element of least degree in D. By exchanging rows and columns we bring it to the $(1, 1)$ position. Let $d_{1j} = d_{11}b_j + r_{1j}$ with $\deg r_{1j} < \deg d_{11}$. Subtract now the first column multiplied by b_j from the jth column. Now either $r_{1j} = 0$ and we repeat the process with the next element in the first row or $r_{1j} \neq 0$ and we move it, by elementary operations, to the $(1, 1)$ position and proceed as before. Next we repeat the same process with rows. Since the degree of the element in the $(1, 1)$ position is decreased with each exchange, it follows that after a finite number of operations we get a matrix of the form

$$
\begin{bmatrix}
e_{11} & 0 & \cdot & \cdot & \cdot & \cdot & \cdot \\
0 & e_{22} & e_{23} & \cdot & \cdot & \cdot \\
\cdot & & & & & \\
\cdot & e_{32} & \cdot & \cdot & \cdot & \\
\cdot & & \cdot & & & \\
\cdot & & \cdot & & & \\
\cdot & & \cdot & & &
\end{bmatrix}
$$

Proceeding inductively we obtain a matrix $\mathrm{diag}(g_1, \ldots, g_r, 0, \ldots, 0)$ equivalent to D, with $g_i \neq 0$. If g_1 does not divide g_i we add the ith row to the first and repeat the process obtaining a diagonal matrix $\mathrm{diag}(h_1, \ldots, h_r, 0, 0, \ldots, 0)$ with $\deg h_1 < \deg g_1$. A finite number of repetitions of this process yields a matrix $\text{diag}(k_1, \ldots, k_r, 0, \ldots, 0)$ equivalent to the original and for which $k_i\,k_{i+1}$, $i = 1, \ldots, r - 1$. By row and column exchange we can reorder the diagonal elements so that the divisibility conditions of the theorem are satisfied.

It remains to prove uniqueness. Suppose $\Delta = \text{diag}(\delta_1, \ldots, \delta_r, 0, \ldots, 0)$ and $\Delta' = \text{diag}(\delta'_1, \ldots, \delta'_r, 0, \ldots, 0)$ are two diagonal matrices equivalent to D and satisfying the conditions of the theorem. By transitivity Δ and Δ' are equivalent. From Lemma 2-20 it follows that the determinant divisors of Δ and Δ' are equivalent in $F[\lambda]$. Now the determinant divisors of Δ are easily computed to be

$$D_0(\Delta) = 1, \qquad D_1(\Delta) = \delta_r, \qquad D_2(\Delta) = \delta_r\delta_{r-1}, \quad \ldots, \qquad D_r(\Delta) = \delta_r \cdots \delta_1$$

and similarly for Δ'. This implies $r = r'$ and $(\delta_r) = (\delta'_r), \ldots, (\delta_1 \cdots \delta_r) = (\delta_1 \cdots \delta_r)$ and hence $(\delta_i) = (\delta'_i)$.

This proof yields an effective way of computing the invariant factors.

Corollary 3-2 Let D be an element of $F[\lambda]^{n \times m}$, $D_i(D)$ its determinant divisors and assume $D_i(D) \neq 0$ for $i = 1, \ldots, r$ and $D_i(D) = 0$ for $i > r$. The invariant factors $\delta_1, \ldots, \delta_r$ of D are given by

$$\delta_1 = D_r(D)/D_{r-1}(D), \ldots, \delta_r = D_1(D)$$

It should be noted that given two polynomials p and q in $F[\lambda]$ the Euclidean algorithm for calculating the g.c.d. of p and q is essentially the reduction of $(p\,q) \in F^{1 \times 2}[\lambda]$ to its Smith canonical form.

4. STRUCTURE OF LINEAR TRANSFORMATIONS

Probably the most elegant approach to the study of linear transformations on a finite dimensional vector space V over a field F is by the study of the $F[\lambda]$-module structure induced by it. The presentation in this section is along these general lines but it stresses the notion of canonical models and the relation between them. The models we are after are either quotient modules of $V[\lambda]$ or submodules of $\lambda^{-1}V[[\lambda^{-1}]]$. To study them we need to have some information about module homomorphism.

The following lemma provides a useful tool for the analysis of the module homomorphism that follows.

Lemma 4-1
(a) Let X, X_1, and Y_2 be modules over the ring R and let $f_1: X \to X_1$ and $f_2: X \to X_2$ be R-homomorphisms of which f_2 is assumed to be surjective. Then there exists a uniquely determined R-homomorphism

$\psi: X_2 \to X_1$ which makes the diagram

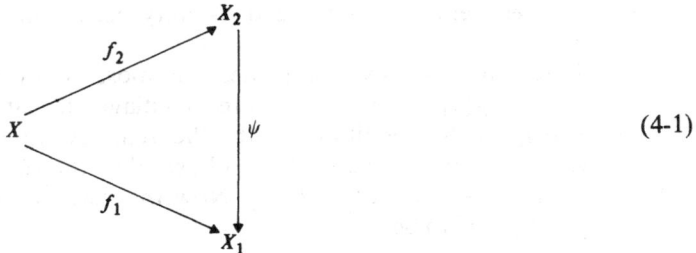

$$(4\text{-}1)$$

commutative if and only if

$$\operatorname{Ker} f_2 \subset \operatorname{Ker} f_1 \qquad (4\text{-}2)$$

Moreover ψ is injective if and only if

$$\operatorname{Ker} f_2 = \operatorname{Ker} f_1 \qquad (4\text{-}3)$$

(b) Let X, X_1, and X_2 be modules over the ring R and let $f_1: X_1 \to X$ and $f_2: X_2 \to X$ be R-homomorphisms of which f_2 is assumed injective. Then there exists a uniquely determined R-homomorphism $\varphi: X_1 \to X_2$ which makes the diagram

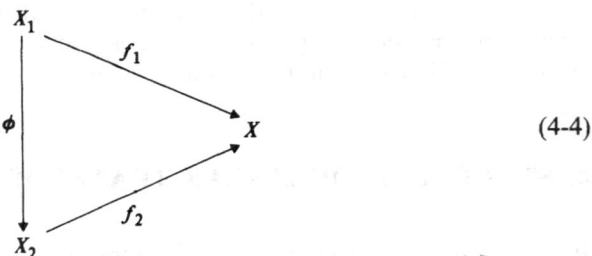

$$(4\text{-}4)$$

commutative if and only if

$$\operatorname{Range} f_1 \subset \operatorname{Range} f_2 \qquad (4\text{-}5)$$

Moreover φ is surjective if and only if

$$\operatorname{Range} f_1 = \operatorname{Range} f_2 \qquad (4\text{-}6)$$

Let us denote by χ the identity polynomial in $F[\lambda]$, that is, $\chi(\lambda) = \lambda$. Define the map $S: V[\lambda] \to V[\lambda]$ by $Sf = \chi f$ or $(Sf)(\lambda) = \lambda f(\lambda)$. S will be called the right shift in $V[\lambda]$. $V[\lambda]$ has a variety of structures associated with it. It is an F-vector space, an $F[\lambda]$-module and a left $(V, V)_F[\lambda]$-module. With each structure there is associated a class of homomorphisms.

Theorem 4-2

(a) A map $\varphi: V[\lambda] \to V[\lambda]$ is an $F[\lambda]$-homomorphism if and only if it is a linear map and the diagram

$$
\begin{array}{ccc}
V[\lambda] & \xrightarrow{\;\varphi\;} & V[\lambda] \\
\downarrow{\scriptstyle S} & & \downarrow{\scriptstyle S} \\
V[\lambda] & \xrightarrow{\;\varphi\;} & V[\lambda]
\end{array}
\tag{4-7}
$$

is commutative.

(b) A map $\varphi: V[\lambda] \to V[\lambda]$ is an $F[\lambda]$-homomorphism if and only if

$$(\varphi f)(\lambda) = \Phi(\lambda) f(\lambda) \tag{4-8}$$

for some $\Phi \in (V, V)_F[\lambda]$.

(c) A map $\varphi: V[\lambda] \to V[\lambda]$ is an $(V, V)_F[\lambda]$-homomorphism if and only if

$$(\varphi f)(\lambda) = p(\lambda) f(\lambda) \tag{4-9}$$

for some $p \in F[\lambda]$.

PROOF

(a) If $\varphi: V[\lambda] \to V[\lambda]$ is an $F[\lambda]$-homomorphism then (4-7) commutes. Conversely if (4-7) commutes then also $\varphi S^n = S^n \varphi$ and by linearity the result follows.

(b) The direct part is obvious. So let $\varphi: V[\lambda] \to V[\lambda]$ be an $F[\lambda]$-homomorphism. Let $e_1, \ldots, e_n$ be a basis of V. Define an element Φ of $(V, V)_F[\lambda]$ by $\Phi(\lambda) e_i = \varphi(e_i)(\lambda)$ and (4-8) follows.

(c) Again the direct part is trivial. Let $\varphi: V[\lambda] \to V[\lambda]$ be an $(V, V)_F[\lambda]$-homomorphism. By (b) $(\varphi f)(\lambda) = \Phi(\lambda) f(\lambda)$. Since φ is by assumption a $(V, V)_F[\lambda]$-homomorphism we have $\Phi(\lambda) D(\lambda) = D(\lambda) \Phi(\lambda)$ for all D in $(V, V)_F[\lambda]$. In particular $\Phi(\lambda) D = D\Phi(\lambda)$ for all $D \in (V, V)_F$ which implies $\Phi(\lambda)$ is a scalar element, that is $\Phi(\lambda) = p(\lambda) I$ for some $p \in F[\lambda]$.

The simple structure of the $F[\lambda]$-homomorphisms of $U[\lambda]$ lets us state the following simple lifting theorem.

Theorem 4-3 Let $\varphi: U[\lambda] \to U_1[\lambda]$ be an $F[\lambda]$-homomorphism and let $j: U[\lambda] \to U((\lambda^{-1}))$ and $j_1: U_1[\lambda] \to U_1((\lambda^{-1}))$ be the canonical injections. Then there exists an $F[\lambda]$-homomorphism $\bar{\varphi}: U((\lambda^{-1})) \to U_1((\lambda^{-1}))$ which makes the diagram

$$
\begin{array}{ccc}
U((\lambda^{-1})) & \xrightarrow{\;\bar{\varphi}\;} & U_1((\lambda^{-1})) \\
\uparrow{\scriptstyle j} & & \uparrow{\scriptstyle j_1} \\
U[\lambda] & \xrightarrow{\;\varphi\;} & U_1[\lambda]
\end{array}
\tag{4-10}
$$

commutative.

PROOF By Theorem 4-2 there exists a $\Phi \in (U, V_1)_F [\lambda]$ such that $\varphi(u) = \Phi u$ for $u \in U[\lambda]$. Define $\bar{\varphi}(u) = \Phi u$ for all $u \in U((\lambda^{-1}))$ then $\bar{\varphi}$ is an $F((\lambda^{-1}))$-homomorphism which makes (4-10) commutative.

One set of the *canonical models* we are after for the representation of finite dimensional linear transformations will be the transformations defined by

$$S(D) f = \pi_D(\chi f) \qquad \text{for} \qquad f \in K_D \tag{4-11}$$

where π_D and K_D are defined by (2-6) and (2-7), respectively.

We recall that two linear transformations A and A_1 acting in F-vector spaces V and V_1, respectively are similar if there exists an invertible F-linear map $R: V \to V_1$ such that $A_1 R = RA$.

To show that our class of models is sufficient for the description, up to similarity, of all linear transformations in finite dimensional vector spaces we prove the following important theorem.

Theorem 4-4 Let A be a linear transformation in a finite dimensional vector space over the field F. Then A is similar to $S(\chi I - A)$ acting in $K_{\chi I - A}$.

PROOF Let $v(\lambda) = \sum \lambda^i v_i$ be an element of the free module $V[\lambda]$. Define a map $\varphi_A: V[\lambda] \to V$ by $\varphi_A v = \sum A^i v_i$. Clearly φ_A is an $F[\lambda]$-homomorphism of $V[\lambda]$ onto V. Thus we have the module isomorphism $V \simeq V[\lambda]/\text{Ker}\,\varphi_A$. Now, by a calculation analogous to the one in Sec. 1, $v \in \text{Ker}\,\varphi_A$ if and only if $\chi I - A|_L v$ or $v(\lambda) = (\lambda I - A) w(\lambda)$ for some $w \in V[\lambda]$. So $\text{Ker}\,\varphi_A = (\chi I - A) V[\lambda]$ and $V \simeq K_{\chi I - A}$ as $F[\lambda]$-modules. This completes the proof.

The introduction of canonical models leads to an extremely simple proof of the Cayley–Hamilton theorem.

As usual given a linear transformation A in a vector space V over F we define the characteristic polynomial d_A of A by $d_A(\lambda) = \det(\lambda I - A)$ where the determinant is computed through any matrix representation of $\lambda I - A$.

Theorem 4-5 (Cayley–Hamilton) Let A be a linear transformation in a finite dimensional vector space V over F and let d_A be its characteristic polynomial. Then $d_A(A) = 0$.

PROOF By similarity it suffices to show that $d_A(S(\chi I - A)) = 0$. This follows, by Cramer's rule, from the inclusion

$$d_A V[\lambda] \subset (\chi I - A) V[\lambda]$$

We proceed now with a more detailed study of the transformation $S(D)$.

Theorem 4-6 A number $\lambda_0 \in F$ is an eigenvalue of $S(D)$ if and only if $\text{Ker}\,D(\lambda_0) \neq \{0\}$. In that case the eigenvectors of $S(D)$ have the form $(\chi - \lambda_0)^{-1} D\xi$ for $\xi \in \text{Ker}\,D(\lambda_0)$.

PROOF Assume $D(\lambda_0) \xi = 0$ and define f by $f = (\chi - \lambda_0)^{-1} D\xi$ then clearly $\pi_{\blacktriangleright} f = f$ that is $f \in K_D$ and

$$(S(D) - \lambda_0 I) f = \pi_D(\chi - \lambda_0) f = \pi_D D\xi = 0$$

that is, f is an eigenvector of $S(D)$ corresponding to the eigenvalue λ_0. Conversely assume f is an eigenvector of $S(D)$ which corresponds to the eigenvalue λ_0 then $\pi_D(\chi - \lambda_0) f = 0$ or $(\chi - \lambda_0) f \in \text{Ker} \, \pi_D = DV[\lambda]$. Therefore $(\chi - \lambda_0) f = Dg$ for some g in $V[\lambda]$, or $f = (\chi - \lambda_0)^{-1} Dg$. It remains to show that g is a constant vector. Since $f \in K_D$ it follows from Lemma 2-15 that $D^{-1}f = (\chi - \lambda_0)^{-1} g$ is proper rational and hence g is necessarily constant.

Corollary 4-7 A number $\lambda_0 \in F$ is an eigenvalue of $S(D)$ if and only if $\chi - \lambda_0$ divides $d = \det D$.

PROOF The polynomial $d = \det D$ is divisible by $\chi - \lambda_0$ if and only if $d(\lambda_0) = 0$ which is equivalent to $\text{Ker} \, D(\lambda_0) \neq \{0\}$.

The above corollary indicates the direction for generalizing Theorem 4-6 and is an instant of a spectral mapping theorem.

Theorem 4-8 Given a polynomial p in $F[\lambda]$ then $p(S(D))$ is invertible if and only if p and $d = \det D$ are coprime.

We omit the direct proof. This theorem follows also as a corollary to the more general result given by Theorem 4-11.

Since we are interested in the relationship between different canonical models it is of importance to characterize the conditions guaranteeing the similarity of two transformations of the form $S(D)$. For this we introduce the notion of intertwining operators. Let K and K_1 be vector spaces over F and let T and T_1 be two linear transformations meeting in K and K_1, respectively. We say that a linear map $X: K \to K_1$ *intertwines* T and T_1 if $XT = T_1 X$. If X happens to be invertible then T and T_1 are similar. In the special case that the spaces are K_D and K_{D_1} and the maps are $S(D)$ and $S(D_1)$, respectively, then a map $X: K_D \to K_{D_1}$ intertwines $S(D)$ and $S(D_1)$ if and only if it is an $F[\lambda]$-module homomorphism. Thus the set of all $F[\lambda]$-module homomorphisms from K_D into K_{D_1} is the one we wish to characterize and in particular the subclass of isomorphisms.

Theorem 4-9 Let D and D_1 be nonsingular elements of $(V, V)_F[\lambda]$ and $(W, W)_F[\lambda]$, respectively. A map $X: K_D \to K_{D_1}$ is an $F[\lambda]$-homomorphism if and only if there exist Ξ and Ξ_1 in $(V, W)_F[\lambda]$ satisfying

$$\Xi D = D_1 \Xi_1 \tag{4-12}$$

and X is defined by

$$Xf = \pi_{D_1} \Xi f \qquad \text{for } f \in K_D \tag{4-13}$$

Before proving Theorem 4-9 we prove the following lemma

Lemma 4-10 Let D_1 be a nonsingular element of $(W, W)_F [\lambda]$. A map $X: V[\lambda] \rightarrow K_{D_1}$ is an $F[\lambda]$-homomorphism if and only if for some Ξ in $(V, W)_F [\lambda]$ X is given by

$$Xf = \pi_{D_1}\Xi f \qquad \text{for} \quad f \in V[\lambda] \tag{4-14}$$

PROOF Assume $X: V[\lambda] \rightarrow K_{D_1}$ is an $F[\lambda]$-homomorphism. Let $e_1, \ldots, e_n$ be a basis of V. Let $Xe_i = \xi_i \in K_{D_1}$. Let Ξ be the element of $(V, W)_F [\lambda]$ defined by $\xi_i(\lambda) = \Xi(\lambda) e_i$. By linearity it follows that $(Xv)(\lambda) = \Xi(\lambda) v$ for $v \in V$. Since X is an $F[\lambda]$-homomorphism we have for any polynomial p in $F[\lambda]$ that $X(pv) = \pi_{D_1}p(\Xi v) = \pi_{D_1}\Xi(pv)$. As all elements of $V[\lambda]$ are sums of elements of the form pv (4-11) follows. The converse is trivial.

PROOF OF THEOREM 4-9 If $X: K_D \rightarrow K_{D_1}$ is defined through (4-13) and (4-12) then it is clearly an $F[\lambda]$-module homomorphism. Conversely let $X: K_D \rightarrow K_{D_1}$ be an $F[\lambda]$-module homomorphism. Thus

$$XS(D) = S(D_1) X \tag{4-15}$$

Right multiplying (4-15) by π_D we obtain

$$XS(D) \pi_D = S(D_1) X\pi_D$$

and this implies

$$(X\pi_D) S = S(D_1)(X\pi_D) \tag{4-16}$$

where $S: V[\lambda] \rightarrow V[\lambda]$ is defined by $Sf = Xf$. Thus $X\pi_D$ satisfies the conditions of Lemma 4-10 and hence

$$X\pi_D f = \pi_{D_1}\Xi f \tag{4-17}$$

for some $\Xi \in (V, W)_F [\lambda]$. Now $X\pi_D$ and X act equally on K_D and hence (4-17) implies (4-13). Also $X\pi_D Dg = 0$ for any $g \in V[\lambda]$ hence $\pi_{D_1}\Xi Dg = 0$ or

$$\Xi DV[\lambda] \subset D_1 V[\lambda] \tag{4-18}$$

But (4-18) implies the existence of a Ξ_1 for which (4-12) holds.

The following theorem characterizes the invertibility properties of the transformations that intertwine two canonical models.

Theorem 4-11 Let D and D_1 be nonsingular elements of $(V, V)_F [\lambda]$ and $(W, W)_F [\lambda]$, respectively, and let $X: K_D \rightarrow K_{D_1}$ be defined by (4-13) with (4-12) holding. Then

(a) X is onto K_{D_1} if and only if $(\Xi, D_1)_L = I$,
(b) X is one-to-one if and only if $(\Xi_1, D)_R = I$.

PROOF

(a) Consider the range of $X = \{\pi_{D_1} \Xi f \mid f \in K_D\}$ which is clearly a submodule of K_{D_1}. X is not onto K_{D_1} if and only if $\{\pi_{D_1} \Xi f \mid f \in K_D\} + D_1 W[\lambda]$ which is equal to $\Xi V[\lambda] + D_1 W[\lambda]$ differs from $W[\lambda]$. Now $\Xi V[\lambda] + D_1 W[\lambda] = \Delta W[\lambda]$ where Δ is a g.c.l.d. of Ξ and D_1. But Δ is not unimodular if and only if $(\Xi, D_1)_L = I$.

(b) Let $f \in K_D$ be in the kernel of X. Since $F \in K_D$ we can write, by Lemma 2-15, $f = Dg$ for some proper rational function g. Now $Xf = 0$ implies $\pi_{D_1} \Xi f = 0$ or $\Xi Dg = D_1 p$ for some $p \in W[\lambda]$. Using (4-12) we obtain $\Xi_1 g = p$. Let us define J_g and I_g by

$$J_g = \{A \in (V, V)_F [\lambda] \mid Ag \in V[\lambda]\}$$

and

$$I_g = \{B \in (V, W)_F [\lambda] \mid Bg \in W[\lambda]\}$$

The representation theorems for ideals and modules proved in Sec. 2 imply that

$$J_g = (V, V)_F [\lambda] \Delta \qquad \text{and} \qquad I_g = (V, W)_F [\lambda] \Delta_1$$

for some Δ and Δ_1 in $(V, V)_F [\lambda]$. Since

$$(V, W)_F [\lambda] J_g \subset I_g \qquad \text{and} \qquad (W, V)_F [\lambda] I_g \subset J_g$$

it follows that Δ and Δ_1 are right associates and without loss of generality can be identified. Now $D \in J_g$ and $\Xi_1 \in I_g$ and hence have a nontrivial common right divisor Δ.

Conversely assume Ξ_1 and D are not right coprime. Let Δ be a g.c.r.d. of Ξ_1 and D. Let g be a proper rational function for which $\Delta g \in V[\lambda]$. Such a g certainly exists. Let $f = D_g$ then $f \in K_D$ and

$$Xf = \pi_{D_1} \Xi f = \pi_{D_1} \Xi Dg = \pi_{D_1} D_1 \Xi_1 g = 0$$

for $\Xi_1 g \in W[\lambda]$ as Δ is a right divisor of Ξ_1. Thus X is not one-to-one.

Since a unimodular matrix U is left or right coprime with any other matrix we get as an easy corollary of the previous theorem the following classical result.

Corollary 4-12 Let A and A_1 be elements of $(V, V)_F$ then A and A_1 are similar if and only if $\chi I - A$ and $\chi I - A_1$ are equivalent.

PROOF That similarity implies equivalence is trivial. Assume equivalence of $\chi I - A$ and $\chi I - A_1$, hence by the previous theorem $S(\chi I - A)$ and $S(\chi I - A_1)$ are similar and by transitivity A and A_1 are also similar.

Theorem 4-11 has important implications inasmuch as, together with the Smith canonical form, it provides the key to the understanding of the structure of finitely generated torsion modules over $F[\lambda]$ and with that to the structure of linear transformations in finite dimensional vector spaces. Before tackling these subjects we wish to indicate the usefulness of another set of canonical models.

For a finite dimensional vector space Y over a field F we note that $Y((\lambda^{-1}))$ is an $F[\lambda]$-module of which $Y[\lambda]$ is a submodule. The quotient module $Y((\lambda^{-1}))/Y[\lambda]$ inherits an induced module structure. We already made the identification of $Y((\lambda^{-1}))/Y[\lambda]$ with $\lambda^{-1}Y[[\lambda^{-1}]]$. Thus $\lambda^{-1}Y[[\lambda^{-1}]]$ is an $F[\lambda]$-module with the action of a polynomial p given by

$$p \cdot y = \pi_-(py) \tag{4-19}$$

for $y \in \lambda^{-1}Y[[\lambda^{-1}]]$, π_- being the canonical projection of $Y((\lambda^{-1}))$ onto $\lambda^{-1}Y[[\lambda^{-1}]]$. As before let π_+ denote the canonical projection of $Y((\lambda^{-1}))$ onto $Y[\lambda]$.

Let now $D \in (Y, Y)_F[\lambda]$ be nonsingular, that is $\det D$ is a nontrivial polynomial in $F[\lambda]$. We define a map $\pi^D: \lambda^{-1}Y[[\lambda^{-1}]] \to \lambda^{-1}Y[[\lambda^{-1}]]$ by

$$\pi^D y = \pi_- D^{-1}\pi_+ Dy \tag{4-20}$$

Obviously π^D is a projection operator in $\lambda^{-1}Y[[\lambda^{-1}]]$ but it is not an $F[\lambda]$-homomorphism. However, L_D defined by

$$L_D = \text{Range } \pi^D \tag{4-21}$$

is a submodule of $\lambda^{-1}Y[[\lambda^{-1}]]$. In fact we have the following counterpart of Theorem 2-12.

Theorem 4-13 A subset M of $\lambda^{-1}Y[[\lambda^{-1}]]$ is a finitely generated torsion submodule if and only if

$$M = L_D = \text{Range } \pi^D \tag{4-22}$$

for some nonsingular $D \in (Y, Y)_F[\lambda]$.

PROOF Let $M = L_D$ for some nonsingular D. By Cramer's rule $DE = (\det D) I$ where E is the cofactor matrix of D. This implies that $\det D$ annihilates all of M, that is, that M is a torsion submodule of $\lambda^{-1}Y[[\lambda^{-1}]]$. As M is a finite dimensional vector space over F it is clearly finitely generated over $F[\lambda]$.

Conversely assume M is a finitely generated torsion submodule of $\lambda^{-1}Y[[\lambda^{-1}]]$. There exists therefore a polynomial $p \in F[\lambda]$ which annihilates all of M. Consider next the set J defined by

$$J = \{A \in (Y, Y)_F[\lambda] \,|\, \pi_-(Ay) = 0 \qquad \text{for all} \quad y \in M\}$$

Then clearly J is a left ideal in $(Y, Y)_F[\lambda]$ and so, by Theorem 2-11, has the form $J = (Y, Y)_F[\lambda] D$ for some $D \in (Y, Y)_F[\lambda]$. Since $p \cdot I \in J$ it follows that D is necessarily nonsingular.

Define now a map $\rho_D: \text{Range } \pi^D \to \text{Range } \pi_D$ by

$$\rho_D y = Dy \tag{4-23}$$

Clearly its inverse is given by $\rho_D^{-1} y = D^{-1}y$ for $y \in K_D$.

Since for every polynomial $p \in F[\lambda]$

$$\rho_D(p \cdot y) = \rho_D \pi_-(py) = D\pi_-(py) = D\pi_-D^{-1}(pDy) = \pi_D p(Dy)$$
$$= \pi_D p(\rho_D y) = p \cdot \rho_D y$$

it follows that ρ_D is an $F[\lambda]$-homomorphism that maps M into a submodule of K_D. But submodules of K_D correspond in a bijective way to left factors of D, hence necessarily $M = \text{Range} \, \pi^D$.

Given any two finitely generated torsion submodules M and M_1 of $\lambda^{-1}Y[[\lambda^{-1}]]$ and $\lambda^{-1}Y_1[[\lambda^{-1}]]$, respectively, we are able now to characterize all $F[\lambda]$-homomorphisms from M into M_1. This is a result analogous to Theorem 4-9.

Theorem 4-14 Let D and D_1 be nonsingular elements of $(Y, Y)_F[\lambda]$ and $(Y_1, Y_1)_F[\lambda]$, respectively. A map $\psi_0 : L_D \to L_{D_1}$ is an $F[\lambda]$-homomorphism if and only if there exist Ψ and Ψ_1 in $(Y, Y_1)_F[\lambda]$ satisfying

$$\Psi D = D_1 \Psi_1 \tag{4-24}$$

and for which

$$\psi_0(y) = \pi_-(\Psi_1 y) \tag{4-25}$$

PROOF Assume there exist Ψ and Ψ_1 satisfying (4-24) and let ψ_0 be defined by (4-25). If $y \in L_D$ then Dy is in $Y[\lambda]$ and from (4-25) we obtain

$$D_1\psi_0(y) = D_1\pi_-(\Psi_1 y) = D_1\pi_-D_1^{-1}D_1\Psi_1 y = \pi_{D_1}(D_1\Psi_1 y) = \pi_{D_1}(Dy)$$

which shows that $\pi_{D_1}(\Psi Dy) \in Y_1[\lambda]$ or that $\psi_0(y) \in L_{D_1}$. To show that ψ_0 is an $F[\lambda]$-homomorphism we note that for $p \in F[\lambda]$

$$\psi_0\big(\pi_-(py)\big) = \pi_-\big(\Psi_1\pi_-(py)\big) = \pi_-(\Psi_1 py) = \pi_-(p\Psi_1 y)$$
$$= \pi_-\big(p\pi_-(\Psi_1 y)\big) = \pi_-\big(p(\psi_0(y))\big)$$

Conversely let $\psi_0 : L_D \to L_{D_1}$ be an $F[\lambda]$-homomorphism. Since $\rho_D : L_D \to K_D$ and $\rho_{D_1} : L_{D_1} \to K_{D_1}$ defined by (4-23) are $F[\lambda]$-homomorphisms then $\psi : K_D \to K_{D_1}$ defined by

$$\psi = \rho_{D_1}\psi_0\rho_D^{-1} \tag{4-26}$$

is also an $F[\lambda]$-homomorphism. Applying Theorem 4-9 which characterizes these homomorphisms we establish the existence of Ψ and Ψ_1 in $(Y, Y_1)_F[\lambda]$ that satisfy (4-24) and for which ψ is given by

$$\psi(u) = \pi_{D_1}(\Psi u) \tag{4-27}$$

As $\psi_0 = \rho_{D_1}^{-1}\psi\rho_D$ we obtain for $y \in L_D$

$$\psi_0(y) = \rho_D^{-1}\psi\rho_{D_1}y = \pi_-D_1^{-1}\pi_{D_1}\Psi Dy = \pi_-D_1^{-1}D_1\pi_-D_1^{-1}\Psi Dy$$
$$= \pi_-(D_1^{-1}\Psi Dy) = \pi_-(\Psi_1 y)$$

by virtue of (4-24).

Since the map $\bar{\psi}_0: \lambda^{-1}Y[[\lambda^{-1}]] \to \lambda^{-1}Y[[\lambda^{-1}]]$ defined by $\bar{\psi}_0(y) = \pi(\Psi_1 y)$ is an $F[\lambda]$-homomorphism we obtain as a corollary the counterpart of Theorem 4-9.

Theorem 4-15 Let $M = L_D$ and $M_1 = L_{D_1}$ be finitely generated torsion submodules of $\lambda^{-1}Y[[\lambda^{-1}]]$ and $\lambda^{-1}Y_1[[\lambda^{-1}]]$, respectively, and let $\psi_0: M \to M_1$ be an $F[\lambda]$-homomorphism. Then there exists an $F[\lambda]$-homomorphism $\bar{\psi}_0: \lambda^{-1}Y[[\lambda^{-1}]] \to \lambda^{-1}Y_1[[\lambda^{-1}]]$ which makes the diagram

$$
\begin{array}{ccc}
\lambda^{-1}Y[[\lambda^{-1}]] & \xrightarrow{\bar{\psi}_0} & \lambda^{-1}Y_1[[\lambda^{-1}]] \\
{\scriptstyle \pi^D}\downarrow & & \downarrow{\scriptstyle \pi^{D_1}} \\
L_D & \xrightarrow{\psi_0} & L_{D_1}
\end{array}
\tag{4-28}
$$

commutative.

Similarly we obtain the dual version of Lemma 4-10.

Corollary 4-16 Let M be a finitely generated torsion submodule of $\lambda^{-1}Y[[\lambda^{-1}]]$ and let $\varphi: M \to \lambda^{-1}Y_1[[\lambda^{-1}]]$ be an $F[\lambda]$-homomorphism. Then there exists an $F[\lambda]$-homomorphism, $\bar{\varphi}: \lambda^{-1}Y[[\lambda^{-1}]] \to \lambda^{-1}Y_1[[\lambda^{-1}]]$ which makes the diagram

$$
\begin{array}{ccc}
\lambda^{-1}Y[[\lambda^{-1}]] & \xrightarrow{\bar{\varphi}} & \lambda^{-1}Y[[\lambda^{-1}]] \\
{\scriptstyle j}\uparrow & \nearrow{\scriptstyle \varphi} & \\
M & &
\end{array}
\tag{4-29}
$$

commutative.

The preceding discussion enables us now to introduce a second class of canonical models. For nonsingular element $D \in (Y, Y)_F[\lambda]$ we define L_D by (4-21) and a map $S^*(D)$ in L_D by

$$
S^*(D)y = \pi_-(\chi y)
\tag{4-30}
$$

$S^*(D)$ is well defined as L_D is a submodule of $\lambda^{-1}Y[[\lambda^{-1}]]$.

Lemma 4-17 For a nonsingular $D \in (Y, Y)_F[\lambda]$ the maps $S(D)$ and $S^*(D)$ are similar, the similarity given by the following commutative diagram.

$$
\begin{array}{ccc}
L_D & \xrightarrow{\rho_D} & K_D \\
{\scriptstyle S^*(D)}\downarrow & & \downarrow{\scriptstyle S(D)} \\
L_D & \xrightarrow{\rho_D} & K_D
\end{array}
\tag{4-31}
$$

PROOF This follows from the following equality

$$
\rho_D^{-1}S(D)\rho_D y = \pi_- D^{-1}\pi_D(\chi D y) = \pi_- D^{-1}D\pi_- D^{-1}(\chi D y) = \pi_-(\chi y) = S^*(D)y
$$

We return now to the study of the structure of finitely generated torsion modules over $F[\lambda]$.

Theorem 4-18 Let M be a finitely generated torsion module over $F[\lambda]$. Then M is isomorphic to a direct sum

$$F[\lambda]/(\delta_1) \oplus \cdots \oplus F[\lambda]/(\delta_r) \qquad (4\text{-}32)$$

with $\delta_{i+1}|\delta_i$. The sequence of ideals (δ_i) is uniquely determined.

PROOF Let M be a finitely generated torsion module over $F[\lambda]$. Then M is isomorphic to a quotient module $F^n[\lambda]/DF^n[\lambda]$ with $\det D \neq 0$. Let $\Delta = \text{diag}(\delta_1, \ldots, \delta_r, 1, \ldots, 1)$ be the Smith canonical form of D then D and Δ are equivalent. Theorem 4-18 implies that $F^n[\lambda]/DF^n[\lambda]$ and $F^n[\lambda]/\Delta F^n[\lambda]$ are $F[\lambda]$-module isomorphic. However, $F^n[\lambda]/\Delta F^n[\lambda]$ is clearly isomorphic to the direct sum $F[\lambda]/(\delta_1) \oplus \cdots \oplus F[\lambda]/(\delta_r)$. Uniqueness follows from the uniqueness of the Smith canonical form.

The previous theorem provides us with the background necessary for the understanding of the structure of linear transformations in finite dimensional vector spaces.

Let V be a finite dimensional vector space over F, and let A be a linear transformation in V. We induce an $F[\lambda]$-module structure in V by way of defining

$$p \cdot x = p(A) x \qquad \text{for } p \in F[\lambda] \quad \text{and} \quad x \in V \qquad (4\text{-}33)$$

Moreover, as a consequence of the Cayley–Hamilton theorem, V as a $F[\lambda]$-module is a finitely generated $F[\lambda]$-torsion module. That this is the case follows also from Theorem 4-4 where the similarity of A and $S(\chi I - A)$ is proved. This similarity is clearly an $F[\lambda]$-module isomorphism between $K_{\chi I - A}$ and V given the naturally induced module structure.

Applying the structure theory developed in Theorem 4-18 we get the following.

Theorem 4-19 Let V be a finite dimensional vector space over F and let A be a linear transformation in V. Let $\delta_1, \ldots, \delta_r$ be the invariant factors of $\chi I - A$ then A is similar to $S(\delta_1) \oplus \cdots \oplus S(\delta_r)$ acting in $K_{\delta_1} \oplus \cdots \oplus K_{\delta_r}$.

PROOF A is similar to $S(\chi I - A)$ and since $\chi I - A$ and $\text{diag}(\delta_1, \ldots, \delta_r, 1, \ldots, 1)$ are equivalent the result follows from Theorem 4-11.

In order to obtain convenient canonical matrix representations we study further the individual summands in the above direct sum.

In the representation (4-32) of the module M the components in the direct sum are cyclic. In fact we have the following.

Theorem 4-20 Let M be a nontrivial cyclic $F[\lambda]$-module. Then either M is isomorphic to $F[\lambda]$ or M is isomorphic to $F[\lambda]/(\delta)$ for some nonzero $\delta \in F[\lambda]$.

PROOF Let g be a generator of M, then the map φ defined by $\varphi(p) = pg$ is a surjective $F[\lambda]$-module homomorphism. Hence $M = F[\lambda]/\text{Ker}\,\varphi$. But $\text{Ker}\,\varphi$ is an ideal in $F[\lambda]$ which is principal so $\text{Ker}\,\varphi = (\delta)$. The result depends on whether δ is zero or not.

To relate the above result to linear transformations we define a linear transformation A in a finite dimensional vector space V to be *cyclic* if V with the $F[\lambda]$-module structure induced by (4-33) is cyclic. A generator $b \in V$ of this module will be called a *cyclic vector* for A. Clearly A is cyclic with b a cyclic vector if and only if $b, Ab, \ldots$ span all of V.

If A is cyclic then by Theorem 4-18, V is isomorphic to $F[\lambda]/(m_A)$ for some $m_A \in F[\lambda]$. The polynomial m_A is the minimal polynomial of A. Clearly we must have $\dim V = \deg m_A$ in case of cyclic transformations. Thus the set $B = \{b, Ab, \ldots, A^{n-1}b\}$ is a basis for V as a vector space. Let the minimal polynomial m_A be given by

$$m_A(\lambda) = \lambda^n + \mu_{n-1}, \lambda^{n-1} + \cdots + \mu_0 \qquad (4\text{-}34)$$

then with respect to the basis B the transformation A has the following matrix representation.

$$\begin{bmatrix} 0 & . & . & . & . & -\mu_0 \\ 1 & & & & & . \\ . & . & & & & . \\ . & & . & & & . \\ . & & & . & & . \\ . & & & . & 1 & -\mu_{n-1} \end{bmatrix} \qquad (4\text{-}35)$$

The matrix (4-35) is called the *companion matrix* of the polynomial m_A. Conversely a matrix of the form (4-35) is clearly cyclic and has m_A as characteristic and minimal polynomials.

Combining this matrix representation of cyclic transformations with Theorem 4-19 we have the following.

Theorem 4-21 Let A be a linear transformation in a finite dimensional vector space V and let

$$\delta_i(\lambda) = \lambda^{r_i} + d_{r_i-1}^{(i)}\lambda^{r_i-1} + \cdots + d_0^{(i)} \qquad (4\text{-}36)$$

be its invariant factors then there exists a basis in V such that with respect to that basis A has the block diagonal form

$$\begin{bmatrix} A_1 & & & \\ & . & & \\ & & . & \\ & & & . \\ & & & & A_r \end{bmatrix} \qquad (4\text{-}37)$$

where A_i is the companion matrix of δ_i. The matrix (4-37) is called the *first canonical form* of A.

Generally a further reduction is possible.

Lemma 4-22 Let $p = p_1 \ldots p_k$ with $p_i \in F[\lambda]$ then $\text{diag}(p, 1, \ldots, 1)$ and $\text{diag}(p_1, \ldots, p_k)$ are equivalent if and only if for $i \neq j$ p_i and p_j are relatively prime.

PROOF Denote by π_i the polynomial $\pi_i = p/p_i$. Assume equivalence of $\text{diag}(p, 1, \ldots, 1)$ and $\text{diag}(p_1, \ldots, p_k)$ then $\text{diag}(p, 1, \ldots, 1)$ is the Smith form of $\text{diag}(p_1, \ldots, p_k)$ and hence, by Corollary 3-2, the g.c.d. of $\pi_1, \ldots, \pi_k$ is 1 which implies the coprimeness of p_i and p_j for $i \neq j$. Conversely if the p_i are pairwise coprime then the g.c.d. of $\pi_1, \ldots, \pi_k$ is 1 which implies that p is the only invariant factor of $\text{diag}(p_1, \ldots, p_k)$ and hence the equivalence.

Corollary 4-23 Let $p \in F[\lambda]$ and let $p = p_1 \ldots p_k$ be a factorization of a polynomial p in $F[\lambda]$ into relative prime polynomials. Then $S(p)$ is similar to $S(p_1) \oplus \cdots \oplus S(p_k)$.

Lemma 4-24 Let $\pi \in F[\lambda]$ be irreducible then there is a basis B in K_{π^r} such that with respect to it $S(\pi^r)$ has the $r \times r$ block matrix representation

$$\begin{bmatrix} P & & & \\ N & P & & \\ & \cdot & \cdot & \\ & & \cdot & \cdot \\ & & & N & P \end{bmatrix} \tag{4-38}$$

where P is the companion matrix of π and N is the matrix

$$\begin{bmatrix} 0 & \cdot & \cdot & \cdot & 0 & 1 \\ \cdot & & & & & 0 \\ \cdot & & & & & \\ \cdot & & & & & \cdot \\ \cdot & & & & & \cdot \\ 0 & & \cdot & \cdot & \cdot & 0 \end{bmatrix} \tag{4-39}$$

PROOF Let $\pi(\lambda) = \lambda^q + p_{q-1}\lambda^{q-1} + \cdots + p_0$. Thus $\dim K_{\pi^r} = rq$. Since the elements of K_{π^r} are all polynomials of degree $\leq qr - 1$ to choose a basis it suffices to choose a set of polynomials of degrees ascending from zero to $qr - 1$.

The set of polynomials

$$\{e_{jq+i} = \chi^{i-1}\pi^j \mid 0 \le j \le r-1, 1 \le i \le q\}$$

satisfies this requirement and hence is a basis for K_{π^r}. With respect to this basis $S(\pi^r)$ has the required form.

Corollary 4-25 If $\pi(\lambda) = \lambda - \alpha$ then $S(\pi^r)$ has a matrix representation of the form

$$\begin{bmatrix} \alpha & & & \\ 1 & . & & \\ & . & . & \\ & & . & . \\ & & 1 & \alpha \end{bmatrix} \tag{4-40}$$

Combining all previous results we have the Jordan canonical form theorem.

Theorem 4-26 Let F be an algebraically closed field and let A be a linear transformation in a finite dimensional vector space V over F. Then for a suitable choice of basis A has a matrix representation of the form

$$\begin{bmatrix} J_1 . & & 0 \\ & . & \\ 0 & & . J_k \end{bmatrix} \tag{4-41}$$

and

$$J_i = \begin{bmatrix} \alpha_i & & & 0 \\ 1 & . & & \\ & . & . & \\ & & . & . \\ 0 & & 1 & \alpha_i \end{bmatrix} \tag{4-42}$$

5. LINEAR SYSTEMS

A *discrete time constant* (*time invariant*) *linear system* Σ consists of three vector spaces U, X, and Y over a field F and a triple (A, B, C) of linear maps $A: X \to X$, $B: U \to X$ and $C: X \to Y$.

The system (A, B, C) is taken to represent the pair of dynamical equations

$$\begin{cases} x_{n+1} = Ax_n + Bu_n \\ y_n = Cx_n \end{cases} \tag{5-1}$$

The space U is referred to as the *input space*, Y as the *output space*, and X as the *state space*. In the rest of this chapter we assume all three spaces are finite dimensional. The dimension of the system Σ, denoted by $\dim \Sigma$, is defined by $\dim \Sigma = \dim X$. If $m = \dim U$ and $p = \dim Y$ then we have an m-input p-output system. Through a choice of basis in U and Y we may, without loss of generality, assume $V = F^m$ and $Y = F^p$. Sometimes the second equation in (5-1) is replaced by $y_n = C x_n + D u_n$. The introduction of the linear map $D: U \to Y$ does not affect the dynamical behaviour of the system and hence will be omitted whenever convenient.

The description of a linear system by way of equations (5-1) is an explicit dynamic description and is referred to as an *internal description*. Given an internal description the *external description* of the system, that is the input/output behaviour of Σ, is easily determined.

To do this right it is convenient to indicate each input choice by the time in which it has been applied. The time axis, which in our case is the set of integers $\mathbb{Z}$, is mapped in a one-to-one way onto all powers of the indeterminate λ such that $k \to \lambda^{-k}$. Thus we will denote by $u_{-k}\lambda^k$ an input u_{-k} that has been applied at time $t = -k$. An output y_j occurring at time $t = j$ will be denoted by $y_j \lambda^{-j}$. This choice is a convention adopted for historical reasons, mainly to get compatibility with the theory of z-transforms.

Since we are interested in sequences of inputs we use $\Sigma u_i \lambda^{-i}$ to denote a sequence of inputs where u_i is applied at time $t = -i$.

Assuming the system to be initially at rest, that is, $x_0 = 0$, then the application of a single input u_0 at time $t = 0$ produces the state evolution $x_n = A^{n-1} B u_0$ and hence a sequence of outputs $y_n = C A^{n-1} B u_0$. Thus to the input $\sum_{-k_i \leq j} u_j \lambda^{-j}$ corresponds the output $\sum_{-k+1 \leq l} y_l u^{-l}$ where

$$y_l = \sum_{0 \leq j} C A^j B u_{l-j-1} \tag{5-2}$$

. Problems of convergence, which have no meaning in this context, do not arise as each sum has at most $1 + k + l$ nonzero terms. Thus, the system Σ induces a map $\bar{f}: U((\lambda^{-1})) \to Y((\lambda^{-1}))$ given by

$$\bar{f} \sum_{-k \leq i} u_i \lambda^{-i} = \sum_{-k+1 \leq l} y_l \lambda^{-l} \tag{5-3}$$

where the y_l are given by Eq. (5-2). The map $\bar{f}$ is called the *input/output map of Σ* or the *result* of Σ.

Lemma 5-1 The result $\bar{f}$ of the system $\Sigma = (A, B, C)$ is an $F[\lambda]$-module homomorphism of $U((\lambda^{-1}))$ into $Y((\lambda^{-1}))$ which satisfies

$$\bar{f}(U[[\lambda^{-1}]]) \subset \lambda^{-1} Y[[\lambda^{-1}]] \tag{5-4}$$

Condition (5-4) is nothing but the expression of the causality of the system Σ. Actually, as we saw in Sec. 2, $U((\lambda^{-1}))$ and $Y((\lambda^{-1}))$ are also $F((\lambda^{-1}))$-modules and it is easily verified that $\bar{f}$ is actually a $F((\lambda^{-1}))$-module homomorphism. Let $j: U[\lambda] \to U((\lambda^{-1}))$ be the inclusion map of $U[\lambda]$ into $U((\lambda^{-1}))$ and π_- the canonical projection of $Y((\lambda^{-1}))$ onto $Y((\lambda^{-1}))/Y[\lambda]$ identified as $\lambda^{-1} Y[[\lambda^{-1}]]$ then the map $f: U[\lambda] \to \lambda^{-1} Y[[\lambda^{-1}]]$ can be defined via the commutative diagram

$$U[\lambda] \xrightarrow{\;f\;} \lambda^{-1}Y[[\lambda^{-1}]]$$

$$i \downarrow \qquad\qquad \uparrow \pi_- \qquad\qquad (5\text{-}5)$$

$$U((\lambda^{-1})) \xrightarrow{\;\bar f\;} {}^{1}Y((\lambda^{-1}))$$

We call f the *restricted input/output map* of Σ. f is clearly an $F[\lambda]$-module homomorphism and together with $\bar f$ it is uniquely determined by Σ.

As a consequence we define a strictly causal input/output map $\bar f$ to be an $F[\lambda]$-module homomorphism of $U((\lambda^{-1}))$ into $Y((\lambda^{-1}))$ which satisfies $\bar f(U[[\lambda^{-1}]] \subset \lambda^{-1}Y[[\lambda^{-1}]]$. Similarly a restricted input/output map f is defined to be an $F[\lambda]$-module homomorphism of $U[\lambda]$ into $\lambda^{-1}Y[[\lambda^{-1}]]$. Given a restricted input/output map f there exists a unique strictly causal input/output map $\bar f$ which makes diagram (5-5) commutative. We simply define

$$\bar f \sum_{-k \le j} u_j \lambda^{-j} = \sum_{-k \le j} \lambda^{-j} f(u_j) \qquad\qquad (5\text{-}6)$$

and the right-hand side is easily seen to be a well-defined element of $Y((\lambda^{-1}))$.

The preceding discussion yields directly to the introduction of transfer functions.

Lemma 5-2 Let $\bar f: D((\lambda^{-1})) \to Y((\lambda^{-1}))$ be an $F((\lambda^{-1}))$-homomorphism then $\bar f$ has a unique representation as multiplication by an element T of $(U, Y)_F((\lambda^{-1}))$, that is

$$\bar f(u) = Tu \qquad \text{for} \qquad u \in U((\lambda^{-1})) \qquad\qquad (5\text{-}7)$$

If $\bar f$ is strictly causal then $T \in \lambda^{-1}(U, Y)_F[[\lambda^{-1}]]$.

If $\bar f$ is the input/output map of the system Σ then its representing multiplier T_Σ is called the *transfer function* of the system Σ. In terms of the transfer function T_Σ the restricted input/output map f is given by

$$f(u) = \pi_-(T_\Sigma u) \qquad \text{for} \qquad u \in U[\lambda] \qquad\qquad (5\text{-}8)$$

For the system (A, B, C) the transfer function is easily computed to be

$$T_\Sigma(\lambda) = \sum_{i=0}^{\infty} CA^iB\lambda^{-i-1} = C(\lambda I - A)^{-1}B \qquad\qquad (5\text{-}9)$$

Finally if $T(\lambda) = \sum_{i=0}^{\infty} T_i\lambda^{-i-1}$ is an element of $\lambda^{-1}(U, Y)_F[[\lambda^{-1}]]$ then we say a system $\Sigma = (A, B, C)$ is a *realization* of T if $T = T_\Sigma$. This is equivalent to

$$T_i = CA^iB \qquad \text{for} \qquad i \ge 0 \qquad\qquad (5\text{-}10)$$

Similarly given an abstract restricted input/output map $f: U[\lambda] \to \lambda^{-1}Y[[\lambda^{-1}]]$ we say $\Sigma = (A, B, C)$ is a realization of f if f coincides with the input/output map of Σ.

6. REACHABILITY, OBSERVABILITY, AND REALIZATIONS

Let $\Sigma = (A, B, C)$ be a finite dimensional constant linear system. We say that Σ is *reachable* if given any state x in X there is a sequence of inputs $u_0, u_1, \ldots, u_k$ which drives the system from $x = 0$ to x. In other words $x = \sum_{i=0}^{k-1} A^{k-i-1} Bu_i$. We say that Σ is *observable* if given any nonzero state x in X at least one of the observations of the free motion $y_k = CA^k x$ is nonzero.

A ternately stated observability of Σ is equivalent to

$$\bigcap_{i \geq 0} \operatorname{Ker} CA^i = \{0\} \tag{6-1}$$

whereas reachability of Σ is equivalent to

$$\bigcap_{i \geq 0} \operatorname{Ker} B^* A^{*i} = \{0\} \tag{6-2}$$

Here $A^* : X^* \to X^*$ and $B^* : X^* \to U^*$ are the maps dual to A and B, respectively.

Define now the *reachability map* $R : U[\lambda] \to X$ and the *observability map* $O : X \to \lambda^{-1} Y[[\lambda^{-1}]]$ by

$$R \sum_{0 \leq i} \lambda^i u_{-i} = \sum_{0 \leq i} A^i Bu_{-i} \tag{6-3}$$

and

$$Ox = \sum_{i=0}^{\infty} \lambda^{-i-1} CA^i x \tag{6-4}$$

respectively.

If we consider X as an $F[\lambda]$-module by way of definition (4-17) then R and O are $F[\lambda]$-module homomorphisms. Clearly the system Σ is reachable if and only if its reachability map R is surjective and observable if and only if its observability map is injective. We say that a realization Σ of an input/output map f is *canonical* if Σ is both reachable and observable.

In terms of the reachability and observability maps the restricted input/output map of Σ can be factored as follows

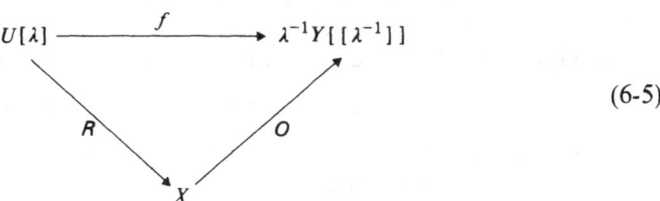

$$\tag{6-5}$$

Conversely given an $F[\lambda]$-module homomorphism $f : U[\lambda] \to \lambda^{-1} Y[[\lambda^{-1}]]$ then any factorization of the form (6-5) with R and O being $F[\lambda]$-module homomorphisms is called canonical if R is surjective and O injective. Assume now that $f = OR$ is a canonical factorization of f then $\operatorname{Ker} f = \operatorname{Ker} R$. Thus we get the $F[\lambda]$-module isomorphism $X \simeq U[\lambda]/\operatorname{Ker} f$. Similarly we get, by surjectivity of

O, the isomorphism $X \simeq \text{Range } f$. Thus each of the $F[\lambda]$-modules $U[\lambda]/\text{Ker } f$ and Range f can serve as the state space of the system.

The preceding discussion serves to define a *realization* of a restricted input/output map $f: U[\lambda] \to \lambda^{-1} Y[[\lambda^{-1}]]$ as a factorization (6-5) $f = OR$ into the product of $F[\lambda]$-module homomorphism.

This definition of a realization is compatible with our previous definition of a realization in terms of triples (A, B, C). Thus, given a factorization $f = OR$ as above we let $A: X \to X$ be the action of λ in X, $B: U \to X$ is defined to be the restriction of R to U as naturally embedded in $U[\lambda]$. Finally we let $C: X \to Y$ be defined by $Cx = (Ox)_{-1}$ where $(Ox)(\lambda) = \sum \lambda^{-n-1}(Ox)_{-n}$. The triple (A, B, C) defined in this manner is a realization of f.

Of course the identification of realizations by triples and realization by factorization allow us to obtain abstract realizations of input/output maps.

In fact let $f: U[\lambda] \to \lambda^{-1} Y[[\lambda^{-1}]]$ be a restricted input/output map. Consider the quotient $F[\lambda]$-module $X = U[\lambda]/\text{Ker } f$ and let R be the canonical projection of $U[\lambda]$ onto X, then obviously R is a surjective $F[\lambda]$-homomorphism. Similarly define a map O from X into $\lambda^{-1} Y[[\lambda^{-1}]]$ by

$$O(Ru) = f(u) \qquad \text{for} \qquad u \in U[\lambda] \tag{6-6}$$

Since R is onto X, O is defined on all of X and is easily checked to be an injective $F[\lambda]$-homomorphism. It is well defined as $\text{Ker } R = \text{Ker } f$. Thus we obtained a canonical factorization of f and hence a canonical realization of f.

The preceding discussion allows us to characterize those input/output maps arising out of finite dimensional realizations.

Theorem 6-1 Let U, Y be finite dimensional vector spaces over F. An element $T \in \lambda^{-1}(U, Y)_F[[\lambda^{-1}]]$ is the transfer function of a finite dimensional constant linear system if and only if it is rational.

PROOF Assume T is the transfer function of the finite dimensional linear system (A, B, C). Let d_A be the characteristic polynomial of A. Then, by Cramer's rule, $d_A T \in (U, Y)_F[\lambda]$ that is T is rational.

Conversely assume T is rational, that is there exists a polynomial d such that $dT \in (U, Y)_F[\lambda]$. It suffices to prove that $U[\lambda]/\text{Ker } f$ is a torsion module. Now for $u \in U[\lambda]$ we have $f(u) = \pi_-(Tu)$. This implies

$$f(du) = \pi_-(Tdu) = \pi_-((dT)u) = 0$$

that is $dU[\lambda] \subset \text{Ker } f$ and hence $U[\lambda]/\text{Ker } f$ is a finitely generated torsion module with d as annihilator.

7. HANKEL MATRICES

Analogous to the matrix representation of a linear transformation we have a matrix representation of input/output maps of constant linear systems.

Given the vector spaces U and Y we let U^* denote the set of finitely nonzero sequences in U and Y^N the set of infinite sequences in Y. U^* and Y^N use given F-linear space structure by the usual definitions of multiplication by scalars and coordinate-wise addition. We induce an $F[\lambda]$-module structure in U^* and Y^N by defining the action of λ to be the right shift σ in U^* and the left shift σ in Y^N, that is

$$\sigma(u_0, u_1, \ldots, u_n, 0, \ldots) = (0, u_0, \ldots, u_n, 0, \ldots) \tag{7-1}$$

and

$$\bar{\sigma}(y_0, y_1, \ldots) = (y_1, y_2, \ldots) \tag{7-2}$$

We define now two maps $\rho: U[\lambda] \to U^*$ and $\rho': \lambda^{-1}Y[[\lambda^{-1}]] \to Y^N$ by

$$\rho \sum_{i=0}^{n} u_i \lambda^i = (u_0, u_1, \ldots, u_n, 0, \ldots) \tag{7-3}$$

and

$$\rho' \sum_{i=0}^{\infty} y_i \lambda^{-i-1} = (y_0, y_1, \ldots) \tag{7-4}$$

and it is easily checked that both ρ and ρ' are $F[\lambda]$-module isomorphisms. Now given a restricted input/output map $f: U[\lambda] \to \lambda^{-1}Y[[\lambda^{-1}]]$ with associated transfer function $T(\lambda) = \sum_{i=0}^{\infty} T_i \lambda^{-i-1}$ we define a map $H_f: U^* \to Y^N$ as the unique $F[\lambda]$-homomorphism which makes the diagram

$$
\begin{array}{ccc}
U[\lambda] & \xrightarrow{\;f\;} & \lambda^{-1}Y[[\lambda^{-1}]] \\
\rho \downarrow & & \downarrow \rho' \\
U^* & \xrightarrow{\;H_f\;} & Y^N
\end{array}
\tag{7-5}
$$

commutative. The fact that H_f is an $F[\lambda]$-homomorphism is equivalent to H_f being linear and satisfying

$$\bar{\sigma}H_f = H_f\sigma \tag{7-6}$$

If $u \in U^*$ and $y \in Y^N$ is given through $y = H_f u$ then a simple computation shows that H_f has the block matrix representation, also denoted by H_f, given by

$$
H_f = \begin{bmatrix}
T_0 & T_1 & T_2 & \cdot & \cdot \\
T_1 & T_2 & & \cdot & \cdot & \cdot \\
T_2 & \cdot & & \cdot & \cdot & \cdot \\
\cdot & \cdot & & \cdot & \cdot & \cdot \\
\cdot & \cdot & & \cdot & \cdot & \cdot
\end{bmatrix}
\tag{7-7}
$$

with $T_i \in (U, Y)_F$.

We call H_f the *Hankel matrix* associated with the input/output map f. In general any block matrix of the form (7-7) is called a Hankel matrix. Clearly there is a bijective correspondence between Hankel matrices and causal input/output maps and hence also with transfer functions.

The isomorphisms ρ and ρ' induce the pair of $F[\lambda]$-isomorphisms $U[\lambda]/\text{Ker} f \simeq U^*/\text{Ker} H_f$ and $\text{Range} f \simeq \text{Range} H_f$. This indicates that the Hankel matrix associated with a given input/output map can be used for realization purposes. This approach will be used extensively in the infinite dimensional setting. In the finite dimensional case the Ho algorithm [10, 11] is one example of realization based on Hankel matrix data.

The characterization of Hankel matrices associated with finite dimensional realization is a direct consequence of Theorem 6-1. We only remark that rank H is defined to be the dimension of the range space.

Theorem 7-1 Let H be the Hankel matrix associated with a transfer function T. Then H has finite rank if and only if T is rational.

8. SIMULATION AND ISOMORPHISM

As we have seen a given system (A, B, C) in state space form determines a unique input/output map. While the converse is not true we can still come up with a great deal of information concerning the relation between different realizations provided extra assumptions are made. The central result of this section is Theorem 8-3 better known as the state space isomorphism theorem.

Let $f: U[\lambda] \to \lambda^{-1}Y[[\lambda^{-1}]]$ and $f_1: U_1[\lambda] \to \lambda^{-1}Y_1[[\lambda^{-1}]]$ be two restricted input/output maps. We say that f is *simulated* by f_1, and write $f | f_1$, if there exists two $F[\lambda]$-module homomorphisms $\varphi: U[\lambda] \to U_1[\lambda]$ and $\psi: \lambda^{-1}Y_1[[\lambda^{-1}]] \to \lambda^{-1}Y_1[[\lambda^{-1}]]$ which make the following diagram commutative.

$$
\begin{array}{ccc}
U[\lambda] & \xrightarrow{\ f\ } & \lambda^{-1}Y[[\lambda^{-1}]] \\
{\scriptstyle\varphi}\downarrow & & \uparrow{\scriptstyle\psi} \\
U_1[\lambda] & \xrightarrow{\ f_1\ } & \lambda^{-1}Y_1[[\lambda^{-1}]]
\end{array}
\tag{8-1}
$$

It is clear that simulation is a transitive relation.

Next we introduce a division relation among transfer functions. Let $T \in \lambda^{-1}(U, Y)_F[[\lambda^{-1}]]$ and $T_1 \in \lambda^{-1}((U_1, Y_1)_F[[\lambda^{-1}]]$ be two transfer functions. Then we say that T *divides* T_1, written $T | T_1$, if there exist polynomial functions Φ, Ψ and Π in $(U, U_1)_F[\lambda]$, $(Y, Y_1)_F[\lambda]$, and $(U, Y)_F[\lambda]$ respectively, for which

$$
T = \Psi T_1 \Phi + \Pi
\tag{8-2}
$$

holds. Both relations, of simulation and division, are reflexive and transitive.

The following theorem relates simulation to the division relation among transfer functions.

Theorem 8-1 Let f and f_1 be two restricted input/output maps having finite dimensional realizations and let T and T_1 be their corresponding transfer functions. Then f is simulated by f_1 if and only if T divides T_1.

PROOF Assume $T|T_1$. As a consequence there exist Φ, Ψ and Π such that (8-2) holds. Define $F[\lambda]$-homomorphisms $\varphi: u[\lambda] \to U_1[\lambda]$ and $\psi: \lambda^{-1}Y_1[[\lambda^{-1}]] \to \lambda^{-1}Y[[\lambda^{-1}]]$ by $\varphi(u) = \Phi u$ and $\psi(y) = \pi_-(\Psi y)$. Then for $u \in U[\lambda]$

$$f(u) = \pi_-(Tu) = \pi_-((\Psi T_1\Phi + \Pi)u) = \pi_-(\Psi\pi_-(T_1\Phi u)) = \Psi f_1\varphi(u)$$

or $f|f_1$.

Conversely assume $f|f_1$ that is $f = \psi f_1\varphi$. Now every $F[\lambda]$-homomorphism $\varphi: U[\lambda] \to U_1[\lambda]$ is of the form $\varphi(u) = \Phi u$ for some $\Phi \in (U, U_1)_F[\lambda]$. As for ψ we restrict it to the range of f_1 which is a finitely generated torsion submodule of $\lambda^{-1}Y_1[[\lambda^{-1}]]$. By Corollary 4-16 there exists an extension $\bar{\psi}: \lambda^{-1}Y_1[[\lambda^{-1}]] \to \lambda^{-1}Y[[\lambda^{-1}]]$ which has the form $\bar{\psi}(y) = \pi_-(\Psi y)$ for some $\Psi \in (Y_1, Y)_F[\lambda]$. Clearly $f = \psi f_1\varphi = \bar{\psi}f_1\varphi$ and so for $u \in U[\lambda]$ we have

$$f(u) = \pi_-(Tu) = \pi_-(\Psi\pi(T_1\Phi u)) = \pi_-(\Psi T_1\Phi u)$$

and this implies (8-2).

Theorem 8-2 Let $f: U[\lambda] \to \lambda^{-1}Y[[\lambda^{-1}]]$ and $f_1: U_1[\lambda] \to \lambda^{-1}Y[[\lambda^{-1}]]$ be two restricted input/output maps having finite dimensional canonical factorizations $f = OR$ and $f_1 = O_1R_1$ through the $F[\lambda]$-modules X and X_1, respectively. Then there exists an injective $F[\lambda]$-homomorphism $\theta: X \to X_1$ which makes the diagram (8-3)

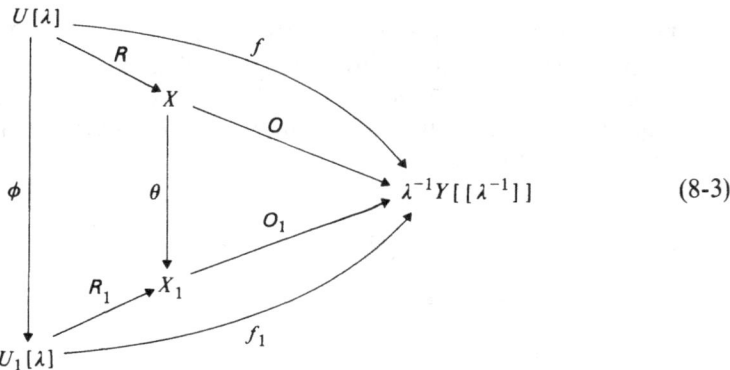

(8-3)

commutative if and only if

$$\text{Range } f \subset \text{Range } f_1 \tag{8-4}$$

PROOF Assume such a homomorphism θ exists. Since the factorizations of f and f_1 are canonical we have $\text{Range } O = \text{Range } f$ as well as $\text{Range } O_1 = \text{Range } f_1$. Since $O = O_1\theta$ it follows that $\text{Range } O \subset \text{Range } O_1$ and so (8-4) follows.

Conversely assume (8-4) holds and consider the homomorphisms $\hat{f}$ and $\hat{f}_1$ induced by f and f_1 in $U[\lambda]/\text{Ker } f$ and $U_1[\lambda]/\text{Ker } f_1$, respectively. Clearly $\hat{f}$ and $\hat{f}_1$ are injective and $\text{Range } \hat{f} \subset \text{Range } \hat{f}_1$. By Lemma 4-1(b) there exists, a necessarily injective, homomorphism $\hat{\varphi}: U[\lambda]/\text{Ker } f \to$

$U_1[\lambda]/\mathrm{Ker}\, f_1$ for which $\hat{f} = \hat{f}_1\hat{\varphi}$ and which, by Theorem 4-9, can be lifted to an $F[\lambda]$-homomorphism $\varphi: U[\lambda] \to U_1[\lambda]$ making the diagram (8-5) commutative.

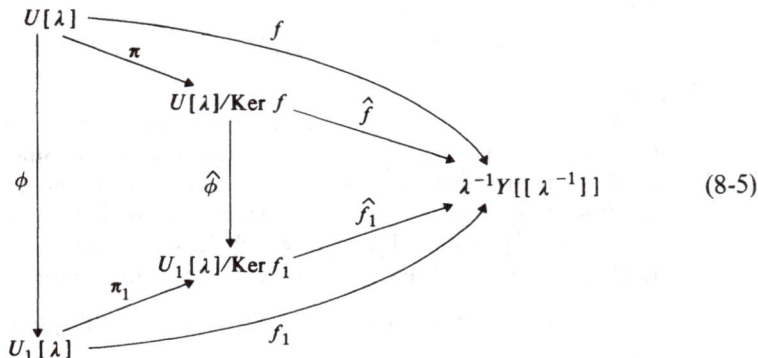

$$(8\text{-}5)$$

By Lemma 4-1 (*a*) there exists a uniquely determined $F[\lambda]$-homomorphism $\theta: X \to X_1$ which satisfies $R_1\varphi = \theta R$. This implies $O_1\theta R = O_1 R_1\varphi = f_1\varphi = f = OR$. As R is surjective we obtain $O_1\theta = O$ and from the injectivity of O it follows that θ too is injective.

The next theorem contains the dual result.

Theorem 8-3 Let $f: U[\lambda] \to \lambda^{-1}Y[[\lambda^{-1}]]$ and $f_1: U[\lambda] \to \lambda^{-1}Y_1[[\lambda^{-1}]]$ be two restricted input/output maps having finite dimensional canonical factorizations $f = OR$ and $f_1 = O_1 R_1$ through the $F[\lambda]$-modules X and X_1, respectively. Then there exists a surjective $F[\lambda]$-homomorphism $\Xi: X_1 \to X$ which makes the diagram (8-6)

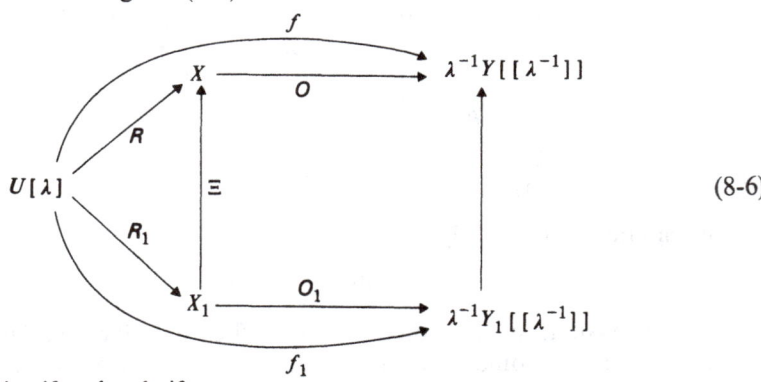

$$(8\text{-}6)$$

commutative if and only if

$$\mathrm{Ker}\, f_1 \subset \mathrm{Ker}\, f \tag{8-7}$$

PROOF Suppose such a homomorphism Ξ exists. Since $\Xi R_1 = R$ it follows that $\mathrm{Ker}\, R_1 \subset \mathrm{Ker}\, R$ which implies (8-7) by our assumption that the factorizations of f and f_1 are canonical.

Conversely assume (8-7) holds. Range f and Range f_1 are finitely generated torsion submodules of $\lambda^{-1}Y[[\lambda^{-1}]]$ and $\lambda^{-1}Y_1[[\lambda^{-1}]]$, respectively. By Lemma 4-1 (a) there exists an $F[\lambda]$-homomorphism $\hat{\psi}$: Range $f_1 \rightarrow$ Range f which satisfies $\psi f_1 = f$. By Theorem 4-15 $\hat{\psi}$ can be lifted to an $F[\lambda]$-homomorphism $\psi: \lambda^{-1}Y_1[[\lambda^{-1}]] \rightarrow \lambda^{-1}Y[[\lambda^{-1}]]$ which still satisfies $\hat{\psi}f_1 = f$. From this we obtain Range $\psi O_1 \subset$ Range O and as O is injective it follows from Lemma 4-1 (b) that there exists an $F[\lambda]$-homomorphism $\Xi: X_1 \rightarrow X$ for which $O\Xi = \psi O_1$. Finally $O\Xi R_1 = \psi O_1 R_1 = \psi f_1 = f = OR$ and by the injectivity of O the equality $\Xi R_1 = R$ follows. This proves the commutativity of diagram (8-6). Finally since R is surjective the equality $\Xi R_1 = R$ shows that Ξ must be surjective too.

As a corollary to the two preceding theorems we obtain the *state space isomorphism theorem* in two equivalent versions.

Theorem 8-4

(a) Let $f = OR$ and $f = O_1R_1$ be two finite dimensional canonical realizations of the restricted input/output map f with state modules X and X_1, respectively. Then there is an $F[\lambda]$-module isomorphism $\theta: X \rightarrow X_1$ which makes the diagram (8-8) commutative.

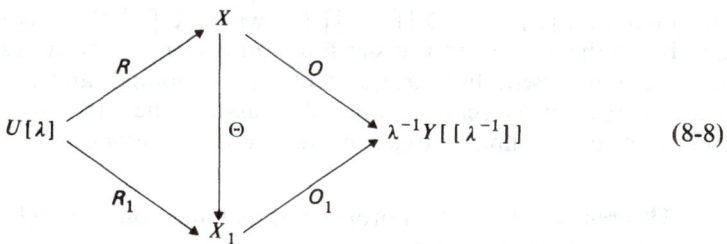

$$(8\text{-}8)$$

(b) Let (A, B, C) and (A_1, B_1, C_1) be two canonical realizations of the restricted input/output map f. Then there exists an invertible linear transformation P which makes the diagram (8-9) commutative.

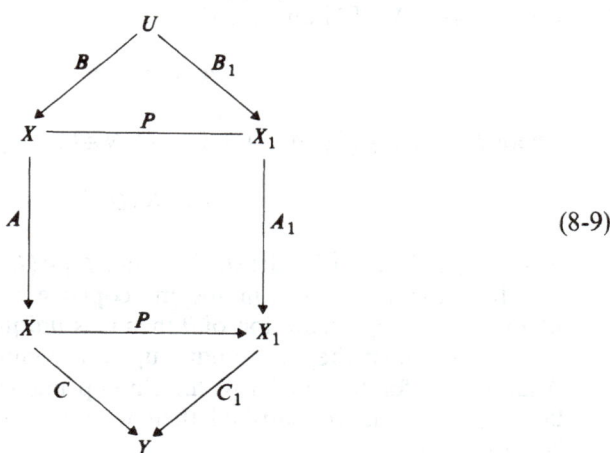

$$(8\text{-}9)$$

Proof

(a) By Theorems 8-2 and 8-3 there exist an injective homomorphism $\theta: X \to X_1$ and a surjective homomorphism $\Xi: X_1 \to X$ which satisfy $\Xi R = R_1$, $O_1\theta = O$, $\Xi R_1 = R$ and $O\Xi = O_1$. It follows from these that $\Xi\theta R = \Xi R_1 = R$ and by the surjectivity of R that $\Xi\theta = I_X$. Similarly $\theta\Xi R_1 = \theta R = R_1$ and so $\theta\Xi = I_{X_1}$. These two relations show that θ and Ξ are actually isomorphisms.

(b) We induce in X and X_1 an $F[\lambda]$-module structure in the normal way. Let R and R_1 be the reachability maps of the two realizations and O and O_1 their observability maps. By (i) there exists an $F[\lambda]$-homomorphism $\theta: X \to X_1$ for which $\theta R = R_1$ and $O = O_1\theta$. Let $P = \theta$ be considered as an F-linear map. The equality $PA = A_1P$ follows from the fact that θ is an $F[\lambda]$-homomorphism. The equalities $PB = B_1$ and $C = C_1P$ follow from $\theta R = R_1$ and $O = O_1\theta$, respectively.

9. TRANSFER FUNCTIONS AND THEIR FACTORIZATIONS

Let U and Y be finite dimensional vector spaces over F. We consider an input/output map $f: U[\lambda] \to \lambda^{-1}Y[[\lambda^{-1}]]$ for which $U[\lambda]/\mathrm{Ker}\, f$ is a torsion module and hence the associated transfer function is rational. Now scalar rational functions have representations as quotients of polynomials and we have the essential uniqueness of such a representation if we assume the numerator and denominator to be coprime. A similar situation exists in the general case.

Theorem 9-1 Let T be a proper rational function in $\lambda^{-1}(U, Y)_F[[\lambda^{-1}]]$ then T has the representations

$$T = \theta/\psi \tag{9-1}$$

where $\theta \in (U, Y)_F[\lambda]$ and $\psi \in F[\lambda]$

$$T = D^{-1}N \tag{9-2}$$

where $D \in (Y, Y)_F[\lambda]$, $\det D \neq 0$, and $N \in (U, Y)_F[\lambda]$, and

$$T = N_1 D_1^{-1} \tag{9-3}$$

where $D_1 \in (U, U)_F[\lambda]$, $\det D_1 \neq 0$, and $N_1 \in (U, Y)_F[\lambda]$.

If we assume ψ to be monic and coprime with the g.c.d. of the elements of any matrix representation of θ then ψ is unique. Similarly, if D and N are left coprime then they are unique up to a common left unimodular factor. Analogously for D_1 and N_1 with right coprimeness assumed. If the coprimeness assumptions are satisfied then we refer to (9-2) and (9-3) as *coprime factorizations*.

PROOF We consider the following sets

$$J = \left\{ \varphi \in F[\lambda] \,\middle|\, \varphi T \in (U, Y)_F[\lambda] \right\}$$

$$J_1 = \left\{ P \in (Y, Y)_F[\lambda] \,\middle|\, PT \in (U, Y)_F[\lambda] \right\}$$

and

$$J_R = \left\{ Q \in (U, U)_F[\lambda] \,\middle|\, TQ \in (U, Y)_F[\lambda] \right\}$$

Obviously J is an ideal in $F[\lambda]$, J_L a left ideal in $(Y, Y)_F[\lambda]$ and J_R a right ideal in $(U, U)_F[\lambda]$. We claim all three ideals are nontrivial. This is essentially equivalent to our definition of rationality. Thus, by principality of $F[\lambda]$, $J = \psi F[\lambda]$ for some ψ which is unique up to a constant factor. Thus $\psi T = \theta$ for some $\theta \in (U, Y)_F[\lambda]$ and hence (9-1). Now ψI belongs to J_L and J_R which are therefore full ideals. By Theorem 2-11 we have $J_L = (Y, Y)_F[\lambda] D$ and $J_R = D_1(U, U)_F[\lambda]$. Since J_L and J_R are full we have $\det D \neq 0$ and $\det D_1 \neq 0$. Thus $DT = N$ for some $N \in (U, Y)_F[\lambda]$ and (9-2) follows. The uniqueness fact follows essentially from Corollary 2-10. The statement about factorization (9-3) is proved analogously.

An alternative approach is to consider the restricted input/output map $f \; U[\lambda] \to \lambda^{-1} Y[[\lambda^{-1}]]$ given by $f(u) = \pi_-(Tu)$ for all $u \in U[\lambda]$. By the rationality of T, $X/\mathrm{Ker}\, f$ is a finitely generated torsion module. Hence by Theorem 2-12 $\mathrm{Ker}\, f = D_1 U[\lambda]$ for some nonsingular $D_1 \in (U, U)_F[\lambda]$. Therefore for each $u \in U[\lambda]$

$$f(D_1 u) = \pi_-(TD_1 u) = 0$$

This implies the existence of $N_1 \in (U, Y)_F[\lambda]$ such that $TD_1 = N_1$ which is equivalent to (9-3). Similarly we can consider Range $f = \{\pi_-(Tu) | u \in U[\lambda]\}$ as a submodule of $\lambda^{-1} Y[[\lambda^{-1}]]$. Now $\lambda^{-1} Y[[\lambda^{-1}]]$ is also left $(Y, Y)_F[\lambda]$-module with the composition $(A, y) \to \pi_-(Ay)$. The set $\bigcap_{y \in \mathrm{Range}\, f} \{A \in (Y, Y)_F[\lambda] \,|\, \pi_-(Ay) = 0\}$ is obviously a left ideal in $(Y, Y)_F[\lambda]$ and hence has a representation as $(Y, Y)_F[\lambda] D$ for some, necessarily nonsingular, D in $(Y, Y)_F[\lambda]$. Thus $\pi_-(DTu) = 0$ for all $u \in U[\lambda]$ which implies $DT = N$ and hence the factorization (9-2).

10. REALIZATION THEORY

While the abstract question of realization has been trivially solved the availability of the canonical models, the factorization of rational transfer functions, and the characterization of intertwining operators for our canonical models allow us to construct some explicit realizations and study their relations.

Thus let $T \in \lambda^{-1}(U, Y)_F[[\lambda^{-1}]]$ be rational and let (9-2) and (9-3) be factorizations of T.

Let K_D and $S(D)$ be defined by (2-7) (4-11), respectively. Define $B: U \to K_D$ by

$$(B\xi)(\lambda) = N(\lambda)\,\xi \tag{10-1}$$

and let $C: K_D \to Y$ be defined by

$$Cf = (D^{-1}f)_1 \qquad \text{for} \qquad f \in K_D \tag{10-2}$$

where $(D^{-1}f)(\lambda) = \sum_{n=1}^{\infty} (D^{-1}f)_n \lambda^{-n}$ is the formal expansion of $D^{-1}f$ which, by Lemma 2-15, belongs to $\lambda^{-1}Y[[\lambda^{-1}]]$.

Theorem 10-1 The system $(S(D), B, C)$ defined above is a realization of the function T which is observable. It is reachable if and only if $(D, N)_L = I$.

PROOF To begin, the map B is actually a map into K_D as $D^{-1}N\xi \in \lambda^{-1}Y[[\lambda^{-1}]]$ for all $\xi \in U$ by another application of Lemma 2-15. Let L denote the set of elements of the form $\sum S(D)^j N\xi_j$. L is the set of reachable states in K_D, it is a submodule of K_D and $L + DY[\lambda]$ is a submodule of $Y[\lambda]$. By Theorem 2-1 we have $L + DY[\lambda] = EY[\lambda]$ for some $E \in (Y, Y)_F[\lambda]$. Since $DY[\lambda] \subset EY[\lambda]$ we have $D = EG$ and as $N\xi \in EY[\lambda]$ for all $\xi \in U$ we have $N = EM$. Thus reachability is equivalent to the left coprimeness of D and N, that is to $(D, N)_L = I$. To show observability assume $f \in K_D$ and $CS(D)^n f = 0$ for all $n \geq 0$. This means that

$$(D^{-1}\pi_D \chi^n f)_1 = (D^{-1}D\pi_- D^{-1}\chi^n f)_1 = (\pi_- \chi^n D^{-1}f)_1 = 0$$

But this implies $(D^{-1}f)_n = 0$ for all n and hence $f = 0$. To show that we have actually a realization let $T(\lambda) = \sum_{i=0}^{\infty} \lambda^{-i-1} T_i$ be the formal expansion of T. It suffices to show that $CS(D)^i B = T_i$. Let $\xi \in U$ then

$$CS(D)^i B\xi = (D^{-1}\pi_D \chi^i N\xi)_1 = (D^{-1}D\pi_- D^{-1}\chi^i N\xi)_1 = (\pi_- d^i D^{-1}N\xi)_1$$

$$= (\pi_- \chi^i T\xi)_1 = T_i\xi$$

which proves the statement.

The second factorization of T gives rise to another realization. So assume $T = N_1 D_1^{-1}$. The equality $N_1 D_1^{-1} = D^{-1}N$ is equivalent to

$$ND_1 = DN_1 \tag{10-3}$$

Since $(D, N)_L = I$ and $(D_1, N_1)_R = I$ it follows from Theorem 4-11 that the map $X: K_{D_1} \to K_D$ given by

$$Xf = \pi_D Nf \qquad \text{for} \qquad f \in K_{D_1} \tag{10-4}$$

is an $F[\lambda]$-module isomorphism. Define now maps $B_1: U \to K_{D_1}$ and

$C_1 : K_{D_1} \to Y$ in such a way that the diagram

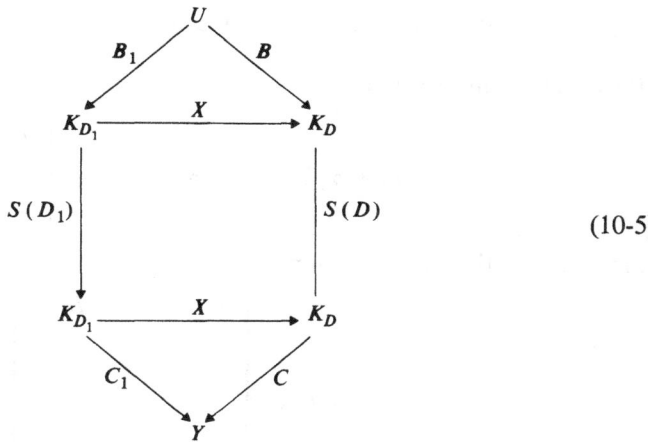

is commutative. The commutativity is equivalent to $XB_1 = B$ and $C_1 = CX$. We check that B_1 is given by

$$B_1 \xi = \pi_{D_1} \xi \tag{10-6}$$

for

$$XB_1 \xi = \pi_D N B_1 \xi = \pi_D N \pi_{D_1} \xi = \pi_D N \xi = N \xi = B \xi$$

Also for every $f \in K_{D_1}$ we have

$$C_1 f = CXf = (D^{-1} Xf)_1 = (D^{-1} \pi_D Nf)_1 = (D^{-1} D \pi_- D^{-1} Nf)_1 = (\pi_- Tf)_1$$

or

$$C_1 f = (\pi_- Tf)_1 \tag{10-7}$$

That $(S(D_1), B_1, C_1)$ is a canonical realization is clear from the invertibility of X and the commutativity of the diagram. This can be verified directly as

$$C_1 S(D_1)^n B_1 \xi = (\pi_- T \pi_{D_1} \chi^n \pi_{D_1} \xi)_1 = (\pi_- \chi^n T \xi)_1 = T_n \xi$$

Summarizing we obtained the following.

Theorem 10-2 The system $(S(D_1), B_1, C_1)$ defined above is a realization of the transfer function T which is reachable. It is observable if and only if $(D_1, N_1)_R = I$.

We consider two special cases. First, let $T = D^{-1} N$ and

$$N(\lambda) = N_0 + N_1 \lambda + \cdots + N_{k-1} \lambda^{k-1}$$

and

$$D(\lambda) = D_0 + D_1 \lambda + \cdots + D_{k-1} \lambda^{k-1} + I \lambda^k$$

It is easily checked that

$$K_D = \pi_D U^n[\lambda] = \{\alpha_0 + \alpha_1\lambda + \cdots + \alpha_{k-1}\lambda^{k-1} \mid \alpha_i \in U^n\}$$

If we make identification

$$\alpha_0 + \alpha_1\lambda + \cdots + \alpha_{k-1}\lambda^{k-1} \leftrightarrow \begin{bmatrix} \alpha_0 \\ \vdots \\ \alpha_{k-1} \end{bmatrix}$$

then we have the representation

$$S(D) \leftrightarrow \begin{bmatrix} 0 & & & & -D_0 \\ I & & & & \cdot \\ & \cdot & & & \cdot \\ & & \cdot & & \cdot \\ & & & \cdot & \cdot \\ & & & I & -D_{k-1} \end{bmatrix}$$

$$B \leftrightarrow \begin{bmatrix} N_0 \\ \vdots \\ N_{k-1} \end{bmatrix} \quad \text{and} \quad C \leftrightarrow (0 \cdots 0 I)$$

and for that reason we call the realization $(S(D), B, C)$ the *standard observable realization*. In the same fashion if $T = N_1 D_1^{-1}$ and

$$N_1(\lambda) = N_0' + \cdots + N_{k-1}'\lambda^{k-1}$$

and

$$D_1(\lambda) = D_0' + \cdots + D_{k-1}'\lambda^{k-1} + I\lambda^k$$

then with the same coordination of K_{D_1} we have

$$S(D_1) \leftrightarrow \begin{bmatrix} 0 & \cdot & \cdot & \cdot & \cdot & & -D_0' \\ I & & & & & & \cdot \\ & \cdot & & & & & \cdot \\ & & \cdot & & & & \cdot \\ & & & \cdot & & & \cdot \\ & & & & I & & -D_{k-1}' \end{bmatrix}$$

$$B_1 \leftrightarrow \begin{bmatrix} I \\ 0 \\ \cdot \\ \cdot \\ \cdot \\ 0 \end{bmatrix} \quad \text{and} \quad C_1 \leftrightarrow (T_0 \cdots T_{k-1})$$

We call the realization $(S(D_1), B_1, C_1)$ the *standard controllable realization*. The above construction should be compared for example with [15].

Given a rational transfer function $T \in \lambda^{-1}(U, Y)_F [[\lambda^{-1}]]$ we define the *McMillan degree* of T, denoted by $\delta(T)$, as the dimension of the state space of any canonical realization of T. The realization results of this section can be used to link the McMillan degree of T with the coprime factorizations of it.

Theorem 10-3 Let $T \in \lambda^{-1}(U, Y)_F [[\lambda^{-1}]]$ be rational and let it have the coprime factorizations (9-2) and (9-3). Then the McMillan degree of T is given by

$$\delta(T) = \deg(\det D) = \deg(\det D_1) \tag{10-8}$$

PROOF The realizations constructed in Theorem 10-1 and Theorem 10-2 are canonical and use K_D and K_{D_1} as state spaces, respectively. The dimensions of K_D and K_{D_1} are given by $\deg(\det D)$ and $\deg(\det D_1)$, respectively.

Since the range of the input/output map induced by T has the same dimension as the range of the Hankel matrix induced by T we immediately obtain a characterization of the McMillan degree in terms of Hankel matrices.

Corollary 10-4 Let $T \in \lambda^{-1}(U, Y)_F [[\lambda^{-1}]]$ be rational then we have the equality

$$\delta(T) = \operatorname{rank} H_T \tag{10-9}$$

We note here two important properties of the McMillan degree.

Theorem 10-5
(a) Let T_1 and T_2 be rational elements of $\lambda^{-1}(U, Y)_F [[\lambda^{-1}]]$. Then

$$\delta(T_1 + T_2) \le \delta(T_1) + \delta(T_2) \tag{10-10}$$

(b) Let T_1 and T_2 be rational elements of $\lambda^{-1}(Y, Z)_F [[\lambda^{-1}]]$ and $\lambda^{-1}(U, Y)_F [[\lambda^{-1}]]$, respectively. Then

$$\delta(T_1 T_2) \le \delta(T_1) + \delta(T_2) \tag{10-11}$$

PROOF If X_1 and X_2 denote the state space of canonical realizations of T_1 and T_2, respectively, then $X_1 \oplus X_2$ can be taken as the state space of a not necessarily canonical realization of $T_1 + T_2$ as well as $T_1 T_2$ simply by joining two canonical realizations of T_1 and T_2 in parallel or in series. Hence the inequalities follow.

11. POLYNOMIAL SYSTEM MATRICES

The object of this section is to make contact with linear system theory as developed by Rosenbrock.

We have seen in Sec. 9 that any rational transfer function $T \in$

$\lambda^{-1}(U, Y)_F[[\lambda^{-1}]]$ has left and right coprime factorizations given by (9-2) and (9-3), respectively. And with each factorization there is a naturally associated state space realization.

We consider now more general factorizations. Assume the transfer function T has a representation of the form

$$T = VD^{-1}W + Q \tag{11-1}$$

where $W \in (U, X)_F[\lambda]$, $D \in (X, X)_F[\lambda]$, $V \in (X, Y)_F[\lambda]$, and $Q \in (U, Y)_F[\lambda^-$. Here X is another finite dimensional vector space over F. No assumptions concerning the left coprimeness of D and W or the right coprimeness of D and V are made. With each factorization (11-1) of a transfer function T we associate a block matrix of the form

$$P = \begin{bmatrix} D & W \\ -V & Q \end{bmatrix} \tag{11-2}$$

and call such a matrix a *polynomial system matrix*.

In analogy with the constructions of the previous sections we use (11-1) as the basis for a state space realization of T. We take K_D as our state space and define operators $B: V \to K_D$ and $C: K_D \to Y$ by

$$B\xi = \pi_D W\xi \qquad \text{for} \qquad \xi \in U \tag{11-3}$$

and

$$Cf = (VD^{-1}f)_1 \qquad \text{for} \qquad f \in K_D \tag{11-4}$$

We claim that $(S(D), B, C)$ is a realization of T. So if $T(\lambda) = \sum_{n=0}^{\infty} T_n \lambda^{-n-1}$ we will show that $T_n = CS(D)^n B$. This follows from the following computation.

$$CS(D)^n B\xi = CS(D)^n \pi_D W\xi = C\pi_D \chi^n \pi_D W\xi$$

$$= (VD^{-1}\pi_D \chi^n \pi_D W\xi) = (VD^{-1}D\pi_- D^{-1}\chi^n W\xi)_1$$

$$= (V\pi_- D^{-1}W\chi^n\xi)_1 = (VD^{-1}W\chi^n\xi)_1 = (\chi^n T\xi)_1$$

$$= T_n\xi$$

We call the realization $(S(D), B, C)$ constructed above the *state space model* associated with the polynomial system matrix P.

From the previous section we know that $(S(D), B, C)$ is reachable if and only if $(D, W)_L = I$ and observable if and only if $(D, V)_R = I$.

Assume now $T = VD^{-1}W + Q = V_1 D_1^{-1}W_1 + Q_1$ are two different representations of the transfer function T. The dimensions of the spaces X and X_1 are not necessarily equal. To the two factorizations we associate two realizations $(S(D), B, C)$ and $(S(D_1), B_1, C_1)$, respectively, where B_1 and C_1 are defined by formulas analogous to (11-3) and (11-4), respectively.

Let us assume now that the two state space realizations are similar and study the effect of this on the relation between the two polynomial system matrices. By similarity there exists an invertible linear map $Z: K_D \to K_{D_1}$ for which $ZS(D) = S(D_1)Z$, $ZB = B_1$ and $C_1 Z = C$ hold. The structure of transformations that intertwine two canonical models is given by Theorem 4-9. Thus there exist M

and M_1 in $L(X, X_1)[\lambda]$ for which

$$MD = D_1 M_1 \qquad (11\text{-}5)$$

and Z is given by

$$Zf = \pi_{D_1} Mf \qquad \text{for} \qquad f \in K_D \qquad (11\text{-}6)$$

Since Z is assumed invertible we must have $(D_1, M)_L = I$ and $(D, M_1)_R = I$.

Next we consider the relation between the respective input and output maps. Since we have $ZB = B_1$ then for every $\xi \in U$

$$\pi_{D_1} W_1 \xi = \pi_{D_1} M \pi_D W \xi = \pi_{D_1} M W \xi \qquad (11\text{-}7)$$

The last equality follows from (11-5) which is equivalent to

$$MDX[\lambda] \subset D_1 X_1[\lambda] \qquad (11\text{-}8)$$

From (11-7) it follows that

$$\pi_{D_1}(W_1 - MW)\xi = 0 \qquad \text{for all} \qquad \xi \in U \qquad (11\text{-}9)$$

and this implies the existence of an $L_1 \in (U, X_1)_F[\lambda]$ such that

$$W_1 - MW = D_1 L_1 \qquad (11\text{-}10)$$

or

$$W_1 = MW + D_1 L_1 \qquad (11\text{-}11)$$

Similarly we have $C_1 Z = C$ and more generally

$$CS(D)^n = C_1 S(D_1)^n Z \qquad (11\text{-}12)$$

Therefore we get for $f \in K_D$

$$(VD^{-1}\pi_D \chi^n f)_1 = (V_1 D_1^{-1}\pi_{D_1}\chi^n \pi_{D_1} Mf)_1$$

or

$$(VD^{-1}\chi^n f)_1 = (V_1 \pi_- D_1^{-1} M\chi^n f)_1$$

which implies in turn

$$((V - V_1 M_1)D^{-1}\chi^n f)_1 = 0 \qquad \text{for all} \qquad n \geq 0$$

Hence $(V - V_1 M_1)D^{-1}$ is necessarily equal to some $K \in (X, Y)_F[\lambda]$. For this K we have

$$V - V_1 M_1 = KD \qquad (11\text{-}13)$$

It is clear that equalities (11-5), (11-11), and (11-13) are equivalent to the matrix equality

$$\begin{bmatrix} M & 0 \\ K & I \end{bmatrix} \begin{bmatrix} D & W \\ -V & Q \end{bmatrix} = \begin{bmatrix} D_1 & W_1 \\ -V_1 & Q_1 \end{bmatrix} \begin{bmatrix} M_1 & -L_1 \\ 0 & I \end{bmatrix} \qquad (11\text{-}14)$$

That $KW + Q = V_1 L_1 + Q_1$ follows from the fact that

$$KW - V_1 L_1 = (V - V_1 M_1) D^{-1} W - V_1 D_1^{-1} (W_1 - MW)$$
$$= VD^{-1}W - V_1 M_1 D^{-1} W - V_1 D_1^{-1} W_1 + V_1 D_1^{-1} MW = Q_1 - Q$$

Here we use the fact that

$$VD^{-1}W + Q = V_1 D_1^{-1} W_1 + Q_1 = T$$

whereas the equality

$$V_1 M_1 D^{-1} W = V_1 D_1^{-1} MW$$

is equivalent to (11-5).

The converse result holds also. If two polynomial system matrices

$$\begin{bmatrix} D & W \\ -V & Q \end{bmatrix} \quad \text{and} \quad \begin{bmatrix} D_1 & W_1 \\ -V_1 & Q_1 \end{bmatrix}$$

are connected via (11-14) with the coprimeness conditions $(M, D_1)_L = I$ and $(M_1, D)_R = I$ holding then the two respective state space models are similar.

This motivates the following definition.

Definition 11-1 Let two polynomial system matrices

$$P = \begin{bmatrix} D & W \\ -V & Q \end{bmatrix} \quad \text{and} \quad P_1 = \begin{bmatrix} D_1 & W_1 \\ -V_1 & Q_1 \end{bmatrix}$$

be given. We say P and P_1 are *strictly system equivalent* if (11-14) holds together with the coprimeness conditions $(M, D_1)_L = I$ and $(M_1, D)_R = I$.

As a direct consequence of the previous discussion and definition we have

Theorem 11-2 Two polynomial matrices are strictly system equivalent if and only if their associated state space models are similar.

We want to remark that as similarity is an equivalence relation it follows that strict system equivalence is also an equivalence relation.

12. GENERALIZED RESULTANT THEOREM

Throughout this chapter the coprimeness of polynomial matrices has played a central role. Thus it seems appropriate to give some effective ways for determining the left or right coprimeness of two polynomial matrices.

A classical result of Sylvester gives a simple criterion, in terms of the nonsingularity of the resultant matrix, for the coprimeness of two polynomials. As motivation for the more general results of this section we review the classical result.

Lemma 12-1 Let $p, q \in F[\lambda]$ then p and q are coprime if and only if

$$F[\lambda]/pqF[\lambda] = p\{F[\lambda]/qF[\lambda]\} + q\{F[\lambda]/pF[\lambda]\} \qquad (12\text{-}1)$$

We identify the quotient ring elements with their unique representative of lowest degree.

PROOF Assume p and q are coprime, then for every $f \in F[\lambda]$ there exist $a, b \in F[\lambda]$ for which $f = ap + bq$. Thus

$$f \bmod (pq) = p(a \bmod q) + p(b \bmod p)$$

or

$$F[\lambda]/pqF[\lambda] \subset p\{F[\lambda]/qF[\lambda]\} + q\{F[\lambda]/pF[\lambda]\}$$

The converse inclusion holds by a dimensionality argument. Conversely, assume now the equality (12-1). In particular there exist polynomials a and b such that $1 = ap + bq$ but this is equivalent to the coprimeness of p and q.
Assume now that

$$p(\lambda) = p_0 p_1 \lambda + \cdots + p_n \lambda^n \quad \text{and} \quad q(\lambda) = q_0 + q_1 \lambda + \cdots + q_m \lambda^m \quad (12\text{-}2)$$

then $F[\lambda]/pF[\lambda]$ is isomorphic to $F_n[\lambda]$ the set of all polynomials of degree less that n with the multiplication being modulo p. Similarly $F[\lambda]/qF[\lambda]$ is isomorphic to $F_m[\lambda]$. The following follows easily from Lemma 12-1.

Corollary 12-2 Let the polynomials p and q in $F[\lambda]$ be given by (12-2). Then p and q are coprime if and only if

$$F_{m+n}[\lambda] = pF_m[\lambda] + qF_n[\lambda]$$

Theorem 12-3 Let p and q be given by (12-2) then p and q are coprime if and only if $\det R(p, q) \neq 0$ where $R(p, q)$ is the resultant matrix

$$R(p, q) = \left.\begin{bmatrix} p_0 \cdot & \cdot & \cdot & p_n \cdot & \cdot & \cdot & 0 \\ & \cdot & & & \cdot & & \\ & & \cdot & & & \cdot & \\ & & & p_0 \cdot & \cdot & \cdot & \cdot & p_n \\ q_0 \cdot & \cdot & \cdot & \cdot & q_m & & \\ & \cdot & & & \cdot & & \\ & & \cdot & & & \cdot & \\ & & \cdot & q_0 \cdot & \cdot & \cdot & \cdot & q_m \end{bmatrix}\right\} \begin{matrix} m \\ \\ \\ \\ n \\ \\ \\ \end{matrix} \qquad (12\text{-}3)$$

PROOF By Corollary 12-2 p and q are coprime if and only if the set

$$B = \{\chi^i p \mid i = 0, \ldots, m - 1\} \cup \{\chi^i q \mid j = 0, \ldots, n - 1\}$$

is a basis for $F_{m+n}[\lambda]$. In terms of the polynomial coefficients this is equivalent to $\det R(p, q) \neq 0$.

A somewhat similar criterion for the coprimeness of two polynomials p and q follows directly from Theorem 4-8.

Theorem 12-4 Given p and q in $F[\lambda]$ then p and q are coprime if and only if $\det p(Q) \neq 0$ where Q is the companion matrix of q.

PROOF By Theorem 4-8 p and q are coprime if and only if $p(S(q))$ is invertible. But $S(q)$ is similar to Q and hence the result.

This result involves calculating a determinant of lower order than the resultant. In view of Theorem 10-1 it has an obvious interpretation in terms of controllability.

We now pass to the generalized result. Let D_1 and D_2 be two nonsingular polynomial matrices in $F^{n \times n}[\lambda]$ and let $M_i = D_i F^n[\lambda]$ be the corresponding full submodules. Define M by $M = M_1 \cap M_2$ so M is also a full submodule and hence has a representation $M = DF^n[\lambda]$ for some nonsingular D. Since $M \subset M_i$ there exist polynomial matrices E_i for which the equalities

$$D = D_1 E_1 = D_2 E_2 \tag{12-4}$$

hold.

Theorem 12-5

(a) The polynomial matrices D_1 and D_2 are left coprime if and only if the equality

$$\det D = \det D_1 \cdot \det D_2 \tag{12-5}$$

holds up to a constant factor on one side. The left coprimeness of D_1 and D_2 implies the right coprimeness of E_1 and E_2 in (12-4).

(b) The equality

$$F^n[\lambda]/DF^n[\lambda] = D_1\{F^n[\lambda]/E_1 F^n[\lambda]\} + D_2\{F^n[\lambda]/E_2 F^n[\lambda]\} \tag{12-6}$$

holds if and only if D_1 and D_2 are left coprime. That this generalizes the resultant theorem is obvious from a comparison with Lemma 12-1.

PROOF Suppose D_1 and D_2 are left coprime. By Theorem 2-8 there exist polynomial matrices G_1 and G_2 such that $I = D_1 G_1 + D_2 G_2$. Therefore every $f \in F^n[\lambda]$ has a representation

$$f = D_1 G_1 f + D_2 G_2 f = D_1 f_1 + D_2 f_2$$

If we apply the projection π_D of $F^n[\lambda]$ onto K_D and use the equalities (12-4) then

$$\pi_D f = D\pi_- D^{-1} f = D_1 E_1 \pi_- E_1^{-1} D_1^{-1} D_1 f_1 + D_2 E_2 \pi_- E_2^{-1} D_2^{-1} D_2 f_2$$

$$= D_1 \pi_{E_1} f_1 + D_2 \pi_{E_2} f_2$$

Therefore we get the inclusion $K_D \subset D_1 K_{E_1} + D_2 K_{E_2}$. To prove the converse inclusion it suffices, by symmetry, to show that $D_1 K_{E_1} \subset K_D$ and hence the equality (12-6) is proved. From the proof it is clear that the inclusion

$$D_1 K_{E_1} + D_2 K_{E_2} \subset K_D \tag{12-7}$$

holds always.

We consider the rational function $D_2^{-1} D_1$ which to begin with we assume to be proper. By Theorem 9-1 there exist polynomial matrices F_1 and F_2 which are right coprime and for which

$$D_2^{-1} D_1 = F_2 F_1^{-1} \tag{12-8}$$

which is equivalent to

$$D_1 F_1 = D_2 F_2 \tag{12-9}$$

Since clearly $D_1 F_1 F''[\lambda] \subset D_1 F''[\lambda]$ and $D_2 F_2 F''[\lambda] \subset D_2 F''[\lambda]$ it follows from (12-9) that

$$D_i F_i F''[\lambda] \subset D_1 F''[\lambda] \cap D_2 F''[\lambda] = DF''[\lambda]$$

Thus for some polynomial matrix G we have

$$D_1 F_1 = D_2 F_2 = DG$$

or

$$DG = D_1 E_1 G = D_2 E_2 G$$

and hence $F_1 = E_1 G$ and $F_2 = E_2 G$. But F_1 and F_2 are assumed to be right coprime and hence necessarily G is unimodular. The unimodularity of G now implies also the right coprimeness of E_1 and E_2. We recall that we assume $D_2^{-1} D_1$ to be a proper rational matrix. We apply now the realization theory developed in Sec. 10 to deduce the similarity of $S(D_2)$ and $S(E_1)$. This in turn implies the equivalence of D_2 and E_1 and hence in particular the equality

$$\det D_2 = \det E_1 \tag{12-10}$$

holds. Using (12-10) and (12-4) the equality (12-5) follows.

To prove the converse half of the theorem we assume D_1 and D_2 to have a nontrivial greatest common left divisor L. L is determined only up to a right unimodular matrix. Thus we have

$$D_1 = LC_1 \quad \text{and} \quad D_2 = LC_2 \tag{12-11}$$

and C_1, C_2 are left coprime. Now

$$D_1 F''[\lambda] \cap D_2 F''[\lambda] = L\{C_1 F''[\lambda] \cap C_2 F''[\lambda]\} = LD'F''[\lambda] = DF''[\lambda]$$

and

$$\det D' = \det C_1 \cdot \det C_2$$

Clearly

$$\det D = \det L \cdot \det D' = \det L \cdot \det C_1 \cdot \det C_2 \neq \det D_1 \cdot \det D_2$$

Similarly the equality (12-6) cannot hold by a dimensionality argument. As linear spaces the dimension of $F^n[\lambda]/Df^n[\lambda]$ is equal to the degree of the polynomial $\det D = \det L \cdot \det D'$ whereas the degree of $D_1\{F^n[\lambda]/E_1 F^n[\lambda]\} + D_2\{F^n[\lambda]/E_2 F^n[\lambda]\}$ is equal to the degree of $\det D'$. Since L is not unimodular there cannot be equality.

We indicate now how to remove the restriction that $D_2^{-1}D_1$ is proper rational. Let π_- be the projection of $F^{n\times n}((\lambda^{-1}))$ onto $\lambda^{-1}F^{n\times n}[[\lambda^{-1}]]$. Then we define Δ_1 by $\Delta_1 = D_2\pi_-D_2^{-1}D_1$. Obviously $D_2^{-1}\Delta_1$ is proper rational and $D_1 = \Delta_1 + D_2R$ for some R in $F^{n\times n}[\lambda]$. Since the conditions $(D_1, D_2)_L = I$ and $(\Delta_1, D_2)_L = I$ are equivalent there exist F_1 and F_2 satisfying $(F_1, F_2)_R = I$ and $D_2^{-1}\Delta_1 = F_2F_1^{-1}$. From the first part of the proof we have $\det D_2 = \det F_1$. Now $\Delta_1 F_1 = D_2 F_2$ implies $D_1 E_1 = D_2 E_2$ where $E_1 = F_1$ and $E_2 = F_2 + RF_1$, and the right coprimeness of E_1 and E_2 follows from that of F_1 and F_2.

13. FEEDBACK

We conclude this chapter by a short study of feedback, feedback equivalence, and canonical forms obtained through the use of feedback.

Let (A, B) be a reachable pair with $A \in (V, V)_F$ and $B \in (U, V)_F$ where U and V are finite dimensional vector spaces over F.

Suppose we augment the dynamical equation

$$x_{\nu+1} = Ax_\nu + Bu_\nu \tag{13-1}$$

by the identity readout map

$$y_\nu = x_\nu \tag{13-2}$$

then the transfer function of the triple (A, B, I) is given by

$$T(\lambda) = (\lambda I - A)^{-1}B \tag{13-3}$$

If we require the input to be a linear combination of a new input and the state, that is, we put

$$u_\nu = Kx_\nu + w_\nu \tag{13-4}$$

then the dynamic equation (13-1) is replaced by

$$x_{\nu+1} = (A + BK)x_\nu + Bw_\nu \tag{13-5}$$

Relation (13-4) is called a *feedback law*. We say that the pair $(A + BK, B)$ has been obtained from (A, B) by state *feedback*. Clearly the applications of feedback form a commutative group. If we enlarge this group to the one generated by similarity transformations in U and V as well as state feedback we obtain the noncommutative feedback group $\mathscr{F}$. Thus an element of $\mathscr{F}$ is a triple of maps

(R, K, P) with $R \in (V, V)_F$ and $P \in (U, U)_F$ nonsingular and $K \in (V, U)_F$. The feedback group $\mathscr{F}$ acts on a pair (A, B) by

$$(A, B) \xrightarrow{(R, K, P)} (R^{-1}AR + R^{-1}BK, R^{-1}BP) \tag{13-6}$$

This implies that the group composition law is

$$(R, K, P) \circ (R_1, K_1, P_1) = (RR_1, PK_1 + KR_1, PP_1) \tag{13-7}$$

and is associative as it can be expressed in terms of matrix multiplications as follows

$$\begin{pmatrix} R & O \\ K & P \end{pmatrix} \begin{pmatrix} R_1 & O \\ K_1 & P_1 \end{pmatrix} = \begin{pmatrix} RR_1 & O \\ KR_1 + PK_1 & PP_1 \end{pmatrix} \tag{13-8}$$

From (13-8) it also follows that

$$(R, K, P)^{-1} = (R^{-1}, -P^{-1}KR^{-1}, P^{-1}) \tag{13-9}$$

which shows that $\mathscr{F}$ is a bona fide group.

From the matrix representation of the feedback group it follows that every element of $\mathscr{F}$ is the product of elements of three basic types, namely

(a) similarity, or change of basis in the state space,
(b) similarity or change of basis in the input space, and finally
(c) pure feedbacks.

This is clear from

$$(R, K, P) = (R, O, I) \circ (I, K, I) \circ (I, O, P) \tag{13-10}$$

The feedback group $\mathscr{F}$ induces a natural equivalence relation in the set of reachable pairs (A, B) with state space and input space given by V and U, respectively. Thus (A, B) and (A_1, B_1) are *feedback equivalent* if there exists an element of $\mathscr{F}$ which transforms (A, B) into (A_1, B_1). It is easily checked that the relation of feedback equivalence is a proper equivalence relation. The equivalence classes are called orbits of the group and we are interested in a characterization of orbits, and of orbit invariants. Moreover we would like to obtain a canonical way of choosing one element in each orbit, a canonical form, which exhibits the orbit invariants.

The situation regarding the question of feedback equivalence of two reachable pairs (A, B) and (A_1, B_1) is analogous to the problem of deciding when two linear transformations A and A_1 are similar. By Corollary 4-12 A and A_1 are similar if and only if $\lambda I - A$ and $\lambda I - A_1$ are equivalent. By Corollary 3-2 and Lemma 2-20 (b) this is equivalent to $\lambda I - A$ and $\lambda I - A_1$ having the same invariant factor. This can be checked by bringing both $\lambda I - A$ and $\lambda I - A_1$ to their Smith canonical forms.

For the methods that will be applied in this section we will have to relax slightly the notion of feedback equivalence. Thus if (A_1, B_1) is another reachable pair with state and input spaces given by V_1 and U_1, respectively, we say that

(A_1, B_1) is feedback equivalent to (A, B) if there exist invertible maps $P: U_1 \rightarrow U$ and $R: V_1 \rightarrow V$ such that (RA_1R^{-1}, RB_1P^{-1}) is feedback equivalent to (A, B).

The feedback group has been introduced through state space formalism. However, we intend to study it through the use of canonical models, polynomial system matrices, and the realization procedures of Sec. 11. To this end we introduce a generalized control canonical form.

Given a reachable pair A, B then in the corresponding transfer function $(\lambda I - A)^{-1} B$ the factorization is left coprime. Associated with it is a right coprime factorization

$$(\lambda I - A)^{-1} B = H(\lambda) D(\lambda)^{-1} \tag{13-11}$$

where the factors H and D are uniquely determined up to a right unimodular factor. By Theorem 11-2 it follows that the pair $(S(D), \pi_D)$ is isomorphic to (A, B) and hence in the study of feedback we may as well start with the former.

The factor D in (13-11) is a nonsingular element of $(U, U)_F [\lambda]$ and hence has the representation

$$D(\lambda) = D_0 + D_1\lambda + \cdots + D_s\lambda^s \tag{13-12}$$

As before we denote by π_+ and π_- the canonical projections of $U((\lambda^{-1}))$ on $U[\lambda]$ and $\lambda^{-1}U[[\lambda^{-1}]]$, respectively. Then we define

$$U_s[\lambda^{-1}] = \pi_- \frac{1}{\lambda^s} U[\lambda] = \left\{ \sum_{j=1}^{s} \frac{\xi_j}{\lambda^j} \middle| \xi_j \in U \right\} \tag{13-13}$$

We clearly have the following direct sum decomposition

$$\lambda^{-1}U[[\lambda^{-1}]] = U_s[\lambda^{-1}] \oplus \frac{1}{\lambda^{s+1}} U[[\lambda^{-1}]] \tag{13-14}$$

For every

$$y \in \frac{1}{\lambda^{s+1}} U[[\lambda^{-1}]]$$

we have, with

$$y(\lambda) = \frac{1}{\lambda^{s+1}} y'(\lambda)$$

and $y' \in U[[\lambda^{-1}]]$ that

$$\pi_+ D(\lambda) y(\lambda) = \pi_+ \frac{D(\lambda)}{\lambda^{s+1}} y'(\lambda) = 0 \tag{13-15}$$

Hence to obtain all vectors in L_D, defined by (4-21), it suffices to consider the linear combinations of the vectors in $U_s[\lambda^{-1}]$.

For $1 \leq j \leq s$

$$\pi_+ \frac{D\xi}{\lambda^j} = (D_j + \cdots + D_s\lambda^{s-j}) \xi \tag{13-16}$$

Let us define now $s + 1$ polynomials in $(U, U)_F [\lambda]$ by

$$E_j(\lambda) = \begin{cases} 0, & j = 0 \\ D_j + D_{j+1}\lambda + \cdots + D_s\lambda^{s-j} & 1 \leq j \leq s \end{cases} \tag{13-17}$$

Equation (13-15) can be rewritten now as

$$\pi_+ \frac{D\xi}{\lambda^j} = E_j(\lambda)\,\xi, \quad 1 \leq j \leq s \tag{13-18}$$

and so

$$\pi^D \frac{\xi}{\lambda^j} = \pi_- D^{-1} E_j \xi \tag{13-19}$$

So for L_D we have the representation

$$L_D = \left\{ \sum_{j=1}^{s} \pi_- D^{-1} E_j \xi_j \,\middle|\, \xi_j \in U \right\} \tag{13-20}$$

Multiplication by D maps L_D onto K_D and, recalling the definition of the projection π_{L}, we obtain

$$K_D = \left\{ \sum_{j=1}^{s} \pi_D E_j \xi_j \,\middle|\, \xi_j \in U \right\} \tag{13-21}$$

We shall call representation (13-21) of K_D the *control representation* of K_D.

The usefulness of the control representation (13-21) of K_D becomes apparent in the study of the operator $S(D)$. Indeed we have the following result.

Theorem 13-1 Let $S(D): K_D \rightarrow K_D$ be defined by (4-11) and let E_j be defined by (13-17). Then

$$S(D)\,\pi_D E_j(\lambda)\,\xi = \pi_D E_{j-1}(\lambda)\,\xi - \pi_D D_{j-1}\xi \tag{13-22}$$

PROOF

$$S(D)\,\pi_D E_j(\lambda)\,\xi = \pi_D \lambda \pi_D E_j(\lambda)\,\xi = \pi_D \lambda E_j(\lambda)\,\xi = \pi_D \big(E_{j-1}(\lambda) - D_{j-1}\big)\xi$$

$$= \pi_D E_{j-1}(\lambda)\,\xi - \pi_D D_{j-1}\xi$$

For $j = 1$ we have of course

$$S(D)\,\pi_D E_1(\lambda) = -\pi_D D_0 \xi \tag{13-23}$$

In order to obtain some feeling for the preceding theorem let us specialize to the case of a degree s monic polynomial D. Thus $D(\lambda) = D_0 + \cdots + D_{s-1}\lambda^{s-1} + I\lambda^s$. In this case

$$\rho_D \pi^D \frac{\xi}{\lambda^j} = \pi_D E_j(\lambda)\,\xi = E_j(\lambda)\,\xi$$

This implies that

$$S(D) E_j(\lambda) \xi = E_{j-1}(\lambda) \xi - D_{j-1}\xi \tag{13-24}$$

Since K_D coincides with all vector polynomials of degree $s - 1$ then each such vector polynomial $u(\lambda)$ can be uniquely expressed in the form $u(\lambda) = \sum_{j=1}^{s} E_j(\lambda)\xi_j$. If we map bijectively K_D onto U^s by mapping $\sum_{j=1}^{s} E_j(\lambda)\xi_j$ into $\begin{pmatrix} \xi_1 \\ \vdots \\ \xi_s \end{pmatrix}$ we obtain for $S(D)$ the block matrix representation

$$\begin{pmatrix} 0 & I & 0 & \cdots & 0 \\ 0 & 0 & I & \cdots & \cdot \\ \cdot & \cdot & \cdot & \cdots & \cdot \\ 0 & 0 & 0 & \cdots & I \\ -D_0 & -D_1 & -D_2 & \cdots & -D_{n-1} \end{pmatrix} \tag{13-25}$$

But this is just the classical control canonical form for $S(D)$.

That D be monic is not necessary for $\pi_D E_j \xi = E_j \xi$ to hold. In fact we have the following.

Lemma 13-2 Let $D \in (U, U)_F [\lambda]$. If D^{-1} is proper then

$$\pi_D E_j \xi = E_j \xi \quad \text{for all} \quad \xi \in U \quad \text{and} \quad 1 \le j \le s \tag{13-26}$$

PROOF

$$\pi_D E_j \xi = D\pi_- D^{-1} E_j \xi = D\pi_- D^{-1} \frac{(D(\lambda) - (D_0 + \cdots + D_{j-1}\lambda^{j-1}))}{\lambda^j} \xi$$

$$= D\pi_- \frac{\xi}{\lambda^j} - D(\lambda)^{-1} \frac{(D_0 + \cdots + D_{j-1}\lambda^{j-1})}{\lambda^j} \xi$$

Now ξ/λ^j and

$$\frac{D_0 + \cdots + D_{j-1}\lambda^{j-1}}{\lambda^j}$$

are strictly proper whereas $D(\lambda)^{-1}$ is proper by assumption. Thus also

$$D(\lambda)^{-1} \frac{(D_0 + \cdots + D_{j-1}\lambda^{j-1})\xi}{\lambda^j}$$

is strictly proper and hence

$$\pi_- \frac{\xi}{\lambda^j} - D(\lambda)^{-1} \frac{(D_0 + \cdots + D_{j-1}\lambda^{j-1})\xi}{\lambda^j}$$

$$= \frac{\xi}{\lambda^j} - D(\lambda)^{-1} \frac{(D_0 + \cdots + D_{j-1}\lambda^{j-1})\xi}{\lambda^j}$$

and this implies (13-26).

The next theorem is a key result in the study of state feedback.

Theorem 13-3 Let $Q, D \in (U, U)_F [\lambda]$ with D nonsingular and QD^{-1} strictly proper. Let $D_1 = D + Q$ and let E_j and E'_j be the polynomials associated with D and D_1, respectively, that are defined by (13-17). Then the map X defined by

$$Xf = \pi_+ D_1 D^{-1} f \qquad \text{for} \qquad f \in K_D \tag{13-27}$$

is an invertible map of K_D onto K_{D_1} that satisfies

$$X\pi_D E_j \xi = \pi_{D_1} E'_j \xi \tag{13-28}$$

PROOF Assume $D(\lambda): D_0 + D_1\lambda + \cdots + D_s\lambda^s$. Since $D_1 D^{-1} = I + QD^{-1}$ with QD^{-1} strictly proper it follows that $D_1(\lambda) = D'_0 + D'_1\lambda + \cdots + D'_s\lambda^s$. Let $D_1 D^{-1}$ have the expansion

$$D_1(\lambda) D(\lambda)^{-1} = I + \frac{\Gamma_1}{\lambda} + \frac{\Gamma_2}{\lambda^2} + \cdots \tag{13-29}$$

or

$$D_1(\lambda) = \left(I + \frac{\Gamma_1}{\lambda} + \frac{\Gamma_2}{\lambda^2} + \cdots \right) D(\lambda) \tag{13-30}$$

By equating coefficients we obtain

$$D'_s = D_s$$

$$D'_{s-1} = D_{s-1} + \Gamma_1 D_s$$

$$\vdots$$

$$D'_0 = D_0 + \Gamma_1 D_1 + \cdots + \Gamma_s D_s \tag{13-31}$$

or in block matrix form

$$\begin{pmatrix} D'_0 \\ \vdots \\ \vdots \\ D'_s \end{pmatrix} = \begin{pmatrix} I & \Gamma_1 & \cdots & & \Gamma_s \\ 0 & I & \Gamma_1 & \cdots & \Gamma_{s-1} \\ & & I & \Gamma_1 & \\ 0 & & & & I \end{pmatrix} \begin{pmatrix} D_0 \\ \vdots \\ \vdots \\ D_s \end{pmatrix} \tag{13-32}$$

$$X\pi_D E_j \xi = \pi_+ D_1 D^{-1} \pi_D E_j \xi = \pi_+ D_1 D^{-1} D \pi_- D^{-1} E_j \xi = \pi_+ D_1 \pi_- D^{-1} E_j \xi$$

$$= \pi_{D_1} \pi_+ D_1 D^{-1} E_j \xi = \pi_{D_1} \pi_+ \left\{ \left(I + \frac{\Gamma_1}{\lambda} + \frac{\Gamma_2}{\lambda^2} + \cdots \right) E_j(\lambda) \xi \right\}$$

$$= \pi_{D_1} \pi_+ \left\{ \left(I + \frac{\Gamma_1}{\lambda} + \frac{\Gamma_2}{\lambda^2} + \cdots \right) (D_j + D_{j+1}\lambda + \cdots + D_s \lambda^{s-j}) \xi \right\}$$

$$= \pi_{D_1} \{ D_s \lambda^{s-i} + (D_{s-1} + \Gamma_1 D_s) \lambda^{s-j-1} + \cdots$$

$$+ (D_j + \Gamma_1 D_{j+1} + \cdots + \Gamma_{s-j} D_s) \} \xi$$

$$= \pi_{D_1} (D'_j + D'_{j+1} + \cdots + D'_s \lambda^{s-j}) \xi = \pi_{D_1} E'_j \xi$$

This shows, by the control representations of K_D and K_{D_1}, that X is a map of K_D onto K_{D_1}. If we define Y on K_{D_1} by

$$Yg = \pi_+ DD_1^{-1} g \qquad g \in K_{D_1} \tag{13-33}$$

then it is easily checked that for $f \in K_D$

$$YXf = \pi_+ DD_1^{-1} \pi_+ D_1 D^{-1} f = \pi_+ DD_1^{-1} (I - \pi_-) D_1 D^{-1} f$$

$$= \pi_+ DD_1^{-1} D_1 D^{-1} f - \pi_+ DD_1^{-1} \pi_- D_1 D^{-1} f$$

$$= f - \pi_+ DD_1^{-1} \pi_- D_1 D^{-1} f = f$$

as $DD_1^{-1} \pi_- D_1 D^{-1} f \in \lambda^{-1} U[[\lambda^{-1}]]$.

We conclude that X is also injective, hence invertible. Necessarily $X^{-1} = Y$.

The map X defined by (13-27) relates also the projections π_D and π_{D_1} in a simple way.

Lemma 13-4 Let D and D_1 be as in Theorem 13-3, and let $X: K_D \to K_{D_1}$ be defined by (13-27). Then for every $P \in (U, U)_F$ and $\xi \in U$ we have

$$X\pi_D P\xi = \pi_{D_1} P\xi \tag{13-34}$$

PROOF

$$X\pi_D P\xi = \pi_+ D_1 D^{-1} \pi_D P\xi = \pi_+ D_1 D^{-1} D\pi_- D^{-1} P\xi$$

$$= \pi_+ D_1 \pi_- D^{-1} P\xi = \pi_{D_1} \pi_+ D_1 D^{-1} P\xi = \pi_{D_1} P\xi$$

As a corollary to Theorem 13-3 we can state the following result.

Theorem 13-5 With the notation of Theorem 4-3 the operator $\bar{X}: U[\lambda] \to U[\lambda]$ defined by

$$\bar{X}f = \pi_+ D_1 D^{-1} f \qquad \text{for} \qquad f \in U[\lambda] \tag{13-35}$$

is an invertible map in $U[\lambda]$.

PROOF We clearly have the direct sum decompositions $U[\lambda] = K_D \oplus DU[\lambda] = K_{D_1} \oplus D_1 U[\lambda]$. We saw that X maps K_D bijectively onto K_{D_1}. Moreover it clearly maps $DU[\lambda]$ bijectively onto $D_1 U[\lambda]$ and hence is invertible.

The following theorem enables us to study the effect of feedback transformation in terms of the coprime factorization of transfer functions and related polynomial system matrices. The result is due to Hautus and Heymann.

Theorem 13-6 Let (A, B), with $A \in (V, V)_F$ and $B \in (U, V)_F$, be a reachable pair and let $H(\lambda) D(\lambda)^{-1}$ be a right coprime factorization of $(\lambda I - A)^{-1} B$. Then a necessary and sufficient condition for a reachable pair (A_1, B_1) to be feedback equivalent to (A, B) is that

$$(\lambda I - A_1)^{-1} B_1 = R H(\lambda) \big(D(\lambda) + Q(\lambda) \big)^{-1} P^{-1} \qquad (13\text{-}36)$$

for some $Q \in (U, U)_F [\lambda]$ for which QD^{-1} is strictly proper and invertible maps R and P in $(V, V)_F$ and $(U, U)_F$, respectively.

PROOF Assume $T(\lambda) = (\lambda I - A)^{-1} B = H(\lambda)^{-1} D(\lambda)$ are coprime factorizations, and let (A_1, B_1) be feedback equivalent to (A, B). Thus there exist invertible maps $R: V \to V$ and $P: U \to U$ such that

$$A_1 = R(A + BK) R^{-1} \qquad \text{and} \qquad B_1 = RBP^{-1}$$

Hence

$$(\lambda I - A_1)^{-1} B_1 = \big(R(\lambda I - A - BK)^{-1} R^{-1} \big) RBP$$

$$= R(\lambda I - A - BK)^{-1} BP^{-1}$$

Now

$$(\lambda I - A - BK)^{-1} B = \big[(\lambda I - A)\big(I - (\lambda I - A)^{-1} BK\big) \big]^{-1} B$$

$$= \big(I - (\lambda I - A)^{-1} BK \big)(\lambda I - A)^{-1} B$$

$$= \big(I - T(\lambda) K \big)^{-1} T(\lambda)$$

But from the equality

$$T(\lambda)\big(I - KT(\lambda) \big) = \big(I - T(\lambda) K \big) T(\lambda)$$

it follows that

$$\big(I - T(\lambda) K \big)^{-1} T(\lambda) = T(\lambda)\big(I - KT(\lambda) \big)^{-1}$$

Hence it follows that

$$T_f(\lambda) = (\lambda I - A_1)^{-1} B_1 = R T(\lambda)\big(I - KT(\lambda) \big)^{-1} P^{-1}$$

$$= R H(\lambda) D(\lambda)^{-1} \big(I - K H(\lambda) D(\lambda)^{-1} \big)^{-1} P^{-1}$$

$$= R H(\lambda) \big(D(\lambda) - K H(\lambda) \big)^{-1} P^{-1}$$

If we put $Q(\lambda) = -KH(\lambda)$ then clearly $T_f(\lambda) = RH(\lambda)(D(\lambda) + Q(\lambda))^{-1} P^{-1}$ and $QD^{-1} = -KT$ is strictly proper.

This proves the necessity part of the theorem.

To prove sufficiency it suffices to show that $H(\lambda)(D(\lambda) + Q(\lambda))^{-1} = (\lambda I - A_1)^{-1} B_1$ for a pair (A_1, B_1) which is feedback equivalent to (A, B). In that case $RH(\lambda)(D(\lambda) + Q(\lambda))^{-1} P^{-1}$ is associated with (RA_1R^{-1}, RE_1P^{-1}).

Let $D_1 = D + Q$ then by Theorem 13-3 the map $X: K_D \to K_{D_1}$ defined by (13-27) is invertible and its inverse $Y = X^{-1}$ is given by (13-33).

The realization procedure of Sec. 11 associates with the factorizations HD^{-1} and HD_1^{-1} realizations in which the state and input operators are $(S(D), \pi_D)$ and $(S(D_1), \pi_{D_1})$, respectively. Thus for our purposes we have to show that $(S(D), \pi_D)$ and $(S(D_1), \pi_{D_1})$ are feedback equivalent, or that for some invertible map $Y: K_{D_1} \to K_D$ and $K: K_D \to U$ we have

$$S(D) - YS(D_1) Y^{-1} = BK \tag{13-37}$$

where $B: U \to K_D$ is given by $B\xi = \pi_D\xi$ for $\xi \in U$. Clearly (13-37) is equivalent to

$$S(D) Y - YS(D_1) = BK_1 \tag{13-38}$$

By Lemma 4-1, applied with vector space structure, it suffices to show that

$$\text{Range}(S(D) Y - YS(D_1)) \subset \text{Range} B \tag{13-39}$$

and it is this we shall prove.

From the control representations of K_{D_1} and K_D we know that they are spanned by vectors of the form $\pi_{D_1} E_j'\xi$ and $\pi_D E_j\xi$, respectively. Thus it suffices to show (13-39) for vectors of this form. Using (13-28), (13-34), and (13-22) we have

$$
\begin{aligned}
(S(D) Y - YS(D_1)) \pi_{D_1} E_j'\xi &= S(D) Y\pi_{D_1} E_j'\xi - YS(D_1) \pi_{D_1} E_j'\xi \\
&= S(D) \pi_D E_j\xi - Y\{\pi_{D_1} E_{j-1}'\xi - \pi_{D_1} D_{j-1}'\xi\} \\
&= \{\pi_D E_{j-1}\xi - \pi_D D_{j-1}\xi\} \\
&\quad - \{\pi_D E_{j-1}\xi - \pi_D D_{j-1}'\xi\} \\
&= -\pi_D(D_{j-1} - D_{j-1}')\xi
\end{aligned}
$$

which proves the assertion.

At this point it will be convenient to fix bases in U and V, respectively. Thus we can identify U and V with F^m and F^n, respectively, and A and B will denote $n \times n$ and $n \times m$ matrices, respectively.

Without loss of generality we will assume B to be injective, that is, of full column rank. Now with the reachable pair $(S(D), \pi_D)$ we associate the polynomial matrix $(D \quad I)$ which is just the upper row of Rosenbrock's polynomial system matrix. Using Theorem 13-6 and previously obtained results an invertible module homomorphism between canonical models, namely, Theorems 4-9 and 4-11, we can state the following

Theorem 13-7 Let D, Q, N, and M be $m \times m$ polynomial matrices with D nonsingular, QD^{-1} strictly proper and N and M unimodular. Let P be an invertible constant $m \times m$ matrix. Then $(D \quad I)$ and $(N(D + Q)M \quad NP)$ are associated with feedback equivalent pairs.

PROOF We consider the following special cases of Theorems 4-9 and 4-11.

$$D_1 = ND \tag{13-40}$$

and

$$D_1 = DM \tag{13-41}$$

Let the maps $X: K_D \to K_{D_1}$ and $Y: K_D \to K_{D_1}$ be defined by

$$Xf = \pi_{D_1} Nf \tag{13-42}$$

and

$$Yf = \pi_{D_1} f \tag{13-43}$$

respectively. Then both maps are invertible as the required coprimeness conditions are trivially satisfied. To see how the input operators are transformed we check that

$$X\pi_D \xi = \pi_{D_1} N\pi_D \xi = \pi_{D_1} N \xi$$

whereas

$$X\pi_D \xi = \pi_{D_1} \pi_D \xi = \pi_{D_1} \xi$$

In terms of polynomial matrices we have the equivalence of $(D \quad I)$ with either $(DM \quad I)$ or $(ND \quad N)$. Combining this with Theorem 13-6 the result follows.

We recall that left (right) multiplication by a unimodular matrix is equivalent to a finite series of elementary row (column) operations. Our next object is to use the freedom of Theorem 13-7 to reduce $(D \quad I)$ to canonical form.

To this end we introduce the notion of column properness. Let $D(\lambda)$ be an $m \times m$ nonsingular polynomial matrix with columns $d^{(1)}(\lambda), \dots, d^{(m)}(\lambda)$. We define the degree of $d^{(i)}(\lambda)$, $\deg d^{(i)}(\lambda)$, to be the degree of the highest degree element in $d^{(i)}(\lambda)$. $D(\lambda)$ is called column proper if $\deg D(\lambda) = \sum_{i=1}^{m} \deg d^{(i)}(\lambda)$.

Theorem 13-8 Let D be an $m \times m$ nonsingular polynomial matrix. Then there exists a unimodular polynomial matrix M such that DM is column proper. If $f^{(i)}$ are the columns of DM we may assume without loss of generality that, with $\kappa_i = \deg f^{(i)}$, $\kappa_1 \geq \kappa_2 \geq \cdots \geq \kappa_m \geq 0$.

PROOF Given a nonsingular $m \times m$ polynomial matrix D we denote by D_h the constant matrix whose ith column $d_h^{(i)}$ consists of the coefficient of the terms of highest degree in the ith column of D. It is clear that D is column proper if and only if $\det D_h \neq 0$.

Now let $d^{(i)}$, $i = 1, \ldots, m$, be the columns of D. If D_h is nonsingular then D is column proper and there is nothing to prove. So we assume $\det D_h = 0$. Let $\delta_i = \deg d^{(i)}$ then clearly $\sum_{i=1}^{m} \delta_i \geq \deg \det D$. We will construct now a unimodular matrix M such that in DM the degree of one column is decreased, the rest remaining unchanged.

To this end we observe that if $\det D_h = 0$ there exist $\alpha_1, \ldots, \alpha_m$ in $\mathbb{F}$, not all zero such that

$$\sum_{i=1}^{m} \alpha_i d_h^{(i)} = 0 \tag{13-44}$$

Let j be an index for which $\alpha_j \neq 0$ and $\delta_i \leq \delta_j$. By dividing (13-44) by α_j we may assume without loss of generality that $\alpha_j = 1$. From equality (13-44) it follows now that $\sum_{i=1}^{m} \alpha_i \lambda^{\delta_j - \delta_i} d_h^{(i)} \lambda^{\delta_i} = 0$. Define a unimodular matrix M by

$$M(\lambda) = \begin{bmatrix} 1 & 0 & . & . & . & \alpha_1 \lambda^{\delta_j - \delta_1} & . & . & . & 0 \\ 0 & 1 & . & . & . & \alpha_2 \lambda^{\delta_j - \delta_2} & . & . & . & 0 \\ & & . & & & . & & & & \\ & & & . & & . & & & & \\ & & & & & 1 & & & & \\ . & . & . & . & & & & . & . & . & 0 \\ 0 & 0 & . & . & . & \alpha_m \lambda^{\delta_j - \delta_m} & . & . & . & 1 \end{bmatrix} \tag{13-45}$$

The columns of DM agree with those of D for $i \neq j$ whereas the jth column of DM is of lower degree than that of D. Since $\det DM = \det D$ we still have $\sum \delta_i^{(1)} \geq \deg \det D$ where $\delta_i^{(1)}$ are the degrees of the columns of DM. We proceed inductively until a column proper matrix is obtained.

As an immediate corollary we obtain the following simple derivation of feedback canonical forms, the first in terms of polynomial system matrices whereas the second is in terms of matrix representations. Both canonical forms are named *Brunovsky canonical forms*.

Theorem 13-9 Let D be a nonsingular $m \times m$ polynomial matrix with $n = \deg \det D$. Then there exist uniquely determined numbers

$$\kappa_1 \geq \cdots \geq \kappa_m \geq 0$$

with $\sum_{i=1}^{m} \kappa_i = n$ such that $(D \quad I)$ and $(\Delta \quad I)$, where

$$\Delta(\lambda) = \operatorname{diag}(\lambda^{\kappa_1}, \ldots, \lambda^{\kappa_m}) \tag{13-46}$$

are associated with the feedback equivalent pairs $(S(D), \pi_D)$ and $(S(\Delta), \pi_D)$, respectively.

PROOF By Theorems 13-7 and 13-8 we may assume without loss of generality that D is column proper with column degrees $\kappa_1 \geq \cdots \geq \kappa_m \geq 0$ and $\sum_{v=1}^{m} \kappa_i = n$. By left multiplication with a constant matrix P we can bring $(D \quad I)$ to $(PD \quad P)$ and PD has the form $\Delta + Q$ and the column degrees of Q are less than the corresponding column degrees of Δ. By the similarity F^{-1} in the input space $(\Delta + Q \quad P)$ is transformed into $(\Delta + Q \quad I)$. Finally, by state feedback $(\Delta + Q \quad I)$ is transformed into $(\Delta \quad I)$.

To prove the uniqueness of the numbers $\kappa_1, \ldots, \kappa_m$ assume $(S(\Delta), \pi_\Delta)$ and $(S(\Delta_1), \pi_{\Delta_1})$ are feedback equivalent with $\Delta_1 = \mathrm{diag}(\lambda^{\delta_1}, \ldots, \lambda^{\delta_m})$. By Theorem 13-6 we have $\Delta_1 = P(\Delta + Q)$ for some invertible matrix P. This in turn implies that $\Delta_1 \Delta^{-1} = P(I + Q\Delta^{-1})$ is proper. Since $\Delta_1 \Delta^{-1} = \mathrm{diag}(\lambda^{\delta_1 - \kappa_1}, \ldots, \lambda^{\delta_m - \kappa_m})$ we have $\kappa_j \geq \delta_j$ for $j = 1, \ldots, m$. Equality now follows by symmetry considerations.

The numbers $\kappa_1, \ldots, \kappa_m$ will be called the *reachability indices* of D. We proceed to show that this definition is in agreement with the common definition of the reachability indices.

Given the pair (A, B) we let $\mathcal{B} = \mathrm{Range}\, B$ and $\langle A | \mathcal{B} \rangle_i = \mathcal{B} + A\mathcal{B} + \cdots + A^{i-1}\mathcal{B}$. Let

$$\alpha_i = \begin{cases} \dim \mathcal{B} & \text{for } i > 1 \\ \dim \langle A | \mathcal{B} \rangle_i / \langle A | \mathcal{B} \rangle_{i-1} & \text{for } i = 1 \end{cases} \tag{13-47}$$

It is clear that, assuming $\mathcal{B}$ to be injective, $m = \alpha_1 \geq \alpha_2 \geq \cdots \geq \alpha_n$. By the Cayley–Hamilton theorem $\alpha_{n+k} = 0$ for $k > 0$. We define now the *reachability indices* of the pair (A, B) to be the set of numbers $\kappa_1, \ldots, \kappa_m$ defined by

$$\kappa_j = \mathrm{Card}\,\{\alpha_i | \alpha_i \geq j\} \tag{13-48}$$

It follows that $\kappa_1 \geq \kappa_2 \geq \cdots \geq \kappa_m$ and $\sum_{j=1}^{m} \kappa_j = n$. Since the α_j are $\mathcal{F}$-orbit invariants so are the κ_j. Now it is easy to check that the reachability indices of Δ, Δ being given by (13-46) coincide with the reachability indices of the pair $(S(\Delta), \pi_\Delta)$.

Corollary 13-10 Let A and B be $n \times n$ and $n \times m$ matrices, respectively. Assume the pair (A, B) is reachable, B injective and the reachability indices of (A, B) are $\kappa_1 \geq \cdots \geq \kappa_m$. Then (A, B) is feedback equivalent to the pair (A_c, B_c) where $A_c = \mathrm{diag}(A_1, \ldots, A_m)$, $B_c = \mathrm{diag}(b_1, \ldots, b_m)$ with

$$A_j = \begin{pmatrix} 0 & 0 & \cdot & \cdot & \cdot & 0 \\ 1 & 0 & & & & 0 \\ & 1 & & & & \cdot \\ & & \cdot & & & \\ & & & \cdot & & \cdot \\ & & & & \cdot & \\ & & & & \cdot & \cdot \\ & & & 1 & 0 & \end{pmatrix}_{\kappa_j \times \kappa_j} \tag{13-49}$$

and

$$b_j = \begin{pmatrix} 1 \\ 0 \\ \vdots \\ 0 \end{pmatrix}_{\kappa_j \times 1} \tag{13-50}$$

PROOF Let $H(\lambda) D(\lambda)^{-1}$ be a right coprime factorization of $(\lambda I - A)^{-1} B$. The pair (A, B) is isomorphic to $(S(D), \pi_D)$ and feedback equivalent to $(S(\Delta), \pi_D)$ where Δ is given by (13-40). Since Δ is diagonal we have

$$K_\Delta = K_{\lambda^{\kappa_1}} \oplus \cdots \oplus K_{\lambda^{\kappa_m}} \tag{13-51}$$

Let $e_1, \ldots, e_m$ be the standard basin in F^m then the vectors $\{\lambda^i e_j \mid j = 1, \ldots, m, i = 0, \ldots, \kappa_j - 1\}$ are a basis for K_Δ. Relative to these bases the pair $(S(\Delta), \pi_\Delta)$ has the matrix representation (A_c, B_c), and this is the *Brunovsky canonical form*.

NOTES AND REFERENCES

For the necessary algebraic background one can consult Jacobson [74, 75], Lang [80], MacLane and Birkhoff [88], or van der Waerden [119]. For matrix theory and polynomial matrices MacDuffee [86] is still useful. Gantmacher [55] is a comprehensive survey. The development of the structure theory for linear transformation in finite dimensional vector spaces follows [51] and is motivated by results in the theory of invariant subspaces that are discussed in the second chapter.

The sections on linear system theory have been written as an effort to bridge the gap between Kalman's stress on seeing linear time invariant systems as $F[\lambda]$-modules [76, 77], state space theory [15], and Rosenbrock's polynomial system matrices approach [100]. The material on linear system theory is fairly standard now. References include besides the previously mentioned books also the path-breaking Zadeh and Desoer [126], Wolovich [123], and Wonham [124]. The concepts of controllability and observability have been introduced by Kalman. Reference [76] contains a historical discussion as well as a comprehensive bibliography. The section on simulation has been motivated by [76] and is based on [53] which contains some applications. Coprime factorizations of rational transfer functions play a dominant role in Rosenbrock's theory. The use of coprimeness in the study of composite systems has been utilized in [19, 46].

The resultant of the two polynomials has been introduced by Sylvester [80, 119]. For generalizations and the use of resultants in system theory we refer to Barnett [9], Rowe [102], and Gohberg and Lerer [60]. A recent series of papers by Gohberg, Lancaster, and Rodman [57–59] contains a large number of results relevant to system theory.

OPERATORS IN HILBERT SPACE

1. GEOMETRY OF HILBERT SPACE

Hilbert space is going to provide the setting for most of the rest of this work. This section provides a quick introduction to the important results concerning the geometry of Hilbert spaces.

We define an *inner product space* to be a complex linear space H with a function $(\, , \,): H \times H \to \mathbb{C}$ that satisfies

(a) $(x, x) \geq 0$ and $(x, x) = 0$ if and only if $x = 0$

(b) $\qquad (\alpha_1 x_1 + \alpha_2 x_2, y) = \alpha_1 (x_1, y) + \alpha_2 (x_2, y)$ for

$\qquad \alpha_1, \alpha_2 \in \mathbb{C}, \quad x_1, x_2, y \in H$ $\hfill (1\text{-}1)$

(c) $(x, y) = (\overline{y, x})$

It follows from (1-1) that the form (x, y) is antilinear in y. A form satisfying (1-1) is also called a *Hermitian form*. Define

$$\|x\| = (x, x)^{1/2} \qquad (1\text{-}2)$$

which is the norm induced by the inner product. It clearly satisfies $\|x\| \geq 0$ for all x and $\|x\| = 0$ if and only if $x = 0$. Also $\|\alpha x\| = |\alpha| \, \|x\|$ for all $\alpha \in \mathbb{C}$ and $x \in H$. The proof of the triangle inequality will follow that of the Schwarz inequality.

Inner product spaces allow us to introduce the important notion of orthogonality. We say two vectors $x, y \in H$ are *orthogonal*, and write $x \perp y$, if $(x, y) = 0$. Given a set M we write $x \perp M$ if $x \perp m$ for all $m \in M$. A set of vectors $\{x_\alpha\}$ is called an *orthogonal set* if $(x_\alpha, x_\beta) = 0$ whenever $\alpha \neq \beta$. A vector x is *normalized* if $\|x\| = 1$. We define an *orthonormal set* as an orthogonal set of normalized vectors. Thus $\{e_\alpha\}$ is an orthonormal set if $(e_\alpha, e_\beta) = \delta_{\alpha\beta}$.

Theorem 1-1 (Pythagorean theorem) Let $\{x_i\}_{i=1}^n$ be an orthogonal set in the inner product space H, then

$$\left\| \sum_{i=1}^n x_i \right\|^2 = \sum_{i=1}^n \left\| x_i \right\|^2$$

PROOF

$$\left\| \sum_{i=1}^n x_i \right\|^2 = \left(\sum_{i=1}^n x_i, \sum_{j=1}^n x_j \right) = \sum_{i=1}^n \sum_{j=1}^n (x_i, x_j) = \sum_{i=1}^n \left\| x_i \right\|^2$$

Theorem 1-2 (Schwarz inequality) For all x, y in H we have

$$|(x, y)| \leq \|x\| \cdot \|y\| \tag{1-3}$$

PROOF For $y = 0$ this is trivial. Assume $y \neq 0$ and let $e = \|y\|^{-1} y$. Noting that $x = (x, e) e + (x - (x, e) e)$ and that $x - (x, e) e \perp e$ the Pythagorean theorem implies

$$\|x\|^2 = \|(x, e) e\|^2 + \|x - (x, e) e\|^2 \geq \|(x, e) e\|^2 = |(x, e)|^2$$

or $|(x, e)| \leq \|x\|$. Substituting $\|y\|^{-1} y$ for e (1-3) is obtained.

Theorem 1-3 (Triangle inequality) For all $x, y \in H$ we have

$$\|x + y\| \leq \|x\| + \|y\| \tag{1-4}$$

PROOF

$$\begin{aligned}
\|x + y\|^2 &= (x + y, x + y) = (x, x) + (x, y) + (y, x) + (y, y) \\
&= \|x\|^2 + 2\mathrm{Re}(x, y) + \|y\|^2 \\
&\leq \|x\|^2 + 2|(x, y)| + \|y\|^2 \\
&\leq \|x\|^2 + 2\|x\| \cdot \|y\| + \|y\|^2 = (\|x\| + \|y\|)^2
\end{aligned}$$

With the proof of the triangle inequality we have proved that the norm defined by (1-2) is a bona fide norm. Thus an inner product space becomes a metric space with the metric ρ defined by $\rho(x, y) = \|x - y\|$. Convergence of a sequence of vectors in this metric is called strong or norm convergence, that is, a sequence x_n converges to x if $\|x_n - x\| \to 0$. We recall that a metric space is called *complete* if every Cauchy sequence converges to an element of the space. A complete inner product space will be called a *Hilbert space*. A subset M of a Hilbert space H is a *linear manifold* if whenever $x, y \in M$ and $\alpha, \beta \in \mathbb{C}$ we have $\alpha x + \beta y \in M$. A linear manifold which is closed is called a *subspace*. Thus a subspace of a Hilbert space is also a Hilbert space.

Theorem 1-4 The inner product is a continuous function in each of its variables.

PROOF Let x_n converge to x then

$$|(x_n, y) - (x, y)| = |(x_n - x, y)| \leq \|x_n - x\| \cdot \|y\|$$

and hence (x_n, y) converges to (x, y).

Actually we can strengthen this result and show that the inner product is simultaneously continuous in both variables. Thus let $\lim x_n = x$ and $\lim y_n = y$ then

$$|(x_n, y_n) - (x, y)| \leq |(x_n, y_n) - (x, y_n) + (x, y_n) - (x, y)|$$

$$\leq |(x_n - x, y_n)| + |(x, y_n - y)|$$

$$\leq \|x_n - x\| \cdot \|y_n\| + \|x\| \cdot \|y_n - y\|$$

Now the sequence $\|y_n\|$ is bounded as a consequence of the uniform boundedness principle and hence we obtain $\lim(x_n, y_n) = (x, y)$.

Corollary 1-5 Given $y \in H$ then $\{x \mid (x, y) = 0\}$ is a subspace.

The norm in a Hilbert space was defined by means of the inner product. It turns out that the inner product can be recovered from the norm. The proof is a simple computation and is omitted.

Theorem 1-6 (Polarization identity) For all vectors $x, y \in H$

$$(x, y) = \tfrac{1}{4}\{\|x + y\|^2 - \|x - y\|^2 + i\|x + iy\|^2 - i\|x - iy\|^2\} \quad (1\text{-}5)$$

Theorem 1-7 (Parallelogram identity) For all vectors $x, y \in H$ we have

$$\|x + y\|^2 + \|x - y\|^2 = 2(\|x\|^2 + \|y\|^2) \quad (1\text{-}6)$$

PROOF Computational.

Theorem 1-8 (Bessel's inequality) Let $\{e_i\}$ be any orthonormal set then for each vector $x \in H$ we have

$$\|x\|^2 \geq \sum |(x, e_i)|^2 \quad (1\text{-}7)$$

PROOF For each finite orthonormal set $\{e_1, \ldots, e_n\}$ we have

$$x = \left(x - \sum_{i=1}^{n} (x, e_i) e_i\right) + \sum_{i=1}^{n} (x, e_i) e_i \quad \text{and} \quad x - \sum_{i=1}^{n} (x, e_i) e_i$$

is orthogonal to $e_1, \ldots, e_n$ and hence to the subspace spanned by them. Using the Pythagorean theorem we obtain

$$\left\|x\right\|^2 = \left\|x - \sum_{i=1}^{n} (x, e_i) e_i\right\|^2 + \left\|\sum_{j=1}^{n} (x, e_i) e_i\right\|^2$$

$$\geq \left\|\sum_{i=1}^{n} (x, e_i) e_i\right\|^2 = \sum_{i=1}^{n} |(x, e_i)|^2$$

If our orthonormal set is infinite we let n go to infinity to obtain the required inequality.

Let y be any vector in the subspace spanned by $e_1, \ldots, e_n$ then $y = \sum_{i=1}^{r} \alpha_i e_i$ for some $\alpha_i \in \mathbb{C}$. Now

$$x - y = x - \sum_{i=1}^{n} (x, e_i) e_i - \sum (\alpha_i - (x, e_i)) e_i$$

and hence as in the proof of Bessel's inequality we have

$$\left\| x - y \right\|^2 = \left\| x - \sum_{i=1}^{n} (x, e_i) e_i \right\|^2 + \sum \left| \alpha_i - (x, e_i) \right|^2 \geq \left\| x - \sum_{i=1}^{n} (x, e_i) e_i \right\|^2$$

Thus $\sum_{i=1}^{n} (x, e_i) e_i$ is the vector in the subspace spanned by $e_1, \ldots, e_n$ which is closest to x. This observation can be used as a basis for the orthogonal decomposition of a Hilbert space with respect to a subspace. To this end define two subspaces M and N to be orthogonal if $(m, n) = 0$ for all $m \in M$ and $n \in N$. Given two orthogonal subspaces M and N we write $M \oplus N$ for the orthogonal direct sum of M and N that is for the subspace $\{m + n \mid m \in M, n \in N\}$.

A set K in a linear space is called *convex* if $x, y \in K$ implies $\alpha x + (1 - \alpha) y \in K$ for all $0 \leq \alpha \leq 1$. Of course any subspace is automatically a convex set. The next theorem, of independent interest, addresses itself to the problem of best approximation by convex sets.

Theorem 1-9 Let K be a closed convex subset of a Hilbert space H and let $x \notin K$. Then there exists a unique vector $k \in K$, such that $\|x - k\| < \|x - k'\|$ for all $k' \in K - \{k\}$.

PROOF Let $d = \inf \{ \|x - k'\| \mid k' \in K \}$ and let k_n be a sequence such that $\lim_{n \to \infty} \|x - k_n\| = d$. The sequence $\{k_n\}$ is a Cauchy sequence for, by the parallelogram identity

$$2(\|x - k_n\|^2 + \|x - k_m\|^2) = 4 \left\| x - \left(\frac{k_n + k_m}{2} \right) \right\|^2 + \|k_n - k_m\|^2$$

or

$$\|k_n - k_m\|^2 = 2(\|x - k_n\|^2 + \|x - k_m\|^2) - 4 \left\| x - \left(\frac{k_n + k_m}{2} \right) \right\|^2$$

But $(k_n + k_m)/2 \in K$ by convexity so

$$\left\| x - \left(\frac{k_n + k_m}{2} \right) \right\| \geq d$$

and

$$\|k_n - k_m\|^2 \leq 2(\|x - k_n\|^2 + \|x - k_m\|^2) - 4d^2$$

Since the right side approaches zero as $n, m \to \infty$ it follows that k_n is a Cauchy sequence. Since K is closed there exists a vector $k \in K$ such that $\lim k_n = k$. By continuity of the norm we have $\|x - k\| = d$. To show uniqueness assume k' is another vector in K for which $\|x - k'\| = d$. Then

$$0 \leq \|k - k'\|^2 \leq 2(\|x - k\|^2 + \|x - k'\|^2) - 4\left\|x - \left(\frac{k + k'}{2}\right)\right\|^2$$

$$= 4d^2 - 4\left\|x - \left(\frac{k + k'}{2}\right)\right\|^2 \leq 0$$

So $k = k'$ and the proof is complete.

Given any set S we define

$$S^{\perp} = \{y \in H \,|\, (y, s) = 0 \qquad \text{for all} \qquad s \in S\}$$

$$= \bigcap_{s \in S} \{y \in H \,|\, (y, s) = 0\}$$

By Corollary 1-5 $S^{\perp}$ is a subspace. For a subspace M, $M^{\perp}$ is called the *orthogonal complement* of M by reason of the following.

Theorem 1-10 Let M be a subspace of a Hilbert space H then $H = M \oplus M^{\perp}$.

PROOF It suffices to show that each vector x in H can be written as $x = y + z$ with $y \in M$ and $z \in M^{\perp}$. Let y be the unique vector in M, whose existence has been established in Theorem 1-9, which is closest to x in M. Put $z = x - y$ then clearly $x = y + z$ and the proof will be complete once we establish $z = x - y$ is orthogonal to M. Let y' be any vector in M then

$$0 \leq \|x - y\| \leq \|x - (y + y')\|^2 = \|(x - y) - y'\|^2$$

$$= \|x - y\|^2 + \|y'\|^2 - 2\operatorname{Re}(x - y, y')$$

Let $y' = \rho e^{i\theta} e$ for some e in M and choose θ so that $e^{-i\theta}(x - y, e) = |(x - y, e)|$. Then $0 \leq |\rho|^2 \|e\|^2 - 2\rho|(x - y, e)|$. Dividing by $\rho > 0$ and letting ρ approach zero we have $0 \leq -|(x - y, e)|$ or $(x - y, e) = 0$. This proves the theorem.

A direct corollary is the Riesz representation theorem for functionals in a Hilbert space. A *linear functional* in a Hilbert space is a linear map F from H into $\mathbb{C}$. A functional is continuous if the function F is continuous.

Theorem 1-11 (Riesz representation theorem) Let F be any continuous linear functional in H. Then there exists a unique vector $y \in H$ such that

$$F(x) = (x, y) \qquad \text{for all} \qquad x \in H \tag{1-8}$$

PROOF If $F = 0$ choose $y = 0$. Otherwise let $M = \operatorname{Ker} F = \{x | F(x) = 0\}$. Since F is linear and continuous M is a subspace. Let $0 \neq z \in M^{\perp}$ and define y by $y = (\overline{F(z)}\, z/\|z\|^2)\, z$ then $F(z) = (z, y)$ and $F(y) = \|y\|^2$. Let $x \in H$ then

$$x = \frac{F(x)}{F(y)} y + \left(x - \frac{F(x)}{F(y)} y \right)$$

where $x - (F(x)/F(y))\, y \in M$ and is therefore orthogonal to y. Taking the inner product of x and y we have

$$(x, y) = \left(\frac{F(x)}{F(y)} y, y \right) = F(x) \frac{\|y\|^2}{F(y)} = F(x)$$

which proves the existence of y.

To prove uniqueness let y' be another vector such that for all $x \in H$ $F(x) = (x, y')$. Then $(x, y - y') = 0$ for all x and choosing $x = y - y'$ we have $\|y - y'\|^2 = 0$ which implies $y = y'$.

We continue with the study of orthonormal sets. An orthonormal set $\{e_\alpha\}$ is an *orthonormal basis* for a Hilbert space H if it is orthonormal and the smallest subspace containing it is H. An orthonormal set $\{e_\alpha\}$ is *closed* if there exists no nonzero vector orthogonal to all e_α. Thus a closed orthonormal set is also a *maximal orthonormal set* in the sense that there exists no orthonormal set which properly includes it. An orthonormal set is complete if for each vector $x \in H$ we have the Parseval identity.

$$\|x\|^2 = \sum_\alpha |(x, e_\alpha)|^2 \tag{1-9}$$

It is a simple consequence of Bellsel's inequality that for an arbitrary orthonormal set at most a countable number of the (x, e_α), that is, of the *generalized Fourier coefficients*, are nonzero.

A Hilbert space H is separable if there exists a countable dense subset in H.

Theorem 1-12 A Hilbert space H is separable if and only if it has a countable orthonormal basis.

PROOF If H has a countable orthonormal basis, say $\{e_i\}_{i=1}^{\infty}$, then, say, the set $\{\sum_{i=1}^{n} \alpha_i e_i | n \geq 0, \operatorname{Re}\alpha_i \text{ and } \operatorname{Im}\alpha_i \text{ rational}\}$ is a countable dense subset. Conversely assume H is separable and let $\{x_i\}_{i=1}^{\infty}$ be a countable dense subset of H. We construct an orthonormal basis by the Gram–Schmidt orthonormalization procedure. If $x_1 \neq 0$ let $e_1 = x_1/\|x_1\|$. Suppose $e_1, \ldots, e_n$ have been defined then we proceed inductively. Let $x_{n'}$ be the first vector in the sequence

$\{x_i\}$ which is not in the subspace spanned by $e_1, \ldots, e_n$. Define e_{n+1} by

$$
e_{n+1} = \frac{x_{n'} - \sum\limits_{i=1}^{n} (x, e_i)\, e_i}{\left\| x - \sum\limits_{i=1}^{n} (x, e_i)\, e_i \right\|}
$$

The resulting orthonormal set $\{e_i\}$ is an orthonormal basis for H as it spans the same subspace as the set $\{x_i\}$.

For simplicity we restrict ourselves from now to the case of separable Hilbert spaces.

Theorem 1-13 Let H be a separable Hilbert space. Then the following statements are equivalent.
(a) $\{e_i\}_{i=1}^{\infty}$ is an orthonormal basis for H.
(b) $\{e_i\}_{i=1}^{\infty}$ is a complete orthonormal set.
(c) $\{e_i\}_{i=1}^{\infty}$ is a closed orthonormal set.

PROOF Assume $\{e_i\}_{i=1}^{\infty}$ is an orthonormal basis, then by Bessel's inequality $\sum_{i=1}^{\infty} |(x, e_i)|^2 < \infty$ and hence $\sum_{i=1}^{\infty}(x, e_i)e_i$ converges. Since $x - \sum_{i=1}^{\infty}(x, e_i)e_i$ is clearly orthogonal to the subspace spanned by the e_i, that is, to all of H, we have $x = \sum_{i=1}^{\infty}(x, e_i)\, e_i$. A simple computation yields $\|x\|^2 = \sum_{i=1}^{\infty} |(x, e_i)|^2$. Thus the Parseval identity holds for all x and $\{e_i\}_{i=1}^{\infty}$ is a complete orthonormal set. Let x be orthogonal to all e_i then from the Parseval identity we have $x = 0$ and the orthonormal set $\{e_i\}_{i=1}^{\infty}$ is closed. Finally assume $\{e_i\}_{i=1}^{\infty}$ is closed. Again $\sum_{i=1}^{\infty}(x, e_i)\, e_i$ converges and $x - \sum_{i=1}^{\infty}(x, e_i)\, e_i$ being orthogonal to all e_i is necessarily zero. So $x = \sum_{i=1}^{\infty}(x, e_i)\, e_i$ for all x and hence $\{e_i\}$ is an orthonormal basis, which completes the proof.

Theorem 1-14 Let H be any separable Hilbert space. Then the cardinalities of two orthonormal bases are equal.

PROOF Since the space is separable the orthonormal basis is at most countable. If H has a finite orthonormal basis $\{e_i, \ldots, e_n\}$ and $\{f_j\}_j$ any other orthonormal basis then by Parseval's identity

$$
\sum_j \|f_j\|^2 = \sum_j \sum_i |(f_j, e_i)|^2 = \sum_i \sum_j |(e_i, f_j)|^2 = \sum_{i=1}^{n} \|e_i\|^2 = n
$$

Since $\|f_j\| = 1$ we have $\sum \|f_j\|^2 = n$ and hence the basis $\{f_j\}$ has also n elements.

We define the *dimension* of a Hilbert space to be the number of elements in any orthonormal basis. By the previous theorem the dimension is well defined. In the same way we define the dimension of a subspace. Given a subspace M the *codimension* of M is the dimension of $M^{\perp}$.

2. BOUNDED OPERATORS IN HILBERT SPACE

Let H_1, H_2 be two Hilbert spaces. A linear operator $T: H_1 \to H_2$ will be called a *bounded operator* if its norm $\|T\|$ defined by

$$\|T\| = \sup_{x \neq 0} \frac{\|Tx\|_2}{\|x\|_1} \tag{2-1}$$

is finite. The set of all bounded linear operators from H_1 into H_2 will be denoted by $B(H_1, H_2)$. We will write $B(H)$ for $B(H, H)$ and note that while $B(H_1, H_2)$ is a Banach space in the operator norm, $B(H)$ is actually a Banach algebra [29, 125]. If $T \in B(H_1, H_2)$ then its *adjoint* T^* is the unique operator in $B(H_2, H_1)$ that satisfies

$$(Tx, y)_2 = (x, T^*y)_1 \tag{2-2}$$

for all $x \in H_1$ and $y \in H_2$. The existence and uniqueness of T^* are a straightforward consequence of the Riesz representation theorem. Moreover a simple computation yields the equality $\|T\| = \|T^*\|$.

An operator T in $B(H)$ will be called *self-adjoint* if $T = T^*$. Self-adjoint operators play in $B(H)$ the role that the reals $\mathbb{R}$ play in the complex number field $\mathbb{C}$. Thus any operator T in $B(H)$ has a unique representation of the form $T = A + iB$ with A and B self-adjoint. The importance of the class of self-adjoint operators stems from their appearance in many applications as well as the fact that they form one of a small class of operators whose structure is completely understood. We will return to the study of self-adjoint operators in Sec. 5. An operator $T \in B(H)$ is called a *contraction* if $\|T\| \leq 1$. Every bounded operator can be rescaled, through multiplication by a scalar, to become a contraction. Thus the study of the structure of contractions yields information about the most general bounded operators.

With each operator T in $B(H_1, H_2)$ we associate two linear manifolds

$$\operatorname{Ker} T = \{x \,|\, x \in H_1, \, Tx = 0\}$$

and

$$\operatorname{Range} T = \{Tx \,|\, x \in H_1\}$$

From the assumption of boundedness it follows that $\operatorname{Ker} T$ is a subspace of H_1 and $\operatorname{Range} T$ is a linear manifold in H_2 which is not necessarily closed. The following is a simple but important result.

Theorem 2-1   Let $T \in B(H_1, H_2)$ then
(a) $H_1 = \operatorname{Ker} T \oplus \overline{\operatorname{Range} T^*}$, and
(b) $H_2 = \operatorname{Ker} T^* \oplus \overline{\operatorname{Range} T}$.

PROOF   From the equality

$$(Tx, y) = (x, T^*y)$$

it follows that $x \in \text{Ker } T$ if and only if it is orthogonal to Range T^* and hence to $\overline{\text{Range } T^*}$. Thus (a); (b) follows by duality.

Let M be a subspace of a Hilbert space H. By Theorem 1-10 we have the direct sum decomposition $H = M \oplus M^{\perp}$. If $x = y + z$ with $y \in M$ and $z \in M^{\perp}$ is the unique representation of x relative to this direct sum decomposition then we define an operator $P \in B(H)$ by $Px = y$. P is linear and satisfies

$$P^2 = P = P^* \tag{2-3}$$

Conversely given any operator $P \in B(H)$ satisfying (2-3) then it induces a direct sum decomposition $H = M \oplus N$ with $M = \text{Range } P = \text{Ker}(I - P)$ and $N = \text{Range}(I - P) = \text{Ker } P$. We call an operator P satisfying $P^2 = P$ a *projection* and an *orthogonal projection* if it satisfies also $P = P^*$. Given a subspace M the orthogonal projection on it is uniquely determined and denoted by P_M.

Let T be a bounded operator in H. A subspace $M \subset H$ is called an *invariant subspace* for T if $TM \subset M$. M is invariant under T if and only if $M^{\perp}$ is invariant under T^*. This is a trivial consequence of (2-2). A subspace M is a *reducing subspace* for T, or reduces T, if it is invariant under both T and T^*. Equivalently M reduces T if and only if both M and $M^{\perp}$ are invariant under T. Since a subspace M and the orthogonal projection on it P_M are in a one-to-one correspondence invariance and reducibility are expressible in terms of the projection P_M.

Theorem 2-2 Let M be a subspace of a Hilbert space H, P the orthogonal projection on it and $T \in B(H)$. Then
(c) M is invariant under T if and only if $PTP = TP$.
(b) M is a reducing subspace for T if and only if $PT = TP$.

PROOF (a) Assume M is invariant under T. For each $x \in H$ we have $Px \in M$ and hence $TPx \in M$ by invariance. Now $M = \text{Ker}(I - P)$ hence $(I - P) TPx = 0$ or $PTPx = TPx$ for all x. Conversely assume $PTP = TP$ and let $x \in M$ then $Px = x$. Since $PTPx = TPx$ we have $Tx = TPx \in \text{Ker}(I - P) = M$ or M is invariant. To prove (b) we note that $M^{\perp}$ is invariant under T if and only if $(I - P) T(I - P) = T(I - P)$ which simplifies to $PTP = PT$. Thus reducibility implies $PT = TP$. Conversely the equality $PT = TP$ implies, by left and right multiplication by P, both $PTP = TP$ and $PTP = PT$ which are equivalent to the invariance of M and $M^{\perp}$ under T.

An operator $T \in B(H_1, H_2)$ is *right* (respectively *left*) *invertible* if there exists an operator R (resp. S) in $B(H_2, H_1)$ such that $TR = I_{H_2}$ (resp. $ST = I_{H_1}$). Clearly right invertibility implies T is onto H_2, that is, surjective, and left invertibility implies T is one-to-one, that is, injective. The converse of the first implication is true, and is a special case of Theorem 7-1, but the other is false. An operator T is *invertible* if it is both left and right invertible. Thus the invertibility of T implies T is one-to-one and onto. The converse is true, but while the existence of an algebraic inverse is trivial the proof of the boundedness of the inverse needs all the power of the closed graph theorem [29, 96].

We denote the inverse of an invertible operator T by T^{-1}. Thus

$$TT^{-1} = T^{-1}T = I$$

Taking adjoints of the previous equality we obtain

$$(T^{-1})^* \, T^* = T^*(T^{-1})^* = I$$

and hence T^* is invertible and

$$(T^*)^* = T^{-1*} \tag{2-4}$$

A relaxation of the notion of invertibility is obtained by introducing quasiaffinities. A bounded map $X : H_1 \to H_2$ which is one-to-one and has dense range is called a *quasiaffinity* or equivalently a *quasi-invertible transformation*.

A different situation occurs when an operator $T : H_1 \to H_2$ has closed range. In that case $T|\{\operatorname{Ker} T\}^{\perp} \to \operatorname{Range} T$ is a boundedly invertible operator. The inverse, originally defined on Range T, can be defined on all of H_2 by letting its restriction to $\{\operatorname{Range} T\}^{\perp} = \operatorname{Ker} T^*$ be zero. The extended operator, uniquely determined by T and denoted by $T^{\#}$, is called the *pseudoinverse* of T. The pseudoinverse of T satisfies

$$TT^{\#}T = T \tag{2-5}$$

Due to the existence of an inner product the notion of isomorphism can be specialized. Generally two spaces H_1 and H_2 are *isomorphic* if there exists an invertible transformation T from H_1 onto H_2. In the case of Hilbert spaces we say two spaces H_1 and H_2 are *isometrically isomorphic*, or *unitarily equivalent*, if there exists an invertible map $U : H_1 \to H_2$ satisfying

$$(Ux, Uy)_2 = (x, y)_1 \tag{2-6}$$

for all $x, y \in H_1$. An invertible map U satisfying (2-6) is called a *unitary* map. From (2-6) it follows that

$$U^*U = I \tag{2-7}$$

and since U is assumed invertible we have also

$$UU^* = I \tag{2-8}$$

which together with (2-7) is equivalent to $U^{-1} = U^*$. An operator $U : H \to H_2$ is called an *isometry* if it satisfies (2-7) and a *coisometry* if (2-8) is satisfied. Contrary to the finite dimensional situation there exist nonunitary isometries in $B(H)$, assuming the dimension of H to be infinite. Thus the set of isometries in $B(H)$ is properly larger than the set of unitary operators. For many purposes it turns out to be useful to introduce a class of operators wider than the class of isometries, namely the partial isometries.

An operator $V \in B(H_1, H_2)$ is called a *partial isometry* if there exists a subspace $M \subset H_1$ such that

$$\| Vx \| = \begin{cases} \|x\| & \text{if} \quad x \in M \\ 0 & \text{if} \quad x \in M^{\perp} \end{cases}$$

The subspace M is called the *initial space* of V whereas Range V, which is a closed subspace of H_2, is called the *final space* of V. The basic properties of partial isometries are given below.

Theorem 2-3

(a) A bounded linear transformation $V : H_1 \to H_2$ is a partial isometry if and only if V^*V is a projection.

(b) V is a partial isometry if and only if V^* is and the initial space of V^* is the final space of V.

(c) If V is a partial isometry then V^*V is the orthogonal projection on the initial space of V whereas VV^* is the orthogonal projection on the final space of V.

(d) V is a partial isometry if and only if $V^* = V^\#$ where $V^\#$ is the pseudo-inverse of V.

PROOF (a) Let $V : H_1 \to H_2$ be a partial isometry with initial space M and let P be the orthogonal projection on M. If $x \in M$ then

$$(V^*Vx, x) = \|Vx\|^2 = \|x\|^2 = (Px, x)$$

on the other hand if $x \in M^\perp$ then $(V^*Vx, x) = \|Vx\|^2 = 0 = (Px, x)$. Thus $(V^*Vx, x) = (Px, x)$ for all x in H hence $V^*V = P$ which proves also the first part of (c).

Conversely assume $V^*V = P$ is a projection, necessarily orthogonal since P is self-adjoint, on a subspace M then for $x \in M$

$$\|Vx\|^2 = (V^*Vx, x) = (Px, x) = \|x\|^2$$

and for $x \in M^\perp$

$$\|Vx\|^2 = (V^*Vx, x) = (Px, x) = 0$$

Thus V is a partial isometry with $M = \text{Range } P$ as initial space.

To prove (b) assume V is a partial isometry and let N be its final space. Let $y \in N$ then $y = Vx$ for some $x \in M$. $\|V^*y\|^2 = (VV^*y, y) = (VV^*Vx, Vx) = (VPx, Vx) = \|Vx\|^2 = \|x\|^2$. If $y \in N^\perp$ then $\|V^*y\|^2 = (VV^*y, y) = (VV^*Vx, y) = (Vx, y) = 0$. Thus $|V^*y\|$ is equal to $\|y\|$ if $y \in N$ and to 0 if $y \in N^\perp$. Thus V^* is a partial isometry with $N = \text{Range } V$ as initial space. It follows from the first part of (c) that VV^* is the projection onto the final space of V.

(d) Assume V a partial isometry then $V^*V = P$ which is equivalent to $V^* = V^\#$. Conversely if $V^* = V^\#$ then $V^*V = P$. From (a) it follows that V is a partial isometry.

While the spectral theorem and the Wold decomposition give us complete information as to the structure of isometries the structure theory for partial isometries is equivalent to the theory of an arbitrary bounded operator in a Hilbert space [64, Problem 103].

The notion of isomorphism of spaces extends to isomorphism of operators.

Given $T_1 \in B(H_1)$ and $T_2 \in B(H_2)$ we say T_1 and T_2 are unitarily equivalent if there exists a unitary operator $U \in B(H_1, H_2)$ for which $UT_1 = T_2U$. T_1 and T_2 are similar if there exists a boundedly invertible operator $R \in B(H_1, H_2)$ for which $RT_1 = T_2R$. We say T_1 is a quasiaffine transform of T_2 if there exists a quasi-affinity $x \in B(H_1, H_2)$ such that $XT_1 = T_2X$. T_1 and T_2 are quasisimilar if each is a quasiaffine transform of the other.

Unitary equivalence, similarity, and quasisimilarity are all equivalence relations while being a quasiaffine transform is a reflexive and transitive but not necessarily symmetric relation.

The object of spectral theory is to study the structure of operators by decomposing them into more elementary ones. This in analogy with the finite dimensional situation summarized by Theorem I 4-21 describing the Jordan canonical form. To this end we introduce the relevant terminology.

We define the *resolvent set* of a bounded operator T as the set $\rho(T)$ of all complex numbers λ such that $\lambda I - T$ has a bounded inverse. We put $R(\lambda, T) = (\lambda I - T)^{-1}$ and call it the *resolvent function*. The complement of $\rho(T)$ is called the *spectrum* of T and denoted by $\sigma(T)$. The spectrum of T can be more finely classified. We denote by $\sigma_p(T)$, the *point spectrum* of T, the set of all eigenvalues of T, that is, the set of all λ where $\lambda I - T$ is not injective. If $\lambda I - T$ is one-to-one but not onto then we assign λ to the *continuous spectrum* $\sigma_c(T)$ if the range of $\lambda I - T$ is dense and to the *residual spectrum* $\sigma_r(T)$ otherwise.

The important basic facts concerning spectrum and resolvent are summarized by the following theorem. The theorem holds just as well in any Banach space.

Theorem 2-4 Let T be a bounded operator then the resolvent set $\rho(T)$ is open and $R(\lambda, T)$ is analytic on $\rho(T)$. The spectrum $\sigma(T)$ is a nonempty compact set and

$$r(T) = \sup\{|\lambda| \; |\lambda \in \sigma(T)\} = \lim_{n \to \infty} \|T^n\|^{1/n} \tag{2-9}$$

The resolvent equation

$$R(\lambda, T) - R(\mu, T) = -(\lambda - \mu) R(\lambda, T) R(\mu, T) \tag{2-10}$$

holds for all $\lambda, \mu \in \rho(T)$.

PROOF We use the fact that if $\|A\| < 1$ then $I - A$ is invertible and the inverse given by the uniformly convergent series $\sum_{i=0}^{\infty} A^i$. Thus if $\lambda \in \rho(T)$ is an arbitrary point of $\rho(T)$ we will show that a full neighbourhood of λ_0 is in $\rho(T)$. Indeed

$$\lambda I - T = (\lambda - \lambda_0) I + (\lambda_0 I - T) = (\lambda_0 I - T)(I + (\lambda - \lambda_0) R(\lambda_0, T))$$

Since $\lambda_0 I - T$ is invertible by assumption $\lambda I - T$ is invertible if

$$(I + (\lambda - \lambda_0) R(\lambda_0, T))$$

is and this is certainly true if

$$\|(\lambda - \lambda_0) R(\lambda_0, T)\| < 1 \quad \text{or if} \quad |\lambda - \lambda_0| < \|R(\lambda_0, T)\|^{-1}$$

Thus $\rho(T)$ is open and moreover we get

$$R(\lambda, T) = \sum_{n=0}^{\infty} (-1)^n (\lambda - \lambda_0)^n R(\lambda_0, T)^{n+1}$$

for all λ such that $|\lambda - \lambda_0| < \|R(\lambda_0, T)\|^{-1}$ which shows the analyticity of the resolvent. Also it follows that if $d(\lambda)$ is the distance of the point $\lambda \in \rho(T)$ from the spectrum then necessarily $\|R(\lambda, T)\| \geq d(\lambda)^{-1}$ which shows that $\lim_{\lambda \to \sigma(T)} \|R(\lambda, T)\| = \infty$ and hence $\rho(T)$ is the natural domain of analyticity of the resolvent function. For $\lambda, \mu \in \rho(T)$ the resolvent equation follows from the equality $(\lambda - \mu) I = (\lambda I - T) - (\mu I - T)$ by multiplication by $R(\lambda, T) R(\mu, T)$. We have for the derivative of the resolvent the formula $dR(\lambda, T)/d\lambda = -R(\lambda, T)^2$. To see that $\sigma(T)$ is bounded we note that for complex λ satisfying $|\lambda| > \|T\|$ then $\lambda \in \rho(T)$ as $(\lambda I - T)^{-1} = \sum_{n=0}^{\infty} (T^n/\lambda^{n+1})$ and the series converges if $\sigma(T)$ were empty then $R(\lambda, T)$ would be an entire function vanishing at infinity, thus necessarily zero. This implies all coefficients of the Laurent expansion $\sum_{n=0}^{\infty} (T^n/\lambda^{n+1})$ vanish, including I which yields the contradiction.

A classical argument about power series shows that the expansion $R(\lambda, T) = \sum_{n=0}^{\infty} (T^n/\lambda^{n+1})$ actually converges for $|\lambda| > \limsup_{n \to \infty} \|T^n\|^{1/n} = \sup\{|\lambda| \,|\, \lambda \in \sigma(T)\}$. Now if $\lambda \in \sigma(T)$ it clearly follows that $\lambda^n \in \sigma(T^n)$ and hence $|\lambda|^n \leq \|T^n\|$. Thus $r(T) = \sup\{|\lambda| \,|\, \lambda \in \sigma(T)\} \leq \|T^n\|^{1/n}$ or $r(T) \leq \liminf \|T^n\|^{1/n}$. This together with the previous inequality implies that $\lim_{n \to \infty} \|T^n\|^{1/n}$ exists and is equal to the *spectral radius* $r(T)$.

If $\lambda \in \rho(T)$ then taking the adjoint of the equality

$$(\lambda I - T) R(\lambda, T) = R(\lambda, T)(\lambda I - T) = I \qquad (2\text{-}11)$$

implies

$$R(\lambda, T)^* = R(\bar{\lambda}, T^*) \qquad (2\text{-}12)$$

If T is a contraction and $\lambda \in \sigma_p(T)$ we do not have necessarily $\bar{\lambda} \in \sigma_p(T)$. More can be said if $|\lambda| = 1$.

Lemma 2-5 Let T be a contraction in a Hilbert space H. If $|\lambda| = 1$ and $Tx = \lambda x$ then $T^*x = \bar{\lambda}x$.

PROOF By considering $\bar{\lambda}T$ instead of T it suffices, without loss of generality to prove that $Tx = x$ implies $T^*x = x$.

Thus we assume $Tx = x$ and hence $(T^*x, x) = (x, Tx) = (x, x) = \|x\|^2$. Now

$$0 \leq \|T^*x - x\|^2 = \|T^*x\|^2 - 2\mathrm{Re}(T^*x, x) + \|x\|^2$$

$$= \|T^*x\|^2 - 2\|x\|^2 + \|x\|^2 = \|T^*x\|^2 - \|x\|^2 \leq 0$$

which implies $\|T^*x - x\| = 0$ or $T^*x = x$.

As in the finite dimensional case, given a polynomial $p(\lambda) = \sum_{i=0}^{n} a_i \lambda^i$ and a bounded operator T, $p(T)$ is defined by

$$p(T) = \sum_{i=0}^{n} a_i T^i \tag{2-13}$$

This is not very useful in the infinite dimensional case due to the fact that the case where $p(T) = 0$ for some nonzero polynomial is the exception rather than the rule. Thus it is of interest to enlarge the class of functions f for which $f(T)$ can be defined.

By a *functional calculus* we mean a homomorphic map of some algebra of functions into the algebra of bounded operators on a Hilbert or a Banach space.

An important instance of a functional calculus is the Riesz–Dunford functional calculus which is the operator theoretic analog of the Cauchy integral formula. Let T be a bounded operator with spectrum $\sigma(T)$, let Ω be an open set that contains $\sigma(T)$ and finally $A(\Omega)$ the set of all analytic functions in Ω. For $f \in A(\Omega)$ we define

$$f(T) = \frac{1}{2\pi i} \int_{\gamma} f(\lambda) R(\lambda, T) \, d\lambda \tag{2-14}$$

where γ is any closed positively oriented path around $\sigma(T)$.

Theorem 2-6 The map $f \to f(T)$ defined by (2-14) is a (multiplicative) homomorphism of $A(\Omega)$ into $B(H)$ that satisfies

$$f(T) = \sum_{n=0}^{\infty} a_n T^n \tag{2-15}$$

for every entire function $f(z) = \sum_{n=0}^{\infty} a_n z^n$.

PROOF   That the map $f \to f(T)$ is linear is obvious. Multiplicativity is a consequence of the resolvent equation (2-10). Indeed let γ_1 and γ_2 be positively oriented paths around $\sigma(T)$ and assume without loss of generality that γ_1 lies inside γ_2, that is the index of each point of γ_1 relative to γ_2 is 1. In that case

$$f(T)g(T) = -\frac{1}{4\pi^2} \left\{ \int_{\gamma_1} f(\lambda) R(\lambda, T) \, d\lambda \right\} \left\{ \int_{\gamma_2} g(\mu) R(\mu, T) \, d\mu \right\}$$

$$= -\frac{1}{4\pi^2} \int_{\gamma_1} \int_{\gamma_2} f(\lambda) g(\mu) R(\lambda, T) R(\mu, T) \, d\mu \, d\lambda$$

$$= -\frac{1}{4\pi^2} \int_{\gamma_1} \int_{\gamma_2} f(\lambda) g(\mu) \frac{R(\lambda, T) - R(\mu, T)}{\mu - \lambda} \, d\mu \, d\lambda$$

$$= \frac{1}{2\pi i} \int_{\gamma_1} f(\lambda) R(\lambda, T) \left\{ \frac{1}{2\pi i} \int_{\gamma_2} \frac{g(\mu)}{\mu - \lambda} \, d\mu \right\} d\lambda$$

$$+ \frac{1}{2\pi i} \int_{\gamma_2} g(\mu) R(\mu, T) \left\{ \frac{1}{2\pi i} \int_{\gamma_1} \frac{f(\lambda)}{\mu - \lambda} \, d\lambda \right\} d\mu$$

$$= \frac{1}{2\pi i} \int f(\lambda) g(\lambda) R(\lambda, T) \, d\lambda = (fg)(T)$$

Finally, assume f is an entire function having a Taylor series $f(z) = \sum_{n=0}^{\infty} a_n z^n$. Expand $R(\lambda, T)$ around infinity then for a contour γ lying in $\{\lambda \mid |\lambda| > \|T\|\}$ we have

$$f(T) = \frac{1}{2\pi i} \int_{\gamma} \sum a_n \lambda^n R(\lambda, T)\, d\lambda = \frac{1}{2\pi i} \sum_{n=0}^{\infty} a_n \int_{\gamma} \lambda^n R(\lambda, T)\, d\lambda$$

$$= \sum_{n=0}^{\infty} a_n \frac{1}{2\pi i} \int_{\gamma} \lambda^n \sum_{k=0}^{\infty} \frac{T^k}{\lambda^{k+1}}\, d\lambda = \sum_{n=0}^{\infty} a_n T^n$$

If $f(T)$ is defined by (2-14) then we expect to recover the spectrum of $f(T)$ from the knowledge of $\sigma(T)$ and the analytic behavior of f. This is the content of the following theorem known as the spectral mapping theorem. It is the analog in this context of Theorem I 4-8.

Theorem 2-7 If $\sigma(T) \subset \Omega$ and $f \in A(\Omega)$ then $\sigma(f(T)) = f(\sigma(T))$.

PROOF It suffices to show that $f(T)$ is invertible if and only if $f(\lambda)$ is different from zero on $\sigma(T)$. Assume $f(\lambda) \neq 0$ for all $\lambda \in \sigma(T)$ then $h(\lambda) = 1/[f(\lambda)]$ is analytic in a neighborhood Ω_1 of $\sigma(T)$. By the multiplicativity property of the map $f \to f(T)$ it follows that $h(T) f(T) = I$. Conversely assume $\lambda \in \sigma(T)$ and $f(\lambda) = 0$. Define g by $g(\mu) = [f(\mu)]/[\lambda - \mu]$ then $g \in A(\Omega)$ and $f(T) = (\lambda I - T) g(T)$. If $f(T)$ were invertible with inverse S so would be $(\lambda I - T)$ with $(\lambda I - T)^{-1} = g(T) S$ in contradiction to the assumption $\lambda \in \sigma(T)$.

3. UNBOUNDED OPERATORS

Unbounded operators will be encountered frequently in the sequel, especially as infinitesimal generators of semigroups. Thus it will be convenient to review some of the basic facts concerning them.

Let T be a linear map whose domain of definition D_T is a linear manifold in a Hilbert space H_1 and whose range is included in a Hilbert space H_2. We define the graph of T as the set $\Gamma(T)$ of all pairs $\{[x, Tx] \mid x \in D_T\}$ in $H_1 \oplus H_2$ the direct sum of H_1 and H_2. The operator T is called *closed* if its graph $\Gamma(T)$ is a closed linear manifold, that is, a subspace, of $H_1 \oplus H_2$. Equivalently stated T is closed if for any sequence x_n in D_T for which $x_n \to x$ and $Tx_n \to y$ we have necessarily $x \in D_T$ and $y = Tx$. Every bounded linear operator T from H_1 to H_2 is closed. The converse is not generally true, however, a closed linear operator whose domain of definition D_T is a Banach space is bounded. This is the content of the closed graph theorem.

If T is injective then T^{-1} is defined on Range T by $T^{-1}y = x$ where $y = Tx$. Clearly T^{-1} is well defined in this case, linear and closed whenever T is closed.

We define the *resolvent set* $\rho(T)$ and the *resolvent function* $R(\lambda, T)$ for an unbounded operator just as for bounded ones. Thus $\lambda \in \rho(T)$ if and only if $R(\lambda, T) =$

$(\lambda I - T)^{-1}$ exists as a bounded operator, that is, $R(\lambda, T)$ is bounded and

$$(\lambda I - T) R(\lambda, T) x = x \qquad \text{for} \qquad x \in H$$

and

$$R(\lambda, T)(\lambda I - T) x = x \qquad \text{for} \qquad x \in D_T$$

While $R(\lambda, T)$ is analytic on $\rho(T)$ the spectrum of T may be unbounded on the one hand or empty on the other hand.

We say that a linear operator S is an *extension* of the linear operator T, and write $T \subset S$, if $D_T \subset D_S$ and $Sx = Tx$ for all $x \in D_T$. Equivalently S is an extension of T if and only if $\Gamma(T) \subset \Gamma(S)$.

A linear manifold $\Gamma \subset H_1 \oplus H_2$ is a graph of a linear operator if and only if $[0, y] \in \Gamma$ implies $y = 0$. If T is not closed $\overline{\Gamma(T)}$, that is, the closure of $\Gamma(T)$ in $H_1 \oplus H_2$ may fail to be a graph. If $\overline{\Gamma(T)}$ is a graph we say T is closable. In that case let $\overline{\Gamma(T)} = \Gamma(\overline{T})$ and we call $\overline{T}$ the *minimal closed extension* of T.

Given a linear operator T with domain D_T in a Hilbert space H_1 and range in a Hilbert space H_2 we consider D the set of vectors $y \in H_2$ for which $\varphi_y(x) = (Tx, y)$ is a continuous linear functional on H_1. By the Riesz representation theorem there exists a vector y^* such that $(Tx, y) = (x, y^*)$. Obviously y^* is uniquely determined if and only if D_T is dense in H_1. Assuming that we define T^*, the adjoint of T, by $D_{T^*} = D$ and $T^*y = y^*$ for $y \in D_{T^*}$. The study of the adjoint operator is facilitated by studying the related graph. To this end consider the map $U: H_1 \oplus H_2 \to H_2 \oplus H_1$ defined by

$$U[x, y] = [iy, -ix] = i[y, -x] \tag{3-1}$$

It is easily checked that U defined by (3-1) is unitary and satisfies also $U^2 = I$. We call such an operator a *conjugation*. In terms of the conjugation U defined by (3-1) we can relate the graphs of T and T^*.

Lemma 3-1 Let T be a densely defined operator in a Hilbert space H then

$$\Gamma(T^*) = \{U\Gamma(T)\}^{\perp} \tag{3-2}$$

the orthogonal complement taken in $H \oplus H$.

PROOF Let $(y, z) \in \{U\Gamma(T)\}^{\perp}$ then for all $x \in D_T$ we have

$$0 = ([iTx, -ix], [y, z]) = i(Tx, y) - i(x, z)$$

or $(Tx, y) = (x, z)$ for all $x \in D_T$. This means that $y \in D_{T^*}$ and $T^*y = z$, or equivalently that $[y, z] \in \Gamma(T^*)$. The converse follows from the same calculation.

Corollary 3-2 Let T be a densely defined operator, then T^* is closed. If T is closed then $T^{**} = T$.

PROOF By the previous lemma $\Gamma(T^*)$ is a closed subspace of $H \oplus H$, so T^* is closed. U being unitary satisfies $U(M^{\perp}) = (UM)^{\perp}$ for each subspace M.

Applying the previous lemma we have

$$\Gamma(T^{**}) = \{U\Gamma(T^*)\}^\perp = \{U\{U\Gamma(T)\}^\perp\}^\perp$$
$$= \{U^2\Gamma(T)\}^{\perp\perp} = \Gamma(T)^{\perp\perp} = \overline{\Gamma(T)}$$

as $U^2 = I$. If T is closed $\Gamma(T) = \overline{\Gamma(T)}$ and $T^{**} = T$.

A densely defined operator A will be called *dissipative* if for all $x \in D_A$ $\text{Im}(Ax, x) \geq 0$. A is *symmetric* if for all $x \in D_A$ we have $\text{Im}(Ax, x) = 0$ which is equivalent to $(Ax, y) = (x, Ay)$ for all $x, y \in D_A$. Stated another way, A is symmetric if and only if $A \subset A^*$, in particular every symmetric operator is closable. A symmetric operator A is called *self-adjoint* if $A = A^*$, so a self-adjoint is automatically closed.

A symmetric operator A is said to be maximally symmetric if it has no proper symmetric extension. That is, if $A \subset A_1$ and A_1 symmetric then necessarily $A = A_1$.

Theorem 3-3 Let A be self-adjoint then A is maximally symmetric.

PROOF Let A be self-adjoint and A_1 a symmetric extension. Since $A \subset A_1$ it follows from Lemma 3-1 that $A_1^* \subset A^*$ which together with the symmetry of A_1 yields

$$A \subset A_1 \subset A_1^* \subset A^* = A$$

and hence $A = A_1 = A_1^*$.

Theorem 3-4 Let A be a dissipative not necessarily densely defined, operator. Then the following statements are true.
(a) $\|(A - iI)x\|^2 \leq \|x\|^2 + \|Ax\|^2 \leq \|(A + iI)x\|^2$ for all $x \in D_A$.
(b) $(A + iI)$ is injective.
(c) A is a closed operator if and only if $\text{Range}(A + iI)$ is closed.

PROOF Let A be dissipative and $x \in D_A$ then

$$\|(A - iI)x\|^2 = \|x\|^2 + \|Ax\|^2 - (ix, Ax) - (Ax, ix)$$

$$= \|x\|^2 + \|Ax\|^2 - 2\,\text{Im}(Ax, x)$$

$$\leq \|x\|^2 + \|Ax\|^2 \leq \|x\|^2 + \|Ax\|^2 + 2\,\text{Im}(Ax, x)$$

$$= \|(A + iI)x\|^2$$

This proves (a) and (b) is an immediate consequence.

Assume now A is closed. Let $(A + iI)x_n$ be a Cauchy sequence in $\text{Range}(A + iI)$. By (a) both x_n and Ax_n are Cauchy sequences converging to x and y, respectively. Since A is closed $x \in D_A$ and $y = Ax$. Thus $(A + iI)x_n$ converges to $(A + iI)x$ and $\text{Range}(A + iI)$ is closed. Suppose conversely that $\text{Range}(A + iI)$ is closed. By (a) we have $\|x\|^2 \leq \|(A + iI)x\|^2$ for all

$x \in D_A$. Since $A + iI$ is injective $(A + iI)^{-1}$ is a well-defined contraction operator on $\text{Range}(A + iI)$. Since the inverse of a bounded injective operator is closed it follows that $A + iI$, and hence also A, is closed.

If one considers operators, bounded or unbounded, on a Hilbert space H to be generalizations of complex numbers then one finds that many simple results concerning complex numbers have nontrivial analogs in the operator theoretic context. One of the most striking and useful results in this vein is the Cayley transform. It is well known that

$$W = \frac{z - i}{z + i}$$

is a fractional linear transformation (Moebius transformation) mapping the upper half plane bijectively onto the unit disc, and the real line onto the unit circle with the point 1 deleted.

Consider now dissipative operators as generalizations of the complex numbers lying in the closed upper half plane, $\{z \mid \text{Im} z \geq 0\}$. With each dissipative operator A we can associate a linear operator in the following manner. Let us define T by

$$Tx = (A - iI)(A + iI)^{-1} x \tag{3-3}$$

for all $x \in \text{Range}(A + iI)$. T is well defined on $\text{Range}(A + iI)$ as $(A + iI)$ is injective. From Theorem 3-4 (a) it follows that T is a contraction. We call T the *Cayley transform* of A.

Since

$$Tx = (A + iI - 2iI)(A + iI)^{-1} x = x - 2i(A + iI)^{-1} x$$

it follows that

$$(T - I)x = -2i(A + iI)^{-1} x \tag{3-4}$$

and therefore $T - I$ is invertible on $\text{Range}(A + iI)$. We can recover A from T as follows. From (3-4) it follows that

$$2(T - I)^{-1} = i(A + iI) = iA - I$$

or

$$A = -i(I + 2(T - I)^{-1}) = -i(T + I)(T - I)^{-1}$$

or

$$A = i(I + T)(I - T)^{-1} \tag{3-5}$$

Lemma 3-5 Every densely defined dissipative operator in H has a maximal dissipative extension which is necessarily closed.

PROOF Let T be the Cayley transform of A. Clearly from (3-3) and (3-5) A has a proper dissipative extension if and only if T has a proper contractive extension. Contractions, however, are easily extended to all of H. For given T which is contractive on D_T we can extend it by continuity to $\overline{D_T}$ and define the extension T_1 to be zero on $D_T^{\perp}$. Next we show that 1 is not an eigenvalue of T_1. If 1 is an eigenvalue of T_1 with an eigenvector y then we have a so, by Lemma 2-5, that $T_1^* y = y$. Let x be an arbitrary vector in D_A then

$$(y, (A + iI) x) = (T_1^* y, (A + iI) x) = (y, T_1 (A + iI) x)$$

$$= (y, T(A + iI) x) = (y, (A - iI) x)$$

This equality implies $(y, x) = 0$ for all x in D_A and since we assumed A to be densely defined $y = 0$ and 1 is not an eigenvalue of T_1. Define now A_1 to be the inverse Cayley transform of T_1 then A_1 is a maximal dissipative extension of A. A_1 is closed by Theorem 3-4 (c) as

$$D_{T_1} = H = \text{Range}(A_1 + iI)$$

The following theorem characterizes maximal dissipative operators.

Theorem 3-6 Let A be a densely defined linear operator in H. The following statements are equivalent.
(a) A is maximal dissipative.
(b) A is dissipative and $\text{Range}(A + iI) = H$.
(c) $A = i(I + T)(I - T)^{-1}$ for some everywhere defined contraction T for which $1 \notin \sigma_p(T)$.

PROOF Let T be the Cayley transform of A. From the previous lemma it is clear that A has a proper dissipative extension if and only if $D_T \neq H$. However, the domain of T is equal to the range of $A + iI$. Thus (a) and (b) are equivalent. From the discussion preceding Lemma 3-5 it follows that the Cayley transform of a dissipative operator is contractive. Also from (3-4) it follows that $I - T$ is invertible on $\text{Range}(A + iI)$. Thus if A is maximal dissipative then $\text{Range}(A + iI) = H$ and $I - T$ is injective or $1 \notin \sigma_p(T)$. So (a) implies (c). Finally assume A is given by (3-5) and $1 \notin \sigma_p(T)$. As all vectors in D_A are of the form $y = (I - T) x$ we have

$$\text{Im}(Ay, y) = \text{Im}\left(i(I + T)(I - T)^{-1} y, y\right)$$

$$= \text{Re}\left((I + T) x, (I - T) x\right) = \|x\|^2 - \|Tx\|^2 \geq 0$$

Thus A is dissipative and maximality follows from the fact that T is everywhere defined.

Since symmetric operators are automatically dissipative all the previous results hold true for this class of operators. However, since an operator A is symmetric together with $-A$ the results can be sharpened.

Theorem 3-7 Let A be a not necessarily densely defined, symmetric operator. Then the following statements are true.

(a) $\|(A - iI)x\|^2 = \|x\|^2 + \|Ax\|^2 = \|(A + iI)x\|^2$ for all $x \in D_A$.
(b) A is closed if and only if $\text{Range}(A \pm iI)$ is closed.
(c) Either $\text{Range}(A + iI) = H$ or $\text{Range}(A - iI) = H$ implies A is maximal symmetric.
(d) The Cayley transform V of a closed symmetric operator A is isometric if and only if $\text{Range}(A + iI) = H$, coisometric if and only if $\text{Range}(A - iI) = H$ and unitary if and only if $\text{Range}(A \pm iI) = H$.

Corollary 3-8 A symmetric operator A has a self-adjoint extension if and only if the codimensions of $\text{Range}(A + iI)$ and $\text{Range}(A - iI)$ are equal.

A class of operators closely related to dissipative operators is the class of accretive operators. A densely defined operator A is accretive if $\text{Re}(Ax, x) \leq 0$ for all $x \in D_A$. Clearly A is accretive if and only if $-iA$ is dissipative. Thus all results concerning dissipative operators can be easily translated to the case of accretive operators. Thus, we have, for example, the following theorem.

Theorem 3-9 Let A be a densely defined linear operator in H. The following statements are equivalent.
(a) A is maximal accretive.
(b) A is accretive and $\text{Range}(A - I) = H$.
(c) $A = (T + I)(T - I)^{-1}$,

for some everywhere defined contraction T for which $1 \notin \sigma_p(T)$.

The maximal accretive operators can be characterized through the resolvent function.

Theorem 3-10 A closed densely defined linear operator A in a Hilbert space H is maximal accretive if and only if

$$\| R(\lambda, A) \| \leq \lambda^{-1} \qquad \text{for all} \qquad \lambda > 0 \qquad (3\text{-}6)$$

PROOF Assume (3-6) holds for all $\lambda > 0$, that is $\|\lambda R(\lambda, A)y\| \leq \|y\|$ for all y. Since $(\lambda I - A)D_A = H$ we have $y = (\lambda I - A)x$ for some $x \in D_A$. So $\|\lambda x\|^2 \leq \|(\lambda I - A)x\|^2$ for all $x \in D_A$. Expanding, we have for $\lambda > 0$

$$|\lambda|^2 \|x\|^2 \leq |\lambda|^2 \|x\|^2 - 2\lambda \,\text{Re}(Ax, x) + \|Ax\|^2$$

from which the inequality

$$\text{Re}(Ax, x) \leq \frac{1}{2\lambda} \|Ax\|^2$$

follows. Letting $\lambda \to \infty$ we have $\text{Re}(Ax, x) \leq 0$, that is A is accretive. Since $(A - I)D_A = H$, A is maximal accretive by Theorem 3-9.

Conversely let A be maximal accretive. Then, as $\mathrm{Re}(Ax, x) \leq 0$ we have

$$|\lambda|^2 \, \|x\|^2 \leq |\lambda|^2 \, \|x\|^2 + \|Ax\|^2 \leq |\lambda|^2 \, \|x\|^2 + \|Ax\|^2$$
$$- 2\lambda \, \mathrm{Re}(Ax, x) = \|(\lambda I - A) \, x\|^2$$

for all $\lambda > 0$ and $x \in D_A$. Since A is maximal $(A - \lambda I) D_A = H$ hence for all $y \in H$ we have $|\lambda|^2 \, \|R(\lambda, A) \, y\|^2 \leq \|y\|^2$ which is equivalent to (3-6).

4. REPRESENTATION THEOREMS

Representation theorems for certain classes of analytic functions have been instrumental in the development of functional analysis in general and the theory of self-adjoint operators in Hilbert space in particular. As we shall see later many of the classical representation theorems have direct interpretations as solutions of special realization problems.

While the current approach to the spectral theorem via the theory of Banach algebras, the Gelfand transform, and the Gelfand–Naimark theorem is unparalleled in power and elegance we will take the classical approach which, though less general, is much closer to those who are application oriented.

Our starting point for the development will be the representation theorem of Herglotz.

A (complex) function u in a domain (open connected set) Ω will be called *harmonic* if it satisfies the Laplace equation

$$\frac{\partial^2 u}{\partial x^2} + \frac{\partial^2 u}{\partial y^2} = 0$$

It is a consequence of the Cauchy–Riemann equations that both the real and the imaginary parts of an analytic function are harmonic. We will be interested primarily in functions harmonic in the open unit disc and without loss of generality we may restrict ourselves to real valued harmonic functions. Clearly the set of all harmonic functions in a given domain is a linear space. This implies that for an analytic function f, its complex conjugate $\bar{f}$, as well as its real and imaginary parts are harmonic. In particular the functions $z^n = r^n e^{in\theta}$ and $\bar{z}^n = r^n e^{-in\theta}$ are harmonic for all $n \in \mathbb{Z}$. It is natural to expect that limits, in a sufficiently strong topology, of harmonic functions, will be harmonic. This is true for uniform convergence on compact subsets of the unit disc D. An important special case is given by

$$P_r(\theta) = \sum_{n=-\infty}^{\infty} r^{|n|} e^{in\theta} \tag{4-1}$$

which is harmonic in D. This can also be verified directly since

$$P_r(\theta) = \mathrm{Re} \, \frac{1 + z}{1 - z} = \mathrm{Re} \, \frac{1 + re^{i\theta}}{1 - re^{i\theta}} = \frac{1 - r^2}{1 - 2r \cos\theta + r^2}$$

We call $P_r(\theta)$ the *Poisson kernel*.

We note the following important properties of the Poisson kernel.

$$P_r(\theta) \geq 0, \qquad \frac{1}{2\pi} \int_{-\pi}^{\pi} P_r(\theta)\, d\theta = 1, \qquad P_r(t) \leq P_r(\delta) \qquad \text{for} \qquad \delta \leq t \leq \pi$$

(4-2)

and

$$\lim_{r \to 1} P_r(\delta) = 0 \qquad \text{for} \qquad 0 < \delta \leq \pi$$

Thus the Poisson kernel is a *summability kernel*, or alternatively an *approximate identity*. The latter name is derived from its role in the commutative Banach algebra $L^1(\mathbb{T})$ where multiplication is defined by convolution. Thus summability kernels play an important role in the derivation of inversion formulas for Fourier series and Fourier transforms. We will return to them in the study of the Fourier transform in Sec. 10. Denote by $C(\mathbb{T})$ the space of all continuous functions on the unit circle.

Theorem 4-1 Let $u \in C(\mathbb{T})$, then the function $u_r(e^{i\theta}) = u(re^{i\theta})$ defined by

$$u(re^{i\theta}) = \frac{1}{2\pi} \int_{-\pi}^{\pi} P_r(\theta - t)\, u(e^{it})\, dt$$

(4-3)

is harmonic in D and u_r converges uniformly to u.

PROOF   Without loss of generality we assume u is real valued. $u(re^{i\theta})$ is harmonic as the real part of the analytic function

$$\frac{1}{2\pi} \int \frac{e^{it} + z}{e^{it} - z}\, u(e^{it})\, dt = \frac{1}{2\pi} \int \frac{e^{it} + re^{i\theta}}{e^{it} - re^{i\theta}}\, u(e^{it})\, dt$$

Now

$$\left| u_r(e^{i\theta}) - u(e^{i\theta}) \right| = \left| \frac{1}{2\pi} \int P_r(t)\left(u(e^{i(\theta-t)}) - u(e^{i\theta}) \right) dt \right|$$

$$\leq \frac{1}{2\pi} \int_{|t| \leq \delta} P_r(t) \left| u(e^{i(\theta-t)}) - u(e^{i\theta}) \right| dt$$

$$+ \frac{1}{2\pi} \int_{\delta < |t| \leq \pi} P_r(t) \left| u(e^{i(\theta-t)}) - u(e^{i\theta}) \right| dt$$

Given $\varepsilon > 0$ we can, by the uniform continuity of u on the unit circle choose $\delta > 0$ so that $\left| u(e^{i(\theta-t)}) - u(e^{i\theta}) \right| < \varepsilon$ for $|t| < \delta$ and all θ. Thus the first integral is bounded by ε. The second integral can be majorized by $2\|u\|_\infty \max_{\delta \leq |t| \leq \pi} P_r(t)$, where $\|u\|_\infty$ is the sup norm of u. Since $\lim_{r \to 1} \max_{\delta < |t| \leq \pi} P_r(t) = 0$ the result follows.

The previous theorem gives a solution to the Dirichlet problem in the unit disc with continuous boundary values, that is the problem of finding a harmonic function in the open disc having preassigned boundary values. The uniqueness of the solution to the Dirichlet problem is a consequence of the following theorem.

Theorem 4-2 Let u be continuous in the closed unit disc and harmonic in the open disc D. If $u(e^{it}) = 0$ on the boundary then $u(re^{it})$ is identically equal to zero in D.

PROOF Without loss of generality we assume u to be real valued. Suppose for some point z_0 of D $u(z_0) > 0$. Choose ε so that $0 < \varepsilon < u(z_0)$ and define v by $v(z) = u(z) + \varepsilon|z|^2$. Then $v(z_0) > \varepsilon$ and $v(z) = \varepsilon$ on $\mathbb{T}$. It follows that v has a local maximum at an interior point z_1 of D. But at a local maximum we must have, since v is twice differentiable, that $v_{xx} \leq 0$ and $v_{yy} \leq 0$. An easy computation yields, however, $v_{xx} = v_{yy} = 2\varepsilon$. Thus $u(z) > 0$ is impossible. Analogously $u(z) < 0$ is impossible. So necessarily $u(z) = 0$ for all $z \in D$.

The Poisson integral can be extended to a much larger space. Specifically, let $M(\mathbb{T})$ be the set of all finite complex Borel measures on $\mathbb{T}$, which by the Riesz representation theorem, is just the dual of $C(\mathbb{T})$. For every measure μ in $M(\mathbb{T})$ we define

$$u(re^{i\theta}) = u_r(e^{i\theta}) = \frac{1}{2\pi} \int_{-\pi}^{\pi} P_r(\theta - t)\, d\mu \qquad (4\text{-}4)$$

By the same reasoning as before u is harmonic in D, and it is a simple consequence of Theorem 4-1 that the measures $u_r\, dt$ converge to μ in the w^*-topology of $M(\mathbb{T})$.

We say a measure μ in $M(\mathbb{T})$ is positive if for each Borel subset σ of $\mathbb{T}$ we have $\mu(\sigma) \geq 0$. This in turn is equivalent to $\int_{-\pi}^{\pi} f\, d\mu \geq 0$ for each nonnegative function f in $C(\mathbb{T})$. The Poisson integrals of positive measures are the key to several important representation theorems. The first is due to Herglotz.

Theorem 4-3 (Herglotz) A function u in D is harmonic and nonnegative if and only if it is the Poisson integral of a positive measure μ in $M(\mathbb{T})$. The measure μ is uniquely determined.

PROOF If $\mu \in M(\mathbb{T})$ is positive and u is defined by (4-4) then clearly u is positive harmonic. To prove the converse let $u_r(e^{i\theta}) = u(re^{i\theta}) \geq 0$ be harmonic in D. Since

$$\frac{1}{2\pi} \int_{-\pi}^{\pi} |u(re^{i\theta})|\, d\theta = \frac{1}{2\pi} \int_{-\pi}^{\pi} u(re^{i\theta})\, d\theta = u(0)$$

it follows that the L^1 normal of u_r are uniformly bounded. Thus we have also the uniform boundedness in $M(\mathbb{T})$ of the absolutely continuous measures $u_r(e^{i\theta})\, d\theta$. But as $M(\mathbb{T}) = C(\mathbb{T})^*$ we have, applying the Banach–Alaoglu theorem [29] that every closed ball in $M(\mathbb{T})$ is w^*-compact. Thus there exists a measure $\mu \in M(\mathbb{T})$ and a subsequence u_{r_i} such that $u_{r_i}\, dt \to \mu$ in the

w^*-topology of $M(\mathbb{T})$. In particular for each nonnegative g in $C(\mathbb{T})$ we have

$$\int_{-\pi}^{\pi} g \, d\mu = \lim_{i \to \infty} \int_{-\pi}^{\pi} u_{r_i} g \, d\mu$$

and, since the integrands on the right are nonnegative, it follows that μ is positive. To complete we note that

$$\frac{1}{2\pi} \int_{-\pi}^{\pi} P_r(\theta - t) \, dt = \lim_{i \to \infty} \frac{1}{2\pi} \int_{-\pi}^{\pi} P_r(\theta - t) u_{r_i}(e^{it}) \, dt$$

$$= \lim_{i \to \infty} u(r_i r e^{i\theta}) = u(re^{i\theta})$$

which is the required representation.

To prove the uniqueness part suppose a harmonic function u has two representations of the form (4-4) corresponding to the measures μ_1 and μ_2. Let $v = \mu_1 - \mu_2$ then $0 = v_r(e^{i\theta}) = \int P_r(\theta - t) \, dv$. Since $v_r \, dt = 0$ converge to v in the w^*-topology we have in particular $\int e^{-int} dv = \lim_{n \to \infty} \int 0 \cdot e^{-int} dt = 0$ so $\int p(e^{it}) \, dv = 0$ for every trigonometric polynomial p and hence for all functions in $C(\mathbb{T})$. Thus $v = 0$ and $\mu_1 = \mu_2$. This completes the proof.

The next result is also associated with Herglotz's name.

Theorem 4-4 A function f analytic in D has a nonnegative real part if and only if it has a representation of the form

$$f(z) = i\beta + \int_{-\pi}^{\pi} \frac{e^{it} + z}{e^{it} - z} \, d\mu \tag{4-5}$$

for some positive measure $\mu \in M(\mathbb{T})$ and real number β. The measure μ in (4-5) is uniquely determined.

PROOF If f has such a representation then

$$\operatorname{Re} f = \operatorname{Re} \int_{-\pi}^{\pi} \frac{e^{it} + z}{e^{it} - z} \, d\mu = \int_{-\pi}^{\pi} P_r(\theta - t) \, d\mu \ge 0$$

Conversely assume $u = \operatorname{Re} f$ is nonnegative. Since u is harmonic the previous theorem guarantees the existence of a positive measure $\mu \in M(\mathbb{T})$ such that $u(re^{i\theta}) = \int_{-\pi}^{\pi} P_r(\theta - t) \, d\mu$. Define g by

$$g(z) = \int_{-\pi}^{\pi} \frac{e^{it} + z}{e^{it} - z} \, d\mu$$

Clearly $f - g$ is analytic in D and its real part is zero. Thus necessarily $f - g$ is a constant which can be taken to be purely imaginary. The uniqueness part follows from the uniqueness part of Theorem 4-3.

Since a function analytic in the unit disc is completely determined by the coefficients of its Taylor expansion at the origin it is natural to expect a characterization of the class of functions analytic in D and having positive real part in terms of the Taylor coefficients. In doing this we make contact with positive definiteness and moment problems.

We say a sequence $\{c_n\}_{n=-\infty}^{\infty}$ is *positive definite* if for each finite set $\xi_1, \ldots, \xi_n$ of complex numbers we have

$$\sum_{i=1}^{n} \sum_{j=1}^{n} c_{i-j} \xi_i \bar{\xi}_j \geq 0 \qquad (4\text{-}6)$$

Theorem 4-5 Let $f(z) = c + \sum_{i=1}^{\infty} c_i z^i$ be an analytic function in D and let $c_c = 2\,\mathrm{Re}\,c, c_{-n} = \bar{c}_n$ for $n > 0$. Then f has positive real part if and only if the sequence $\{c_n\}_{n=-\infty}^{\infty}$ is positive definite.

PROOF Assume f has positive real part. By Theorem 4-4

$$f(z) = i\beta + \int_{-\pi}^{\pi} \frac{e^{it} + z}{e^{it} - z} \, d\mu'$$

for some positive measure μ'. By expanding the kernel of the integral, and using the absolute and uniform convergence of the related series on closed subsets of D, we obtain $f(z) = i\beta + \int d\mu' + 2\sum_{n=1}^{\infty} \int_{-\pi}^{\pi} e^{-int} \, d\mu' \cdot z^n$. If we define μ by $\mu = 2\mu'$ we readily obtain

$$c_n = \int_{-\pi}^{\pi} e^{-int} \, d\mu \qquad \text{for all} \qquad n \in \mathbb{Z} \qquad (4\text{-}7)$$

Now let $\xi_1, \ldots, \xi_n$ be any finite set of complex numbers then

$$\sum_{i=1}^{n} \sum_{j=1}^{n} c_{i-j} \xi_i \bar{\xi}_j = \int_{-\pi}^{\pi} \left| \sum_{k=1}^{n} \xi_k e^{ikt} \right|^2 \, d\mu \geq 0 \qquad (4\text{-}8)$$

and hence the sequence $\{c_n\}_{n=-\infty}^{\infty}$ is positive definite.

Conversely assume $\{c_n\}_{n=-\infty}^{\infty}$ is a positive definite sequence. Choosing $\xi_0 = 1$ we have from (4-6) that $c_0 \geq 0$. Choosing $\xi_0 = 1, \xi_1 = \cdots = \xi_{k-1} = 0$ we have from (4-6) that $c_0 |\xi_0|^2 + c_k \bar{\xi}_0 \xi_k + c_{-k} \xi_0 \bar{\xi}_k + c_0 |\xi_k|^2 \geq 0$ so that $c_k \bar{\xi}_k + c_{-k} \bar{\xi}_k$ is real for all complex ξ_k. Since $c_k \xi_k + \bar{c}_k \bar{\xi}_k$ is also real we have necessarily that $c_{-k} = \bar{c}_k$. Finally applying (4-6) again with $\xi_0 = 1$ and $\xi_k = \rho e^{\theta}$ the choice of $\theta = \arg c_k$ implies $c_0 \rho^2 + 2|c_k| \rho + c_0 \geq 0$ for all real ρ. The discriminant of this equation has to be negative and we obtain $|c_k| \leq c_0$ for all k.

Define now a function f in D by $f(z) = \frac{1}{2} c_0 + \sum_{n=1}^{\infty} c_n z^n$. The inequality $|c_n| \leq c_0$ implies the analyticity of f in D. We will show that f has positive real part. Indeed for $|z| < 1$

$$\frac{2\,\mathrm{Re}\,f(z)}{1 - |z|^2} = \frac{f(z) + \overline{f(z)}}{1 - z\bar{z}} = \sum_{m=0}^{\infty} z^m \bar{z}^m \left\{ \sum_{k=0}^{\infty} c_k z^k + \sum_{k=1}^{\infty} c_k \bar{z}^k \right\}$$

$$= \sum_{m=0}^{\infty} \sum_{k=0}^{\infty} c_k z^{k+m} \bar{z}^m + \sum_{m=0}^{\infty} \sum_{k=1}^{\infty} c_{-k} z^m \bar{z}^{k+m}$$

$$= \sum_{m=0}^{\infty} \sum_{n=m}^{\infty} c_{n-m} z^n \bar{z}^m + \sum_{m=0}^{\infty} \sum_{n=m+1}^{\infty} c_{m-n} z^m \bar{z}^n$$

$$= \sum_{m=0}^{\infty} \sum_{n=m}^{\infty} c_{n-m} z^n \bar{z}^m + \sum_{n=0}^{\infty} \sum_{m=n+1}^{\infty} c_{n-m} z^n \bar{z}^m$$

$$= \sum_{n=0}^{\infty} \sum_{m=0}^{\infty} c_{n-m} z^n \bar{z}^m = \lim_{N \to \infty} \sum_{n=0}^{N} \sum_{m=0}^{N} c_{n-m} z^n \bar{z}^m \geq 0$$

and the proof is complete.

The proof of Theorem 4-5 actually contains a solution of the trigonometric moment problem which is to characterize the Fourier coefficients of positive measures.

Theorem 4-6 Given an infinite sequence $\{c_n\}_{n=0}^{\infty}$ of complex numbers, there exists a positive Borel measure μ on $\mathbb{T}$ such that

$$c_n = \int_{-\pi}^{\pi} e^{-int}\, d\mu \qquad \text{for all} \qquad n \in \mathbb{Z} \tag{4-9}$$

if and only if $\{c_n\}_{n=-\infty}^{\infty}$ is a positive definite sequence. The measure μ is uniquely determined.

PROOF Assume μ is a positive measure and (4-9) holds then the sequence $\{c_n\}_{n=-\infty}^{\infty}$ is positive definite by (4-8). Conversely if $\{c_n\}_{-\infty}^{\infty}$ is a positive definite sequence then by Theorem 4-5 the function $f(z) = \frac{1}{2}c_0 + \sum_{n=1}^{\infty} c_n z^n$ is analytic in D, has positive real part and $\mathrm{Im}\, f(0) = 0$. Thus it has the representation

$$f(z) = \frac{1}{2} \int_{-\pi}^{\pi} \frac{e^{it} + z}{e^{it} - z}\, d\mu$$

for some positive measure μ. This representation clearly implies (4-9). The uniqueness part follows, by the use of the Poisson integral, from the w^*-convergence of $P_r^* \mu$ to μ.

There is an alternative way of introducing positive definiteness. Given a sequence $\{c_n\}_{n=-\infty}^{\infty}$ we define a functional C on the set of all trigonometric polynomials $\sum_{k=-N}^{N} \eta_k e^{ikt}$, $N \geq 0$ by

$$C \sum \eta_k e^{ikt} = \sum c_k \eta_k \tag{4-10}$$

The functional C is called a *positive functional* if $Cy \geq 0$ whenever $y(e^{it}) = \sum \eta_k e^{ikt} \geq 0$. It is not surprising that the two notions of positivity coincide. Indeed we have the following theorem.

Theorem 4-7 A necessary and sufficient condition for a sequence $\{c_k\}_{k=-\infty}^{\infty}$ to be positive definite is that the functional C defined by (4-10) be a positive functional.

PROOF To prove necessity assume $\{c_k\}_{k=-\infty}^{\infty}$ is a positive definite sequence. By Theorem 4-6 $c_k = \int e^{-ikt} d\mu$ for some positive Borel measure μ. Let $y(e^{it}) = \sum_{k=-n}^{n} \eta_k e^{ikt}$ be a positive trigonometric polynomial then so is $y(e^{-it}) = \sum_{k=-n}^{n} \eta_k e^{-ikt}$. Hence

$$\sum_{k=-n}^{n} c_k \eta_k = \sum_{k=-n}^{n} \eta_k \int e^{-ikt} d\mu = \int \sum_{k=-n}^{n} \eta_k e^{-ikt} d\mu \geq 0$$

and C is a positive functional.

Conversely assume C is a positive functional and let $\xi_0, \ldots, \xi_n$ be any set of complex numbers. Define a trigonometric polynomial y by $y(e^{it}) = |p(e^{it})|^2$ where p is the analytic polynomial $p(e^{it}) = \sum_{k=0}^{n} \xi_k e^{ikt}$. From this we get the following representation for y.

$$y(e^{it}) = \left| \sum_{k=0}^{n} \xi_k e^{ikt} \right|^2 = \sum_{k=0}^{n} \xi_k e^{ikt} \sum_{l=0}^{n} \bar{\xi}_l e^{-ilt}$$

$$= \sum_{k=0}^{n} \sum_{l=0}^{n} \xi_k \bar{\xi}_l e^{i(k-l)t} = \sum_{i=-n}^{n} \eta_\nu e^{i\nu t}$$

where

$$\eta_\nu = \sum_{k=\nu}^{n} \xi_k \bar{\xi}_{k-\nu} \qquad \text{for} \qquad \nu \geq 0 \tag{4-11}$$

and $\eta_\nu = \bar{\eta}_{-\nu}$ for $\nu < 0$. Since C is a positive functional it follows that $\sum_{\nu=-n}^{n} c_\nu \eta_\nu \geq 0$. But $\sum_{\nu=-n}^{n} c_\nu \eta_\nu = \sum_{k=0}^{n} \sum_{l=0}^{n} c_{k-l} \xi_k \bar{\xi}_l$ which proves the positive definiteness of the sequence $\{c_k\}_{k=-\infty}^{\infty}$.

The necessity part could be proved without recourse to the solution of the trigonometric moment problem by applying a factorization theorem of Fejer and Riesz which is of interest in itself.

Theorem 4-8 A necessary and sufficient condition for a trigonometric polynomial $y(e^{it}) = \sum_{k=-n}^{n} \eta_k e^{ikt}$ to have a representation

$$y(e^{it}) = |p(e^{it})|^2 \tag{4-12}$$

for some analytic polynomial $p(e^{it}) = \sum_{k=0}^{n} \xi_k e^{ikt}$ is that $y(e^{it}) \geq 0$. In the factorization (4-12) of a nonnegative trigonometric polynomial y we may choose p to have no zeros in the open unit disc.

PROOF The necessity part is trivial. So we assume y is nonnegative and therefore

$$\bar{\eta}_{-k} = \eta_k \qquad \text{for} \qquad k = 0, \ldots, n \tag{4-13}$$

For simplicity assume $\eta_k \neq 0$. Define now a polynomial Y by $Y(z) = \sum_{k=-n}^{n} \eta_k z^{n+k}$, which is of degree $2n$. In $\mathbb{C}_{2n}[\lambda]$ the space of all polynomials of degree $\leq 2n$ we introduce an involution by defining $q^*(z) = z^{2n}\overline{q(1/\bar{z})}$. Condition (4-13) guarantees that $Y^* = Y$. But from this we infer immediately that if α_1 is a zero of Y so is $\bar{\alpha}_1^{-1}$. Now $h_1(z) = (z - \alpha_1)(1 - \bar{\alpha}_1 z)$ is a polynomial with zeros at α_1 and $\bar{\alpha}_1^{-1}$ which satisfies $h_1 = h_1^*$ and for which $\bar{z}h_1(z)$ is nonnegative on $\mathbb{T}$. Since h_1 divides Y we have $Y = hY_1$ and Y_1 satisfies $Y_1^* = Y_1$ and $\bar{z}^{(n-1)}Y_1(z)$ is nonnegative on $\mathbb{T}$. Repeating this process we can remove all zeros of Y which lie off the unit circle and write

$$Y(z) = \left\{ \prod_{j=1}^{s} (z - \alpha_j)(1 - \bar{\alpha}_j z) \right\} X(z)$$

where $X = X^*$ is a polynomial all of whose zeros lie on the unit circle and for which $\bar{z}^{n-s}X(z)$ is nonnegative there. Next we show that every zero of Y on the unit circle is of even order. If we let $y(z) = \sum_{k=-n}^{n} \eta_k z^k$ then $Y(z) = z^n y(z)$. For a positive number α $y(z) + \alpha = [Y(z) + \alpha z^n]/z^n$ which shows that on $\mathbb{T}$ $y(z) + \alpha$ and $Y(z) + \alpha z^n$ have the same zeros, but $y(z) + \alpha$ has none. Since $Y(z) + \alpha z^n$ is invariant with respect to the previously defined involution its zeros come in pairs of numbers conjugate with respect to the unit circle. Choosing α small enough an application of a theorem of Hurwitz shows that the zeros of Y on $\mathbb{T}$ are of even order. Now for β of absolute value one the polynomial g defined by $g(z) = (z - \beta)(1 - \bar{\beta}z) = -\bar{\beta}(z - \beta)^2$ satisfies $g^* = g$ and $\bar{z}g(z) \geq 0$ on $\mathbb{T}$. Everything put together shows that Y has the representation

$$Y(z) = A \left\{ \prod_{i=1}^{s} (z - \alpha_i)(1 - \bar{\alpha}_i z) \right\} \left\{ \prod_{j=1}^{n-s} (z - \beta_j)(1 - \bar{\beta}_j z) \right\} \tag{4-14}$$

for some positive number A, α_i in D and β_j of absolute value one. Define a polynomial p by

$$p(z) = A^{1/2} \left\{ \prod_{i=1}^{s} (1 - \bar{\alpha}_i z) \right\} \left\{ \prod_{j=1}^{n-s} (1 - \bar{\beta}_j z) \right\} \tag{4-15}$$

then a simple calculation yields (4-11) and moreover p has no zeros in D. Clearly we could have chosen p to have no zeros outside the closed unit disc.

We give now another proof of the necessity part of Theorem 4-7. Let $\{c_k\}_{k=-\infty}^{\infty}$ be a positive definite sequence and let $y(e^{it}) = \sum_{v=-n}^{n} \eta_v e^{ivt}$ be a positive trigonometric polynomial. By the Fejér–Riesz theorem $y(e^{it}) = |p(e^{it})|^2$ for some $p(e^{it}) = \sum_{k=0}^{n} \xi_k e^{ikt}$ and η_ν is given by (4-11). So

$$Cy = \sum_{\nu=-n}^{n} c_\nu \eta_\nu = \sum_{k=0}^{n} \sum_{l=0}^{n} c_{k-l} \xi_k \bar{\xi}_l \geq 0$$

and C is a positive functional.

Theorem 4-5 can be used as a starting point for obtaining other related representation theorems. The following one is due to Nevanlinna.

Theorem 4-9 A function F defined in the open upper half plane $\Pi_+ = \{w \mid \operatorname{Im} w > 0\}$ admits a representation

$$F(w) = \alpha + \gamma w + \int_{-\infty}^{\infty} \frac{1 + tw}{t - w} \, d\sigma \tag{4-16}$$

where α and $\gamma \geq 0$ are real constants and σ a finite positive Borel measure on $\mathbb{R}$ if and only if it is analytic in Π_+ and has a nonnegative imaginary part. The measure σ is uniquely determined by F.

PROOF Assume F is analytic in Π_+ and has a nonnegative imaginary part. The fractional linear transformation $W = i[(1 + z)/(1 - z)]$ maps the unit disc D onto Π_+. Define a function f on D by $f(z) = -iF\{i[(1 + z)/(1 - z)]\}$. Then f is analytic in D and has a nonnegative real part. By Theorem 4-4 $f(z) = \beta + \int (e^{it} + z)/(e^{it} - z) \, d\mu$ for some real number β and a positive measure μ on $\mathbb{T}$. Let $\gamma = \mu(\{0\})$ and define μ' on $\mathbb{T}$ by $\mu' = \mu - \gamma\delta$ where δ is the Dirac measure of the point 1. Thus we have

$$f(z) = i\beta + \gamma \frac{1 + z}{1 - z} + \int \frac{e^{it} + z}{e^{it} - z} \, d\mu' \tag{4-17}$$

Since $z = (w - i)/(w + i)$ it follows that

$$\frac{e^{it} + z}{e^{it} - z} = \frac{(w + i)e^{it} + (w - i)}{(w + i)e^{it} - (w - i)} = \frac{w(e^{it} + 1) + i(e^{it} - 1)}{w(e^{it} - 1) + i(e^{it} + 1)} = \frac{-iw\,ctg(t/2) + i}{w + ctg(t/2)}$$

Define a measure σ on $\mathbb{R}$ such that for each Borel subset Δ of $\mathbb{R}$, $\sigma(\Delta) = \mu'[\varphi^{-1}(\Delta)]$ where $\varphi(t) = ctg(1/2)$. Then with the change of variable $s = ctg(t/2)$ we obtain

$$f(z) = i\beta + \gamma \frac{1 + z}{1 - z} + \frac{1}{i} \int_{-\infty}^{\infty} \frac{1 + sw}{s - w} \, d\sigma$$

which in turn implies that

$$F(w) = if(z) = -\beta + \gamma w + \int_{-\infty}^{\infty} \frac{1 + sw}{s - w} \, d\sigma$$

which reduces to (4-16) by putting $\alpha = -\beta$.

Conversely if F has the representation (4-16) then it is clearly analytic in Π_+ and it is easily checked that

$$\operatorname{Im} F(w) = \gamma \operatorname{Im} w + \int_{-\infty}^{\infty} \frac{(1 + t^2) \operatorname{Im} w}{|t - w|^2}\, d\sigma \geq 0$$

The uniqueness of the measure σ in (4-16) follows from the uniqueness of part of Theorem 4-4. This completes the proof.

Up to this point our representation theorems were all of the Poisson type. To get a Cauchy-type representation much stronger growth conditions have to be satisfied. A typical result is the following.

Theorem 4-10 A function F defined in the open upper half plane Π_+ admits a representation of the form

$$F(w) = \int \frac{d\mu}{t - w} \tag{4-18}$$

for some finite positive Borel measure μ on $\mathbb{R}$ if and only if F is analytic, has nonnegative imaginary part and satisfies

$$\sup_{y > 0} \left| y F(iy) \right| < \infty \tag{4-19}$$

The measure μ is uniquely determined by F.

PROOF If μ is a finite positive measure and F is defined by (4-18) then F is clearly analytic in Π_+ and for $w \in \Pi_+$

$$\operatorname{Im} F(w) = \int \operatorname{Im} \frac{1}{t - w}\, d\mu = \int \frac{\operatorname{Im} w}{|t - w|^2}\, d\mu \geq 0$$

Finally

$$\left| y F(iy) \right| = \left| y \int \frac{d\mu}{t - iy} \right| \leq \int \frac{|y|}{|t - iy|}\, d\mu \leq \int d\mu = \|\mu\|$$

which implies (4-19).

To prove the converse we apply Theorem 4-9 to obtain the representation (4-16). For $w = iy$ we have

$$y F(iy) = \alpha y + i\gamma y^2 + \int y \frac{(1 + isy)}{s - iy}\, d\sigma$$

and by assumption (4-19)

$$\left| \alpha y + i\gamma y^2 + \int \frac{y(1 + isy)}{s - iy}\, d\sigma \right| \leq M$$

for some $M > 0$ and all $y > 0$. The boundedness of the previous expression is equivalent to the separate boundedness of both the real and imaginary parts. So for $y > 0$ we have

$$\left| \alpha y + \int \frac{ys(1 - y^2)}{s^2 + y^2} \, d\sigma \right| \leq M \tag{4-20}$$

and

$$\left| \gamma y^2 + \int \frac{y^2(q + s^2)}{s^2 + y^2} \, d\sigma \right| \leq M \tag{4-21}$$

Since the integral in (4-21) is positive we must have $|\gamma y^2| \leq M$ for $y > 0$ which forces $\gamma = 0$. As

$$\lim_{y \to \infty} \frac{y^2}{s^2 + y^2} = 1$$

it follows that $\int (1 + s^2) \, d\sigma \leq M$. Define a measure μ by $d\mu = (1 + s^2) \, d\sigma$ then μ is a finite Borel measure on $\mathbb{R}$. From (4-20) we have

$$\alpha = -\lim_{y \to \infty} \int \frac{s(1 - y^2)}{s^2 + y^2} \, d\sigma = \int s \, d\sigma$$

and upon substituting back in (4-16) we obtain

$$F(w) = \alpha + \int \frac{1 + sw}{s - w} \, d\sigma = \int s \, d\sigma + \int \frac{1 + sw}{s - w} \, d\sigma = \int \frac{(1 + s^2)}{s - w} \, d\sigma = \int \frac{d\mu}{s - w}$$

Once again the uniqueness of the measure μ in the representation (4-18) follows from the uniqueness of the measure σ in the representation (4-16).

5. THE SPECTRAL THEOREM

The representation theorems obtained in the previous section form the basis and analysis of the structure of self-adjoint operators. There is an advantage in using this approach inasmuch as there is no need to assume that the self-adjoint operator under study is bounded.

Let us review first the finite dimensional situation concerning self-adjoint operators. In this case there exists an orthonormal basis $\{e_i\}_{i=1}^n$ consisting of eigenvectors of the self-adjoint operator A corresponding to the real eigenvalues λ_i. Since $\{e_i\}_{i=1}^n$ is an orthonormal basis we have $x = \sum_{i=1}^n (x, e_i) e_i$ and hence $Ax = \sum_{i=1}^n \lambda_i (x, e_i) e_i$. Now the operator E_i defined by $E_i x = (x, e_i) e_i$ is an orthogonal projection and hence A has the representation $A = \sum_{i=1}^n \lambda_i E_i$. If we pass to the infinite dimensional situation we expect the finite sum to be replaced either by an infinite sum or by an integral. This in fact is the situation. An examination of the simplest concrete self-adjoint operators in infinite dimensional Hilbert spaces indicates that we cannot expect generally the spectrum to consist of eigenvalues. Thus if we consider $L^2(0, 1)$ the space of all (equivalence classes) of Lebesgue square integrable functions on $(0, 1)$ and the operator A defined by

$$(Af)(\lambda) = \lambda f(\lambda)$$

then A is a bounded self-adjoint operator with $\sigma(A) = \sigma_c(A) = [0, 1]$ and $\sigma_r(A) = \varnothing$. Still in some sense $L^2(0, 1)$ decomposes into reducing subspaces of A where A acts as multiplication. To make this more precise we introduce the notion of spectral measures.

We define a *spectral measure* on $\mathbb{R}$ to be a map from the σ-algebra of all Borel sets in $\mathbb{R}$ into the set of orthogonal projections in H which has the following properties.

(a) $E(\mathbb{R}) = I$
(b) If $\{S_i\}$ is a countable set of pairwise disjoint Borel sets then for each $x \in H$,
$E\left(\bigcup_{i=1}^{\infty} S_i\right) x = \sum_{i=1}^{\infty} E(S_i) x$.

We shall use the spectral measure to construct self-adjoint operators in H. Let B be the set of all bounded Borel measurable functions on $\mathbb{R}$. In particular simple functions, that is functions of the form $\sum \alpha_i \chi_{\delta_i}$ where χ_{δ_i} is the characteristic function of a Borel set δ_i, are in B, and are dense there in the norm $\|f\| = \sup_{\lambda \in \mathbb{R}} |f(\lambda)|$.

For a simple function $\varphi(\lambda) = \sum_i \alpha_i \chi_{\delta_i}(\lambda)$ we define

$$\int \varphi(\lambda) E(d\lambda) = \sum \alpha_i E(\delta_i)$$

It is easily checked that the integral of simple functions with respect to a spectral measure is well defined in the sense that it is independent of the particular representations of φ used in defining it. If x, y are vectors in H then $\mu_{x,y} = (E(\cdot) x, y)$ is a complex Borel measure in R. It follows that

$$\left| \int \varphi(\lambda) (E(d\lambda) x, y) \right| = \left| \int \varphi(\lambda) d\mu_{x,y} \right| \leq \|\varphi\| \cdot \|\mu_{x,y}\|$$

where $\|\varphi\|$ is the norm of φ as an element of B and $\|\mu_{x,y}\|$ is the total variation of $\mu_{x,y}$. Using the decomposition of complex measures [29] we have

$$\|\mu_{x,y}\| \leq 4 \sup |\mu_{x,y}(\delta)|$$

the supremum taken over all Borel subsets of $\mathbb{R}$. Now $|\mu_{x,y}(\delta)| = |(E(\delta) x, y)| \leq \|E(\delta)\| \cdot \|x\| \cdot \|y\|$ and hence

$$\left| \int \varphi(\lambda) (E(d\lambda) x, y) \right| \leq 4 \|\varphi\| \cdot \|x\| \cdot \|y\| \tag{5-1}$$

From this inequality it is clear that if $\{\varphi_n\}$ is a sequence in B converging to φ then we can define $\int \varphi(\lambda) (E(d\lambda) x, y) = \lim \int \varphi_n(\lambda) (E(d\lambda) x, y)$ and the inequality

(5-1) is still satisfied. By the Riesz representation theorem there exists a bounded operator $A(\varphi)$ such that

$$\big(A(\varphi)\,x,\,y\big) = \int \varphi(\lambda)\,\big(E(d\lambda)\,x,\,y\big) \tag{5-2}$$

Using the properties of the spectral measure it follows that the map $\varphi \to A(\varphi)$ is not only linear but also multiplicative, that is $A(\varphi\psi) = A(\varphi)\,A(\psi)$, that is, it is a homomorphism of B into $B(H)$. Since for a simple function $\varphi(\lambda) = \sum \alpha_i \chi_{\delta_i}(\lambda)$ we have $\bar{\varphi}(\lambda) = \overline{\varphi(\lambda)} = \sum \bar{\alpha}_i \chi_{\delta_i}(\lambda)$ it follows that

$$A(\bar\varphi) = \sum \bar\alpha_i E(\delta_i) = \big(\sum \alpha_i E(\delta_i)\big)^* = A(\varphi)^*$$

and by a continuity argument this holds for all $\varphi \in B$. Thus the homomorphism $\varphi \to A(\varphi)$ is actually a $*$-homomorphism. It is clear that $A(\varphi)$ is self-adjoint if and only if φ is real valued almost everywhere with respect to the spectral measure E. Next we note that for simple functions φ we have

$$\left\| A(\varphi)\,x \right\|^2 = \int \big|\varphi(\lambda)\big|^2 \big(E(d\lambda)\,x,\,x\big) \tag{5-3}$$

and again by continuity this holds for all $\varphi \in B$.

If the spectral measure E has compact support then $A = \int \lambda E(d\lambda)$ is a well-defined bounded self-adjoint operator. The case in which the spectral measure does not have compact support is different and we no longer expect a bounded operator as the outcome of the integral $\int \lambda E(d\lambda)$. We state the result as the following.

Theorem 5-1 Let E be in spectral measure on $\mathbb{R}$. Define an operator A by

$$D_A = \left\{ x \ \middle| \ \int \lambda^2 \big(E(d\lambda)\,x,\,x\big) < \infty \right\} \tag{5-4}$$

and

$$Ax = \lim_{n \to \infty} \int_{-n}^{n} \lambda E(d\lambda)\,x \tag{5-5}$$

then A is a densely defined self-adjoint operator.

PROOF Let $\varphi_n(\lambda) = \lambda$ for $|\lambda| \le n$ and zero otherwise then

$$A(\varphi_n)\,x = \int \varphi_n(\lambda)\,E(d\lambda)\,x = \int_{-n}^{n} \lambda E(d\lambda)\,x$$

Since $\varphi_n \in B$ we have $\|A(\varphi_n)\,x\|^2 = \int_{-n}^{n} |\lambda|^2\,(E(d\lambda)\,x,\,x)$, hence if $x \in H$ is such that $\lim \int_{-n}^{n} \lambda E(d\lambda)\,x$ exists then necessarily $\int \lambda^2 \big(E(d\lambda)\,x,\,x\big) < \infty$. Conversely if $x \in D_A$ then for $n > m$

$$\|A(\varphi_n)\,x - A(\varphi_m)\,x\|^2 = \int_{m \le |\lambda| \le n} \lambda^2 \big(E(d\lambda)\,x,\,x\big)$$

which shows that $A(\varphi_n) x$ is a Cauchy sequence and hence there exists $Ax = \lim\limits_{n \to \infty} A(\varphi_n) x$. To note that the operator A is densely defined we note that for each $x \in H$ $\lim\limits_{n \to \infty} E((-n, n)) x = x$ and hence $\bigcup_{n=1}^{\infty} E((-n, n)) H$ is dense in H. But for each n $E((-n, n)) H \subset D_A$ since with $\delta_n = (-n, n)$

$$\int \lambda^2 (E(d\lambda) E(\delta_n) x, E(\delta_n) x) = \int_{\delta_n} \lambda^2 (E(d\lambda) x, x) < \infty$$

To conclude we will prove that A is self-adjoint. Let x and y be in D_A then

$$(Ax, y) = \int \lambda (E(d\lambda) x, y) = \overline{\int \lambda (E(d\lambda) y, x)} = (\overline{Ay, x}) = (x, Ay)$$

which shows that A is symmetric, that is, $A \subset A^*$. It remains to show that $D_{A^*} \subset D_A$. Let $y \in D_{A^*}$ then with the previous definitions of δ_n and $A(\varphi_n)$ we have, since $A(\varphi_n)$ is bounded and self-adjoint

$$(x, A(\varphi_n) y) = (A(\varphi_n) x, y) = (AE(\delta_n) x, y) = (x, E(\delta_n) A^* y)$$

From this equality it follows that for $y \in D_{A^*}$ we have $A^* y = \lim E(\delta_n) A^* y = \lim A(\varphi_n) y$. Since $\| A(\varphi_n) y \|^2 = \int_{-n}^{n} \lambda^2 (E(d\lambda) y, y)$ it follows that $D_{A^*} \subset D_A$ and the proof is complete.

Our next concern is to show that the self-adjoint operators constructed by integration of spectral measures are not special but represent the most general self-adjoint operators. To show this we start with the study of the spectrum of a self-adjoint operator as well as the analytic properties of its resolvent function.

Theorem 5-2 Let A be a self-adjoint operator with domain $D(A)$ dense in the Hilbert space H. Then the spectrum of A is real and the resolvent of A satisfies

$$R(\zeta, A)^* = R(\bar{\zeta}, A) \tag{5-6}$$

and

$$\| R(\zeta, A) \| \leq |\mathrm{Im}\,\zeta|^{-1} \quad \text{for} \quad \mathrm{Im}(\zeta) \neq 0 \tag{5-7}$$

PROOF Let $x \in D(A)$ and $\zeta = \rho + i\sigma$ then

$$\| (\zeta I - A) x \|^2 = ((\zeta I - A) x, (\zeta I - A) x)$$

$$= \| (\rho I - A) x \|^2 + \| \sigma x \|^2 \geq \sigma^2 \| x \|^2$$

which implies

$$\frac{\| (\zeta I - A) x \|}{|\mathrm{Im}\,\zeta|} \geq \| x \| \quad \text{for} \quad x \in D(A) \tag{5-8}$$

The above inequality shows that $(\zeta I - A)^{-1}$ exists as a bounded operator with norm bounded by $|\text{Im}\,\zeta|^{-1}$. To show that ζ is in the resolvent set of A we have to show that the domain of $(\zeta I - A)^{-1}$ coincides with H. Since self-adjoint operators are automatically closed and the inverse of a closed operator is closed it follows that $(\zeta I - A)^{-1}$ is closed, and being bounded its domain is closed. To show density of the domain of $(\zeta I - A)^{-1}$ which is the same as the range of $(\zeta I - A)$ let us note that y is orthogonal to the range of $\zeta I - A$ if and only if for all $x \in D(A)$ $0 = ((\zeta I - A)\,x, y) = (x, (\bar{\zeta} I - A)\,y)$ and since $D(A)$ is dense in H this holds only if $(\bar{\zeta} I - A)\,y = 0$. However, $\text{Im}\,\bar{\zeta} = -\text{Im}\,\zeta \neq 0$ hence from inequality (5-8) it follows that y is necessarily zero. The equality (5-7) follows from (5-8).

From Theorem 5-2 it follows that $R(\zeta, A)$ is defined in the open upper and lower half planes.

Lemma 5-3 The resolvent function of A satisfies the resolvent equation

$$R(z, A) - R(\zeta, A) = -(\zeta - z)\,R(z, A)\,R(\zeta, A) \tag{5-9}$$

for all $z, \zeta \in \rho(A)$.

PROOF We observe that for all z in $\rho(A)$ the range of $R(z, A)$ coincides with $D(A)$. Thus if $z, \zeta \in \rho(A)$ we have

$$R(z, A)(z - \zeta)\,R(\zeta, A) = R(z, A)(z - A + A - \zeta)\,R(\zeta, A)$$
$$= R(\zeta, A) - R(z, A)$$

which is equivalent to (5-9).

As an immediate corollary we have

Corollary 5-4 The resolvent function of a self-adjoint operator A is analytic in the open lower and upper half planes and for each $x \in D_A$

$$\text{Im}(R(z, A)\,x, x) = -\left\| R(z, A)\,x \right\|^2 (\text{Im}\,z) \tag{5-10}$$

PROOF For all $x, y \in H$ and z, ζ nonreal we have

$$\frac{(R(z, A)\,x, y) - (R(\zeta, A)\,x, y)}{z - \zeta} = -(R(z, A)\,R(\zeta, A)\,x, y) \tag{5-11}$$

Letting ζ approach z we have

$$\frac{d(R(z, A)\,x, y)}{dz} = -(R(z, A)^2\,x, y)$$

and $R(z, A)$ is weakly analytic. However, the various types of analyticity are equivalent [29] which implies that $[dR(z, A)]/dz = -R(z, A)^2$ actually holds in the norm operator topology. To prove (5-10) we substitute in (5-11) $\zeta = \bar{z}$ and use the fact that $R(\bar{z}, A)^* = R(z, A)$.

Theorem 5-5 (Spectral theorem) A densely defined operator A in a Hilbert space H is self-adjoint if and only if there exists a spectral measure E defined on the Borel subsets of the real line such that

$$D_A = \left\{ x \;\middle|\; \int \lambda^2 (E(d\lambda)\, x, x) < \infty \right\}$$

and

$$Ax = \int \lambda E(d\lambda)\, x \tag{5-12}$$

where the last integral is defined in the strong topology.

PROOF Let A be self-adjoint. For each vector $u \in H$ we consider the function $\varphi_u(z) = (R(z, A)\, u, u)$ defined and analytic off the spectrum of A, in particular in the upper and lower half planes. By (5-10) $\varphi_u(z)$ has nonpositive imaginary part in the upper half plane and from (5-7) it follows that it satisfies $\sup_{y > 0} |y\varphi_u(iy)| < \infty$. Thus, making allowance for the required change in sign, Theorem 4-10 implies the existence of a finite positive Borel measure μ_x such that

$$(R(z, A)\, x, x) = \int \frac{1}{z - \lambda} \, d\mu_x \tag{5-13}$$

for all z in the upper half plane. Since $R(z, A)^* = R(\bar{z}, A)$ we have

$$(R(\bar{z}, A)\, x, x) = (\overline{R(z, A)\, x, x}) = \int \frac{1}{\bar{z} - \lambda} \, d\mu_x$$

and hence the representation (5-13) holds actually for all nonreal complex numbers.

We apply now the polarization identity by which for all x, y in H

$$(R(z, A)\, x, y) = \tfrac{1}{4}\{ (R(z, A)(x + y), x + y) - (R(z, A)(x - y), x - y)$$
$$+ i((R(z, A)(x + iy), x + iy) - (R(z, A)(x - iy), x - iy)) \}$$

If we define a complex measure $\mu_{x,y}$ by

$$\mu_{x,y} = \tfrac{1}{4}\{ \mu_{x+y} - \mu_{x-y} + i\mu_{x+iy} - i\mu_{x-iy} \}$$

then $\mu_{x,y}$ is a finite complex Borel measure and

$$(R(z, A)\, x, y) = \int \frac{1}{z - \lambda} \, d\mu_{x,y} \tag{5-14}$$

By the uniqueness of the representing measure we must have that $\mu_{x,y}$ is linear in x and antilinear in y. From (5-7) we have

$$|\eta(R(i\eta, A)\, x, x)| \leq \|x\|^2$$

and hence

$$\left| \int \frac{\eta}{i\eta - \lambda} d\mu_{x,x} \right| \leq \|x\|^2$$

By the Lebesgue dominated convergence theorem it follows, as $\lim_{n \to \infty} (\eta/i\eta - \lambda) = -i$ that $\|\mu_{x,x}\| \leq \|x\|^2$. This implies the boundedness of the Hermitian bilinear form $\int_\sigma d\mu_{x,y} = \mu_{x,y}(\sigma)$ for each Borel set σ. We apply now the Riesz representation theorem to obtain the existence of a self-adjoint operator $E(\sigma)$ of norm less than or equal to one such that

$$(E(\sigma)x, y) = \mu_{x,y}(\sigma) \tag{5-15}$$

and moreover the operator $E(\sigma)$ is uniquely determined. Applying the resolvent identity

$$(R(z, A)x, R(\bar{\zeta}, A)y) = (R(\zeta, A)R(z, A)x, y)$$

$$= -\frac{1}{\zeta - z}((R(\zeta, A) - R(z, A))x, y)$$

and hence

$$\int \frac{1}{z - \lambda}(E(d\lambda)x, R(\bar{\zeta}, A)y) = \frac{1}{\zeta - z}\int \left(\frac{1}{\zeta - \lambda} - \frac{1}{z - \lambda}\right)(E(d\lambda)x, y)$$

$$= \int \frac{1}{(z - \lambda)(\zeta - \lambda)}(E(d\lambda)x, y)$$

By the uniqueness part of Theorem 4-10 we have for each Borel set σ

$$(E(\sigma)x, R(\bar{\zeta}, A)y) = \int_\sigma \frac{1}{\zeta - \lambda}(E(d\lambda)x, y) = \int \frac{1}{\zeta - \lambda}\chi_\sigma(\lambda)(E(d\lambda)x, y)$$

But

$$(E(\sigma)x, R(\bar{\zeta}, A)y) = \int \frac{1}{\zeta - \lambda}(E(\sigma)x, E(d\lambda)y)$$

So we have the equality

$$\int \frac{1}{\zeta - \lambda}(E(\sigma)x, E(d\lambda)y) = \int \frac{1}{\zeta - \lambda}\chi_\sigma(\lambda)(E(d\lambda)x, y)$$

We apply now once more the uniqueness part of Theorem 4-10 to obtain

$$\int_\rho (E(\sigma)x, E(d\lambda)y) = \int_\rho \chi_\sigma(\lambda)(E(d\lambda)x, y) \tag{5-16}$$

for each Borel subset ρ of R. Equality (5-16) is equivalent to the following one

$$(E(\sigma)x, E(\rho)y) = (E(\sigma \cap \rho)x, y) \tag{5-17}$$

for all Borel sets σ and ρ and all vectors x, y in H. Since $E(\rho)$ is self-adjoint we have finally

$$E(\sigma)\, E(\rho) = E(\sigma \cap \rho) \tag{5-18}$$

In particular $E(\sigma)$ is orthogonal projection valued or $E(\cdot)$ is a spectral measure.

The spectral measure reduces the resolvent function of A, that is for each Borel set σ and all nonreal ζ we have

$$E(\sigma)\, R(\zeta, A) = R(\zeta, A)\, E(\sigma) \tag{5-19}$$

To see this we note that using (5-18)

$$
\begin{aligned}
(E(\sigma)\, R(\zeta, A)\, x, x) = (R(\zeta, A)\, x, E(\sigma)\, x) &= \int \frac{1}{\zeta - \lambda}\, (E(d\lambda)\, x, E(\sigma)\, x) \\
&= \int \frac{\chi_\sigma(\lambda)}{\zeta - \lambda}\, (E(d\lambda)\, x, x) = \int \frac{\chi_\sigma(\lambda)}{\zeta - \lambda}\, (x, E(d\lambda)\, x) \\
&= \int \frac{1}{\zeta - \lambda}\, (E(\sigma)\, x, E(d\lambda)\, x) = (E(\sigma)\, x, R(\bar{\zeta}, A)\, x)
\end{aligned}
$$

from which (5-19) follows.

We proceed with the characterization of the domain of A. Let λ be any nonreal complex number. Then the range of $R(\lambda, A)$ coincides with the domain of $\lambda I - A$ and hence with that of A. If x is in D_A then there exists a vector y such that $x = R(i, A)\, y$. Now

$$
\begin{aligned}
\| R(i, A)\, y \|^2 &= (R(i, A)\, y, R(i, A)\, y) \\
&= (R(-i, A)\, R(i, A)\, y, y) \\
&= -\frac{1}{2i}\, ((R(i, A) - R(-i, A))\, y, y) \\
&= -\frac{1}{2i} \int \left(\frac{1}{i - \lambda} - \frac{1}{-i - \lambda} \right) (E(d\lambda)\, y, y) \\
&= \int \frac{1}{1 + \lambda^2}\, (E(d\lambda)\, y, y)
\end{aligned}
$$

Using the reducibility of $R(\zeta, A)$ by the spectral measure we have

$$
\begin{aligned}
(E(\sigma)\, x, x) &= (E(\sigma)^2\, R(i, A)\, y, R(i, A)\, y) \\
&= \| R(i, A)\, E(\sigma)\, y \|^2 = \int_\sigma \frac{1}{1 + \lambda^2}\, (E(d\lambda)\, y, y)
\end{aligned}
$$

and therefore

$$\int \lambda^2 (E(d\lambda) x, x) = \int \frac{\lambda^2}{1 + \lambda^2} (E(d\lambda) y, y) \le \int (E(d\lambda) y, y) = \| y \|^2$$

Thus we have proved that $D_A \subset \{x | \int \lambda^2 (E(d\lambda) x, x) < \infty\}$. To show that we have actually equality we define an operator A_1 by

$$D_{A_1} = \left\{ x \middle| \int \lambda^2 (E(d\lambda) x, x) < \infty \right\}$$

and

$$A_1 x = \lim_{n \to \infty} \int_{-n}^{n} \lambda E(d\lambda) x \qquad \text{for} \qquad x \in D_{A_1}$$

By Theorem 5-1 A_1 is a densely defined self-adjoint operator. We will show that A and A_1 coincide. For arbitrary $u \in H$ we have

$$(R(i, A)(iI - A_1) x, u) = \int \frac{1}{i - \lambda} (E(d\lambda)(iI - A_1) x, u)$$

$$= \int \frac{i - \lambda}{i - \lambda} (E(d\lambda) x, u) = (x, u)$$

Since u is arbitrary we obtain

$$R(i, A)(iI - A_1) x = x \qquad \text{for all} \qquad x \in D_{A_1} \tag{5-20}$$

Thus Range $R(i, A) = D_{A_1}$ which together with Range $R(i, A) = D_{iI-A} = D_A$ shows that $D_A = D_{A_1}$. Now (5-20), together with $R(i, A)(iI - A) x = x$ for all $x \in D_A$, implies that $R(i, A)((iI - A) x - (iI - A_1) x) = 0$. Since $R(i, A)$ is injective it follows that $(iI - A) x = (iI - A_1) x$ for all $x \in D_A$ and hence that $A = A_1$. This completes the proof.

As a straightforward consequence of the spectral theorem we can construct a much more powerful functional calculus than the Riesz–Dunford calculus described in Sec. 2. This calculus is based on integration of bounded Borel functions with respect to spectral measures.

Let A be a self-adjoint operator which is not necessarily bounded. For $f \in B$ we define $f(A)$ by

$$f(A) = \int f(\lambda) E(d\lambda) \tag{5-21}$$

where E is the spectral measure of A.

Theorem 5-6 The map $f \to f(A)$ is an algebra homomorphism of B into $B(H)$ mapping the function 1 onto I and satisfies $\bar{f}(A) = f(A)^*$ for $\bar{f}(\lambda) =$

$\overline{f(\lambda)}$, and

$$\| f(A) x \|^2 = \int |f(\lambda)|^2 (E(d\lambda) x, x) \qquad \text{for each} \qquad x \in H \qquad (5\text{-}22)$$

If $\{f_n\}$ is a uniformly bounded sequence of functions converging pointwise to f then $f_n(A)$ converges to $f(\lambda)$ strongly.

PROOF That the map of $f \to f(A)$ is a homomorphism has been established in the beginning of this section. That $\bar{f}(A) = f(A)^*$ is trivial for simple functions and extends by continuity to all of B. Since $\| f(A)x \|^2 = (f(A)x, f(A)x) = (f(A) \bar{f}(A) x, x)$ (5-22) follows. Finally, since

$$\| f(A) x - f_n(A) x \|^2 = \int |f(\lambda) - f_n(\lambda)|^2 (E(d\lambda) x, x)$$

the last statement is a consequence of the Lebesgue dominated convergence theorem.

If the operator A is bounded we can replace the algebra B by the algebra $B(\sigma(A))$ of all Borel measurable functions defined and bounded on $\sigma(A)$. In this case all polynomials are in $B(\sigma(A))$ and if $p(\lambda) = \sum_{i=0}^{n} a_i \lambda^i$ we have

$$\sum \int \sum_{i=0}^{n} a_i \lambda^i E(d\lambda) = \sum_{i=0}^{n} a_i A^i$$

thus the various definitions of $p(A)$ coincide. In the next section we will extend the present functional calculus even further.

An important consequence of the spectral theorem is the existence of square roots of positive operators. An operator A in a Hilbert space H is called positive if for each $x \in H$ $(Ax, x) \geq 0$. Since the Hilbert space is assumed to be complex a positive operator is self-adjoint. For real Hilbert spaces the self-adjointness has to be postulated.

Theorem 5-7 A bounded operator A is positive if and only if it has a positive square root. The positive square root of a positive operator is unique.

PROOF Let P be a positive square root of A then $(Ax, x) = (P^2 x, x) = \| Px \|^2 \geq 0$ and A is positive. Conversely assume A is positive then its spectrum lies in $[0, \infty)$. The positive function $\lambda^{1/2}$ is in $B(\sigma(A))$ and hence $P = \int \lambda^{1/2} E(d\lambda)$ is a well-defined positive operator. By the properties of the functional calculus $P^2 = A$. To prove uniqueness assume Q is another positive square root of A. The operators Q and A commute as $QA = Q^3 = AQ$ and consequently, P being a limit in norm of polynomials in A, P and Q commute. Let now x be an arbitrary vector in H and put $y = (P - Q) x$. Then if $P^{1/2}$ and $Q^{1/2}$ are arbitrary positive square roots of P and Q, respectively

$$\| P^{1/2} y \|^2 + \| Q^{1/2} y \|^2 = (Py, y) + (Qy, y) = ((P + Q)(P - Q) x, y)$$

$$= ((P^2 - Q^2) x, y) = 0$$

So $P^{1/2}y = Q^{1/2}y = 0$ and hence $Py = Qy = 0$. Now $\|(P - Q)x\|^2 = ((P - Q)^2 x, x) = ((P - Q)y, x) = 0$ and we obtain $Px = Qx$. As x was arbitrary $P = Q$ and the proof is complete.

We saw already that bounded operators on a Hilbert space have properties resembling those of the complex numbers. An important instance is the generalization of the fact that every complex number z can be represented uniquely in the polar form $z = re^{i\theta}$ where $r > 0$ and $0 \le \theta < 2\pi$.

Theorem 5-8 (Polar decomposition)

(a) Every bounded operator T in a Hilbert space H can be written in the form

$$T = VP \tag{5-23}$$

where P is positive and V a partial isometry. The decomposition (5-23) is unique if we require

$$\operatorname{Ker} V = \operatorname{Ker} P \tag{5-24}$$

(b) Every operator $T \in B(H)$ can be written in the form

$$T = QW \tag{5-25}$$

where Q is positive and W a partial isometry. The decomposition (5-25) is unique if we require

$$\operatorname{Ker} W^* = \operatorname{Ker} Q \tag{5-26}$$

PROOF (a) Define $P = (T^*T)^{1/2}$ which exists by Theorem 5-7 as T^*T is clearly a positive operator. Now for each $x \in H$

$$\|Px\|^2 = (P^2 x, x) = (T^*Tx, x) = \|Tx\|^2 \tag{5-27}$$

Define an operator V on Range P by $VPx = Tx$. Equality (5-27) shows that V is isometric on its domain of definition and hence can be extended by continuity to an isometry on $\overline{\operatorname{Range} P}$. Extend the domain of definition of V to all of H by letting $V | \{\operatorname{Range} P\}^{\perp} = V | \operatorname{Ker} P = 0$. So V is a partial isometry with Range P as initial space and $\overline{\operatorname{Range} T}$ as final space. So (5-23) is proved and (5-24) satisfied. To prove uniqueness assume $T = WQ$ is another decomposition of the same type with $\operatorname{Ker} W = \operatorname{Ker} Q$. By Theorem 2-3 W^*W is a projection on the initial space of W which is equal to $\{\operatorname{Ker} W\}^{\perp} = \{\operatorname{Ker} Q\}^{\perp} = \overline{\operatorname{Range} Q}$. Consequently $W^*WQx = Qx$ for each $x \in H$ and hence $T^*T = QW^*WQ = Q^2$ or $Q = (T^*T)^{1/2} = P$ by the uniqueness part of Theorem 5-7. This in turn yields the equality $VP = WP$ or V and W are equal on $\overline{\operatorname{Range} P}$. Now $\{\operatorname{Range} P\}^{\perp} = \operatorname{Ker} P = \operatorname{Ker} V = \operatorname{Ker} W$ and hence V and W are equal.

To prove (b) apply (a) to T^*. We note that necessarily $Q = (TT^*)^{1/2}$ in this case.

Generally similarity of two operators does not imply their unitary equivalence, but for self-adjoint operators even more is true.

Theorem 5-9 Let A and A_1 be two self-adjoint operators acting in the Hilbert spaces H and H_1, respectively. Let $X: H \to H_1$ intertwine A and A_1, that is, $XA = A_1 X$. Then

(a) If X has range dense in H_1 there exists a coisometry V such that $VA = A_1 V$
(b) If X is one-to-one there exists an isometry W such that $WA = A_1 W$
(c) If X is one-to-one and has range dense in H_1 (in particular if X is boundedly invertible) then there exists a unitary U such that $UA = A_1 U$

PROOF From $XA = A_1 X$ it follows by taking adjoints that $AX^* = X^* A_1$ and hence $AX^*X = X^* A_1 X = X^* XA$ or $A(X^*X) = (X^*X) A$ and analogously $A_1(XX^*) = (XX^*) A_1$. By a standard approximation argument it follows that

$$A(X^*X)^{1/2} = (X^*X)^{1/2} A \tag{5-28}$$

and

$$A_1(XX^*)^{1/2} = (XX^*)^{1/2} A_1 \tag{5-29}$$

Now assume X has range dense in H_1. Since $\{0\} = \{\text{Range }X\}^{\perp} = \text{Ker }X^* = \text{Ker}(XX^*)^{1/2} = \{\text{Range}(XX^*)^{1/2}\}^{\perp}$ it follows that also $(XX^*)^{1/2}$ has range dense in H_1. From the equality $\|X^*y\| = \|(XX^*)^{1/2} y\|$ it follows that if we define V by

$$VX^*y = (XX^*)^{1/2} y$$

then V can be extended by continuity for an isometry from $\overline{\text{Range }X^*}$ onto H_1. Extend V to all of H by defining $V|\text{Ker }X = 0$ and V becomes a coisometry satisfying $VX^* = (XX^*)^{1/2}$. By our assumption $(XX^*)^{1/2}$ has dense range hence $(XX^*)^{-1/2}$ is a closed densely defined operator. Thus $VX^*(XX^*)^{-1/2} y = y$ for all y in $\text{Range}(XX^*)^{1/2}$. Since V is isometric on $\text{Range }X^*$ we have $X^*(XX^*)^{-1/2}$ is isometric on its domain of definition hence extendible by continuity to an isometry on H_1 which has to coincide with V^*. So we have

$$V = (XX^*)^{-1/2} X \tag{5-30}$$

Since from (5-29) it follows that $A_1(XX^*)^{-1/2} = (XX^*)^{-1/2} A_1$ we have

$$VA = (XX^*)^{-1/2} XA = (XX^*)^{-1/2} A_1 X = A_1(XX^*)^{-1/2} X = A_1 V$$

which proves (a). Part (b) follows by duality considerations. Finally, if X is one-to-one and has dense range then both $X^*(XX^*)^{-1/2}$ and $X(X^*X)^{-1/2}$ are isometric. Now from the equality $X(X^*X) = (XX^*) X$ it follows that $X(X^*X)^{1/2} = (XX^*)^{1/2} X$ and hence that $(XX^*)^{-1/2} X = X(X^*X)^{-1/2}$. This means that V given by (5-30) is also isometric and therefore unitary.

6. SPECTRAL REPRESENTATIONS

The spectral theorem for self-adjoint operators proved in the previous section while stating that diagonalization of these operators is possible does not yield much insight into their structure.

In the spirit of Sec. I-4 we would like to describe a general self-adjoint operator in terms of operators of simple type. Essentially given a self-adjoint operator A we look for a model of it, that is a unitarily equivalent operator, acting in a function space.

Consider a positive Borel measure μ on $\mathbb{R}$. For each $\varphi \in B$, that is, for each bounded Borel measurable function on $\mathbb{R}$ we define a multiplication $M_{\varphi,\mu}$ acting in the Hilbert space $L^2(\mu)$ by

$$M_{\varphi,\mu}f = \varphi f \tag{6-1}$$

We could replace B by $B(\Omega)$ the set of all bounded Borel measurable functions on the closed set Ω which we assume contains the support of μ. Algebraically we have induced in $L^2(\mu)$ a B, or $B(\Omega)$, module structure. We note that the map $\varphi \to M_{\varphi,\mu}$ is an algebra homomorphism of B, or $B(\Omega)$, into $B(L^2(\mu))$ which satisfies

$$M_{\varphi,\mu}^* = M_{\bar{\varphi},\mu} \tag{6-2}$$

and

$$\|M_{\varphi,\mu}\| \leq \|\varphi\|_\infty \tag{6-3}$$

where $\|\varphi\|_\infty$ is the sup norm of φ. If the support of μ is compact then with $\chi(\lambda) = \lambda$ $M_{\chi,\mu}$ is a bounded self-adjoint operator, and so is a direct sum of such operators $M_{\chi,\mu_1} \oplus \cdots \oplus M_{\chi,\mu_n}$ acting in $L^2(\mu_1) \oplus \cdots \oplus L^2(\mu_n)$. Actually if $\{\mu_\alpha\}$ is any family of positive measures with a uniformly bounded support then $\oplus_\alpha M_{\chi,\mu_\alpha}$ acting in $\oplus_\alpha L^2(\mu_\alpha)$ is a bounded self-adjoint operator. If we want the Hilbert space under consideration to be separable then necessarily the family $\{\mu_\alpha\}$ has to be countable.

Given a self-adjoint operator A in a Hilbert space H a unitary map $\Phi: H \to \oplus_\alpha L^2(\mu_\alpha)$ is called a *spectral representation* of A if for each $\varphi \in B$ we have

$$(\Phi\varphi(A)x)_\alpha = M_{\varphi,\mu_\alpha}(\Phi x)_\alpha \tag{6-4}$$

for all $x \in H$.

A moment's reflection brings us to the conclusion that if there exists a spectral representation it is not unique. The questions of existence and uniqueness, assuming some extra conditions, of spectral representations are central in this section.

To simplify matters as much as possible we will not discuss the most general self-adjoint operator but rather restrict ourselves to the case of finite multiplicity. The general case can be handled in similar fashion.

Given a self-adjoint operator A with the associated spectral measure E we induce a B-module structure on H by letting

$$\varphi \cdot x = \varphi(A)x = \int \varphi(\lambda) E(d\lambda) x \tag{6-5}$$

A self-adjoint operator A is called *cyclic* if there exists a vector x in H such that the set of vectors $\{\varphi(A)x \mid \varphi \in B\}$ is dense in H. For a bounded operator this is equivalent to the density of the set $\{p(A)x \mid p \in \mathbb{C}[\lambda]\}$. More generally a set of vectors $\{x_\alpha\}$ in H is a *set of generators* for A if the set of all finite sums $\sum \varphi_\alpha(A) x_\alpha$ with $\varphi_\alpha \in B$ is dense in H. A self-adjoint operator has *finite multiplicity* if there exists a finite set of generators for it. *A minimal set of generators* is a set of generators of smallest possible cardinality. The *multiplicity* of A is the cardinality of a minimal set of generators.

Let $\{x_1, \ldots, x_r\}$ be a fixed set of generators for a self-adjoint operator A and let B^r be the cartesian product of r copies of B. Clearly B^r is a B-module. We define the map $\rho: B^r \to H$ by

$$\rho(\varphi_1, \ldots, \varphi_r) = \sum_{i=1}^{r} \varphi_i(A) x_i \tag{6-6}$$

where $x_1, \ldots, x_r$ is the fixed set of generators for A. The map ρ is, by elementary properties of the functional calculus a B-module homomorphism, and by our assumption that $x_1, \ldots, x_r$ is a set of generators it follows that ρ has range which is dense in H.

Computing the norm of $\rho(\varphi_1, \ldots, \varphi_r)$ we obtain

$$\left\| \sum \varphi_i(A) \cdot x_i \right\|^2 = \sum_i \sum_j (\varphi_i(A) x_i, \varphi_j(A) x_j) = \sum_i \sum_j (\bar{\varphi}_j(A) \varphi_i(A) x_i, x_j)$$

$$= \sum_i \sum_j \int \overline{\varphi_j(\lambda)} \varphi_i(\lambda) (E(d\lambda) x_j, x_j)$$

Define now the (complex) measures μ_{ij} by

$$\mu_{ij}(\sigma) = (E(\sigma) x_i, x_j) \tag{6-7}$$

for all Borel sets σ and let $\mathbf{M}$ be the matrix whose i, j entry is μ_{ij}. We call such an object a *matrix measure* [29]. We say a matrix measure is a positive matrix measure if for each Borel set σ, $\mathbf{M}(\sigma)$ is a nonnegative definite Hermitian matrix. It is easily checked that the matrix measure $\mathbf{M}$ constructed in (6-7) is a positive matrix measure. Indeed let σ be a Borel subset of $\mathbb{R}$ and let $\alpha_1, \ldots, \alpha_r$ be complex numbers then with $a = (\alpha_1, \ldots, \alpha_r)$

$$(\mathbf{M}(\sigma) a, a) = \sum_i \sum_j \mu_{ij} \alpha_i \bar{\alpha}_j = \sum_i \sum_j (E(\sigma) x_i, x_j) \alpha_i \bar{\alpha}_j$$

$$= \left(E(\sigma) \sum_i \alpha_i x_i, \sum_j \alpha_j x_j \right) = \left\| E(\sigma) \sum_i \alpha_i x_i \right\|^2 \geq 0$$

In terms of the matrix measure introduced we have

$$\left\| \rho F \right\|^2 = \left\| \sum_{i=1}^{r} f_i(A) x_i \right\|^2 = \int (d\mathbf{M}F, F) \tag{6-8}$$

where $F \in B^r$ is the vector function whose components are $f_1, \ldots, f_r$. Equality (6-8) indicates that if we define properly the L^2 space of a matrix measure $\mathbf{M}$ which we will denote naturally by $L^2(\mathbf{M})$ then the map $\rho; B^r \to H$ will have a natural extension to a unitary map of $L^2(\mathbf{M})$ onto H. Moreover, such a map satisfies

$$\rho(\varphi F) = \varphi(A)(\rho F) \qquad \text{for all} \qquad \varphi \in B \tag{6-9}$$

Also for any vector x in the domain of A we have

$$[\rho^{-1}(Ax)](\lambda) = \lambda \cdot (\rho^{-1}x)(\lambda) \tag{6-10}$$

Thus in the functional representation A acts like multiplication by λ.

We note that $\mathbf{M}$ has a convenient description in terms of the spectral measure $E(\cdot)$ that is associated with A. If $J: \mathbb{C}^r \to H$ is the map sending $(\alpha_1, \ldots, \alpha_r)$ onto $\sum_{i=1}^r c_i x_i$ then for each Borel set σ we have

$$\mathbf{M}(\sigma) = J^* E(\sigma) J \tag{6-11}$$

To define $L^2(\mathbf{M})$ we proceed as follows. We denote by $L_0^2(\mathbf{M})$ the set of all r-tuples $(f_1, \ldots, f_r)$ of Borel measurable functions for which

$$\|F\|^2 = \int_{-\infty}^{\infty} (d\mathbf{M} F, F) = \int_{-\infty}^{\infty} \sum_{i=1}^r \sum_{j=1}^r f_i(\lambda) \, \overline{f_j(\lambda)} \, d\mu_{ij} < \infty \tag{6-12}$$

and define $L^2(\mathbf{M})$ as the set of all equivalence classes in $L_0^2(\mathbf{M})$ modulo the set of null function, a null function being one for which $\|F\| = 0$. With the inner product in $L^2(\mathbf{M})$ defined by

$$(F, G) = \int (d \, \mathbf{M} F, G) = \int \sum_{i=1}^r \sum_{j=1}^r f_i(\lambda) \, \overline{g_j(\lambda)} \, d\mu_{ij} \tag{6-13}$$

$L^2(\mathbf{M})$ becomes a pre-Hilbert space and the only open question is that of completeness. There is one class of matrix measures for which $L^2(\mathbf{M})$ is clearly complete, namely the class of positive diagonal measures, that is, those for which $i \neq j$ implies $\mu_{ij} = 0$ and the diagonal elements are positive measures. If $\mu_1, \ldots, \mu_r$ are the diagonal elements of a diagonal matrix measure then in this case

$$\|F\|^2 = \int \sum_{i=1}^r |f_i(\lambda)|^2 \, d\mu_i = \sum_{i=1}^r \int |f_i(\lambda)|^2 \, d\mu_i = \sum_{i=1}^r \|f_i\|^2$$

where $\|f_i\|$ is the norm of f_i and an element of $L^2(\mu_i)$. Hence in this case $L^2(\mathbf{M})$ is clearly equal to the direct sum $L^2(\mu_1) \oplus \cdots \oplus L^2(\mu_r)$ which is a complete space. We will use this observation to show completeness of $L^2(\mathbf{M})$ by exhibiting a unitary map that diagonalizes $\mathbf{M}$.

As a first step we simplify the problem by replacing matrix measures by density matrices and one scalar measure. We choose a positive measure μ such that all μ_{ij} are absolutely continuous with respect to μ, $\mu_{ij} \ll \mu$. One candidate is the sum of the total variation of all the μ_{ij}. A better choice turns out later to be the trace of $\mathbf{M}$. If $m_{ij} = d\mu_{ij}/d\mu$ is the Radon–Nikodym derivative of μ_{ij} with respect to μ then we introduce the density matrix

$$M(\lambda) = m_{ij}(\lambda) \tag{6-14}$$

Lemma 6-1 If $M(\lambda)$ is the density matrix of a matrix measure $\mathbf{M}$ with respect to a scalar measure μ then $M(\lambda)$ is nonnegative definite μ-a.e.

PROOF Observe first that the set $\Lambda_0 = \{\lambda | (m_{ij}(\lambda))$ is nonnegative definite$\}$ is a measurable set, for it is the intersection of all sets $\{\lambda | \sum_i \sum_j m_{ij}(\lambda) \xi_i \bar{\xi}_j \geq 0\}$ where $(\xi_1, \ldots, \xi_n)$ vary over all vectors with rational coordinates. If $M(\lambda)$ is not nonnegative definite μ-a.e. then there exists a set Λ of positive μ-measure in the complement of Λ_0, and a rational vector $(\xi_1, \ldots, \xi_n)$ such that for λ in Λ $\sum_i \sum_j m_{ij}(\lambda) \xi_i \bar{\xi}_j < 0$. This would imply

$$(M(\Lambda) x, x) = \int \sum_i \sum_j m_{ij}(\lambda) \xi_i \bar{\xi}_j \, d\mu < 0$$

contradicting our assumption that M is a positive matrix measure.

Consider next the set of all positive matrix measures on $\mathbb{R}$. We say that M *divides* N, and write $M | N$, if there exists a Borel matrix function H such that

$$dM = H^* dN \, H \tag{6-15}$$

Two matrix measures M and N are *equivalent* and we write $M \sim N$ if $M | N$ and $N | M$.

The division relation is clearly reflexive and transitive and hence induces a partial order in the set of all matrix measures. Relation (6-15) is a generalization of the concept of absolute continuity as applied to matrix measures. Heuristically the matrix function H has the interpretation of a "square root" of a generalized Radon–Nikodym derivative of M with respect to N. We point out that M and N do not have to be necessarily of the same size. In that case H will not be a square matrix. For scalar measures μ and ν we have of course $\mu | \nu$ if and only if $\mu \ll \nu$.

The partial order in the set of positive matrix measures is reflected in the corresponding $L^2(M)$ spaces. Given two matrix measures M and N we say that a map $U: L^2(M) \to L^2(N)$ is an *embedding* if it is an injective B-homomorphism. If U is also an isometry we say U is an *isometric embedding*.

The next lemma provides a large class of isometric embeddings.

Lemma 6-2 Let M and N be positive matrix measures and assume that $M | N$. Then there exists an isometric embedding of $L^2(M)$ into $L^2(N)$.

PROOF Since $M | N$ there exists a measurable matrix function H such that (6-15) holds. Define $U_M^N: L^2(N)$ by

$$U_M^N F = HF \qquad \text{for} \qquad F \in L^2(M) \tag{6-16}$$

then clearly

$$\|U_M^N F\|^2 = \int (dN \, HF, HF) = \int (H^* dN \, HF, F) = \int (dM \, F, F) = \|F\|^2$$

So U_M^N is an isometry and it is easily checked that it is a B-homomorphism.

We note that the set of isometries U_M^N is a *coherent set of isometries* [18] in the sense that if $M | N$ and $N | S$ then we have

$$U_M^S = U_N^S U_M^N \tag{6-17}$$

The equivalence of two matrix measures can be described also in terms of their density matrices with respect to a common scalar measure. To this end we define a notion of equivalence between measurable matrix functions. Let M and N be Borel measurable $n \times m$ matrix functions defined on a subset of $\mathbb{R}$, and let σ be a positive measure on $\mathbb{R}$. We say that M and N are σ-*equivalent* if there exist σ-a.e. invertible measurable $n \times n$ and $m \times m$ matrix functions P and R such that

$$M(\lambda) = P(\lambda) N(\lambda) R(\lambda) \qquad \sigma\text{-a.e.} \tag{6-18}$$

If M and N are square matrix functions we say that M and N are *unitarily σ-equivalent* if there exists a measurable σ-a.e. unitary matrix function P such that

$$M(\lambda) = P(\lambda)^* N(\lambda) P(\lambda) \qquad \sigma\text{-a.e.} \tag{6-19}$$

It is clear that both relations are bona fide equivalence relations and unitary σ-equivalence implies σ-equivalence. Also if ν is a positive measure and $\nu \ll \sigma$ then σ-equivalence implies ν-equivalence.

We prove now a lemma which is the main technical result needed for the proof of completeness. In terms of the equivalence notions introduced we can state it as follows.

Lemma 6-3 Let **M** be a positive matrix measure and let M be its density matrix with respect to a positive measure μ that satisfies $m_{ij} \ll \mu$. Then there exists a diagonal matrix function D such that M and D are unitarily μ-equivalent.

Alternately stated there exists a measurable matrix function H such that

$$H(\lambda)^* H(\lambda) = I \tag{6-20}$$

and

$$M(\lambda) = H(\lambda)^* D(\lambda) H(\lambda) \tag{6-21}$$

hold μ-a.e.

PROOF We note that μ-a.e. $M(\lambda)$ is a nonnegative definite matrix and hence can be diagonalized by a unitary matrix. The content of the lemma is that the pointwise diagonalizations can be fitted together in a globally measurable way.

Observe first that if there exists a sequence of sets e_n and μ-measurable matrix functions $H^{(n)}(\lambda)$ and $D^{(n)}(\lambda)$ which are unitary and diagonal, respectively, and such that $M(\lambda) = H^{(n)}(\lambda)^* D^{(n)}(\lambda) H^{(n)}(\lambda)$ holds μ-a.e. on e_n then $H(\lambda)$ and $D(\lambda)$ defined by $H(\lambda) = H^{(n)}(\lambda)$ and $D(\lambda) = D^{(n)}(\lambda)$ for $\lambda \in e_n - \cup_{i=1}^{n-1} e_i$ satisfy (6-20) and (6-21) on $\cup_{i=1}^{\infty} e_i$.

By a theorem of Lusin [29] given any $\varepsilon > 0$ the m_{ij} are actually continuous on a set whose complement has at most μ-measure ε. Taking a sequence of ε_i converging to zero, and recalling that μ is a finite measure, then, but for a set of μ-measure zero, $\mathbb{R}$ is the union of measurable sets where all

m_{ij} are continuous. By the preceding remark we only need to prove the lemma on a measurable set e where all m_{ij} are continuous.

Now if the $m_{ij}(\lambda)$ are continuous on a set e so are the eigenvalues of $M(\lambda)$, that is, the zeros of $\det[M(\lambda) - \zeta I]$, for they are the zeros of a polynomial with continuously varying coefficient. However, in general it is impossible to choose a continuously varying set of corresponding eigenvectors. The trouble can occur at points where the multiplicity of an eigenvalue changes. To avoid this difficulty we let $N(\lambda)$ denote the number of distinct eigenvalues of λ. Then the sets $\{\lambda \in e \,|\, n(\lambda) \geq s\}$ are open in the relative topology of e. It follows that $e_s = \{\lambda \in e \,|\, n(\lambda) = s\} = \{\lambda \in e \,|\, n(\lambda) \geq s\} - \cup_{k=1}^{\infty} \{\lambda \in e \,|\, n(\lambda) \geq s + 1/k\}$ is a Borel set. Thus e decomposes into the union of $e_1, \ldots, e_n$ which are disjoint Borel sets. Thus it suffices to construct $H(\lambda)$ and $D(\lambda)$ on any of the e_k. Let $\lambda_0 \in e_k$ and let $\hat{\varphi}_1(\lambda_0), \ldots, \hat{\varphi}_k(\lambda)$ be the distinct eigenvalues of $M(\lambda_0)$. Then there exists by continuity of the eigenvalues a unique enumeration $\hat{\varphi}_1(\lambda), \ldots, \hat{\varphi}_k(\lambda)$ of the eigenvalues of $M(\lambda)$ such that the $\hat{\varphi}_i(\lambda)$ are continuous on e_k.

Let $\varepsilon > 0$ be such that no two distinct eigenvalues of $M(\lambda_0)$ are closer than ε. Let γ_i be a positively oriented circular path with center at $\hat{\varphi}_i(\lambda_0)$ and radius less than $\varepsilon/2$. Let $E_{M(\lambda_0)}$ and $E_{M(\lambda)}$ be the spectral measures of $M(\lambda_0)$ and $M(\lambda)$, respectively. For λ in a sufficiently small neighborhood N_1 of λ_0 $\hat{\varphi}_i(\lambda)$ will be within the circle γ_i. In that case we have

$$E_i(\lambda_0) = E_{M(\lambda_0)}(\{\hat{\varphi}_i(\lambda_0)\}) = \frac{1}{2\pi i} \int_{\gamma_i} R(\zeta, M(\lambda_0))\, d\zeta$$

and

$$E_i(\lambda) = E_{M(\lambda)}(\{\hat{\varphi}_i(\lambda)\}) = \frac{1}{2\pi i} \int_{\gamma_i} R(\zeta, M(\lambda))\, d\zeta$$

Since $M(\lambda)$ varies continuously with $\lambda \in N_1$ so does $R(\zeta, M(\lambda))$ and therefore $E_i(\lambda)$ is a continuous function of $\lambda \in N_1$.

Choose an orthonormal basis $v_1, \ldots, v_n$ of $\mathbb{C}^n$ consisting of eigenvectors of $M(\lambda_0)$ ordered so that $E_i(\lambda_0) v_j = v_j$ for $n_{i-1} < j \leq n_i$, $0 = n_0 < n_1 < \cdots < n_k = n$. Define $\hat{v}_j(\lambda) = E_i(\lambda) v_j$ for $n_{i-1} < j \leq n_i$. The $\hat{v}_j(\lambda)$ depend continuously on $\lambda \in N_1$ and as $E_i(\lambda) \hat{v}_j(\lambda) = \hat{v}_j(\lambda)$ they form a basis of eigenvectors of $M(\lambda)$, but it may fail to be an orthonormal basis. Since eigenvectors corresponding to distinct eigenvalues of a self-adjoint operator are orthogonal we can produce an orthonormal basis for $\mathbb{C}^n$ consisting of eigenvectors of $M(\lambda)$ by applying the Gram–Schmidt process to each of the sets $\hat{v}_{n_{i-1}+1}(\lambda), \ldots, \hat{v}_{n_i}(\lambda)$. If $v_{n_{i-1}+1}(\lambda), \ldots, v_{n_i}(\lambda)$ is the resultant orthonormal basis for the range of $E_i(\lambda)$ then $v_1(\lambda), \ldots, v_n(\lambda)$ is an orthonormal basis for $\mathbb{C}^n$ and $M(\lambda) v_i(\lambda) = \varphi_i(\lambda) v_i(\lambda)$, $i = 1, \ldots, n$. Let $e_1, \ldots, e_n$ be the standard orthonormal basis of $\mathbb{C}^n$. Let $H(\lambda)$ be a matrix function such that $H(\lambda) e_i = v_i(\lambda)$ $i = 1, \ldots, n$. As $H(\lambda)$ transforms one orthonormal basis into another it is necessarily unitary. Moreover from $M(\lambda) v_i(\lambda) = \varphi_i(\lambda) v_i(\lambda)$ it follows that $M(\lambda) H(\lambda) e_i = \varphi_i(\lambda) H(\lambda) e_i$ or (6-21) holds with

$$D(\lambda) = \text{diag}\big(\varphi(\lambda), \dots, \varphi_n(\lambda)\big)$$

This completes the proof of the lemma.

Theorem 6-4 If M is a positive measure on $\mathbb{R}$ then $L^2(M)$ is a Hilbert space.

PROOF Let μ, H, and D be as in the previous lemma and let $d\mathbf{D} = D d\mu$. The map $U_\mathbf{M}^\mathbf{D} \colon L^2(\mathbf{M}) \to L^2(\mathbf{D})$ given by (6-16) is an isometric embedding. However, since $D(\lambda) = H(\lambda) M(\lambda) H(\lambda)^*$, it is invertible and we have

$$(U_\mathbf{M}^\mathbf{D})^{-1} = U_\mathbf{D}^\mathbf{M} = (U_\mathbf{M}^\mathbf{D})^* \tag{6-22}$$

where $U_\mathbf{D}^\mathbf{M} G = H^* G$. Thus $U_\mathbf{M}^\mathbf{D}$ is a unitary map and

$$L^2(\mathbf{D}) = L^2(\delta_1) \oplus \cdots \oplus L^2(\delta_r) \tag{6-23}$$

where δ_i are the measures defined by $\delta_i(\sigma) = \int d_i(\lambda) \, d\mu$. Thus $L^2(\mathbf{D})$ is complete and so is $L^2(\mathbf{M})$.

For $\varphi \in B$ we define the operator on multiplication by φ in $L^2(\mathbf{M})$ by

$$M_{\varphi,\mathbf{M}} F = \varphi F \qquad \text{for} \qquad F \in L^2(\mathbf{M}) \tag{6-24}$$

In terms of the identity function χ, $\chi(\lambda) = \lambda$ we can summarize the previous results and exhibit a functional representation for the Hilbert space H and the self-adjoint operator A acting in it.

Theorem 6-5 Any operator A in a Hilbert space H is unitarily equivalent to an operator $M_{\chi,\mathbf{M}}$ in $L^2(\mathbf{M})$ for some positive matrix measure $\mathbf{M}$ on the real line if and only if A is a finitely generated self-adjoint operator.

The combination of Lemma 6-3 and Theorem 6-4 which is now available to us poses naturally the question of canonical forms. Our aim is to simplify a matrix density by transformations of the form (6-19) for σ-a.e. unitary measurable matrix P, and this simplification will be reflected in a simpler spectral representation for the corresponding self-adjoint operator. This problem of canonical forms for self-adjoint operators is a classical one, first resolved by Hellinger. Our approach uses only simple matrix manipulation. The price for that is the loss of generality involved by assuming finite multiplicity.

Lemma 6-6 Let $\mathbf{L} = (\lambda_{ij})$ be a positive matrix measure and let σ be a positive measure such that $\mathbf{L} | \sigma I$. Then there exists a diagonal matrix measure $\mathbf{M}$ with diagonal entries $\mu_1, \dots, \mu_p$ such that $d\mu_i = m_i \, d\sigma$ and the following statements hold
(a) $\mu_1 \gg \mu_2 \gg \cdots \gg \mu_p$ and
(b) $\mathbf{L}$ and $\mathbf{M}$ are unitarily σ-equivalent.

Moreover if **N** is another diagonal matrix measure with diagonal entries $v_1, \ldots, v_p$ such that $dv_i = n_i \, d\rho$ and the statements

(*a'*) $v_1 \gg v_2 \gg \cdots \gg v_p$ and
(*b'*) **L** and **N** are unitarily ρ-equivalent hold then **M** and **N** are unitarily τ-equivalent where $\tau = \rho \wedge \sigma$ is the infimum of the measures ρ and σ [62].

PROOF By Lemma 6-3 it suffices to show that given a diagonal matrix measure **L** it can be reduced to canonical form. Thus without loss of generality we let **L** be diagonal with diagonal elements $\lambda_1, \ldots, \lambda_p$ where, by assumption, $\lambda_i \ll \sigma$. Let $d\lambda_i = l_i \, d\sigma$, that is, l_i is the Radon–Nikodym derivative of λ_i with respect to σ. For simplicity of notation we assume $p = 2$. Let $\lambda_2 = \lambda_2' + \lambda_2''$ be the Lebesque decomposition of λ_2 with respect to λ_1, assuming $\lambda_2' \ll \lambda_1$ and $\lambda_2'' \perp \lambda_1$. Let $l_2 = l_2' + l_2''$ with l_2' and l_2'' the respective Radon–Nikodym derivatives of λ_2' and λ_2'' with respect to σ. Let $E_2 = \{\lambda \mid l''(\lambda) \neq 0\}$ and $F_2 = \{\lambda \mid l''(\lambda) = 0\}$ and let χ_{E_2} and χ_{F_2} be the corresponding characteristic function of the two sets. Define a 2×2 matrix function $H(\lambda)$ by

$$H(\lambda) = \begin{pmatrix} \chi_{F_2}(\lambda) & \chi_{E_2}(\lambda) \\ \chi_{E_2}(\lambda) & \chi_{F_2}(\lambda) \end{pmatrix}$$

A simple calculation yields the equality

$$L(\lambda) = H(\lambda)^* \, L'(\lambda) \, H(\lambda) \tag{6-25}$$

where

$$L(\lambda) = \begin{pmatrix} l_1(\lambda) & 0 \\ 0 & l_2(\lambda) \end{pmatrix} \quad \text{and} \quad L'(\lambda) = \begin{pmatrix} l_1(\lambda) + l_2''(\lambda) & 0 \\ 0 & l_2'(\lambda) \end{pmatrix}$$

$$\tag{6-26}$$

which proves the statement for $p = 2$. The necessary modifications needed to make the proof work for $p > 2$ are obvious. Thus we proved the existence of the canonical diagonalization.

To prove the uniqueness part we note the obvious fact that if **L** and **M** are unitarily σ-equivalent they are also unitarily σ'-equivalent for any $\sigma' \ll \sigma$. It follows that if we form the infimum $\tau = \rho \wedge \sigma$ of the measures ρ and σ transitivity **M** and **N** are unitarily τ-equivalent. Thus τ-a.e. the diagonal matrices

$$\begin{pmatrix} m_1'(\lambda) & \cdots & 0 \\ & \cdot & \\ & \cdot & \\ & \cdot & \\ 0 & \cdots & m_p'(\lambda) \end{pmatrix} \quad \text{and} \quad \begin{pmatrix} n_1'(\lambda) & \cdots & 0 \\ & \cdot & \\ & \cdot & \\ 0 & \cdots & n_p'(\lambda) \end{pmatrix}$$

are unitarily equivalent. Here m_i' and n_i' are the Radon–Nikodym derivatives of μ_i and ν_i with respect to τ. Since assumptions (a) and (a') imply $m_{i+1}'(\lambda) = 0$ whenever $m_i'(\lambda) = 0$ it follows that the zero sets of m_i' and n_i' are equal τ-a.e. This is equivalent to $\mu_i \approx \nu_i$.

As an immediate corollary we obtain the *ordered spectral representation* for a finitely generated self-adjoint operator A. The integer p is referred to as the *multiplicity* of A.

Theorem 6-7 Let A be a finitely generated self-adjoint generator in a Hilbert space H. Then there exists a finite sequence of positive measures $\mu_1 \gg \mu_2 \gg \cdots \gg \mu_p$ such that A is unitarily equivalent to

$$M_{\chi,\mu_1} \oplus \cdots \oplus M_{\chi,\mu_p} \tag{6-27}$$

acting in

$$L^2(\mu_1) \oplus \cdots \oplus L^2(\mu_p) \tag{6-28}$$

The sequence $\mu_1, \ldots, \mu_p$ is determined by A up to equivalence of measures.

There is another representation associated with a self-adjoint operator which is closely related to the ordered spectral representation.

Theorem 6-8 Let A be a finitely generated self-adjoint operator in a Hilbert space H. Then there exists a finite sequence of mutually singular positive measures $\nu_1, \ldots, \nu_p$ such that A is unitarily equivalent to

$$M_{\chi,\mathbf{N}_1} \oplus \cdots \oplus M_{\chi,\mathbf{N}_p} \tag{6-29}$$

acting in

$$L^2(\mathbf{N}_1) \oplus \cdots \oplus L^2(\mathbf{N}_p) \tag{6-30}$$

where $N_j = \nu_j I_j$, I_j being the $j \times j$ identity matrix. The sequence of measures $\nu_1, \ldots, \nu_p$ is determined by A up to equivalence of measures.

The representation (6-29) is referred to as the *canonical spectral representation* of A.

The passage from Theorem 6-7 to Theorem 6-8 is straightforward using repeatedly the Lebesque decomposition theorem for measures. We omit the details.

We remark that another alternative way of writing the canonical spectral representation is to define the matrix measure by

$$\mathbf{N} = \begin{pmatrix} \nu_1 + \cdots + \nu_p & & & & \\ & \nu_2 + \cdots + \nu_p & & & \\ & & \ddots & & \\ & & & \ddots & \\ & & & & \nu_p \end{pmatrix} \tag{6-31}$$

then A is unitarily equivalent to the operator

$$M_{\chi,\mathbf{N}} \tag{6-32}$$

acting in $L^2(\mathbf{N})$.

Having obtained a spectral representation for a self-adjoint operator it is easy to extend the functional calculus constructed in Sec. 5.

Let $\Phi: H \to \oplus_{i=1}^p L^2(\mu_i)$ be the ordered spectral representation of a finitely generated self-adjoint operator A. Since the spectral representation is ordered we have $\mu_p \ll \mu_{p-1} \ll \cdots \ll \mu_1 = \mu$. Each space $L^2(\mu_i)$ is a module over $L^\infty(\mu)$, that is, multiplication by $L^\infty(\mu)$ functions makes sense and the usual module axioms are satisfied. Thus also the direct sum $\oplus_{i=1}^p L^2(\mu_i)$ becomes a module over $L^\infty(\mu)$. Using the map Φ we induce an $L^\infty(\mu)$ module structure in H by defining for each $x \in H$ and $\varphi \in L^\infty(\mu)$

$$\varphi(A)x = \Phi^{-1}(\varphi \cdot \Phi x) \tag{6-33}$$

It is easily checked that for functions $\varphi \in B$ the new definition coincides with the previous one. We summarize the result.

Theorem 6-9 The map $\varphi \to \varphi(A)$ defined by (6-33) is an isometric algebra isomorphism of $L^\infty(\mu)$ into $B(H)$ that satisfies $\bar{\varphi}(A) = \varphi(A)^*$.

Suppose now we are given two self-adjoint operators A_1 and A_2 acting in H_1 and H_2, respectively. By our functional calculus each of the spaces H_1 and H_2 becomes a B-module. Given a map $X: H_1 \to H_2$ that intertwines A_1 and A_2, that is, for which

$$XA_1 = A_2 X \tag{6-34}$$

then it follows easily that for each function $\varphi \in B$ we have

$$X\varphi(A_1) = \varphi(A_2)X \tag{6-35}$$

and this means that $X: H_1 \to H_2$ is a B-module homomorphism. Our object is to study these module homomorphisms in terms of the spectral representations of A_1 and A_2.

As a consequence of Theorem 6-5 the study of operators intertwining two (finitely generated) self-adjoint operators reduces to those intertwining two operators of the form $M_{\chi,\mathbf{M}}$. In the set of all matrix measures we single out the set of all *scalar type measures* which are the matrix measures of the form σI, diagonal matrix measures with all diagonal elements being equal to σ.

Given a matrix measure $\mathbf{M}$, a subspace K of $L^2(\mathbf{M})$ is called an *invariant subspace* if

$$M_{\varphi,\mathbf{M}}K \subset K \tag{6-36}$$

for all $\varphi \in B$, that is, if it is invariant under all multiplication operators by bounded measurable functions.

It is important to have a characterization of invariant subspaces and this is given by the next theorem.

Theorem 6-10 A subspace K of $L^2(\sigma I)$ is an invariant subspace if and only f $K = PL^2(\sigma I)$ where P is a measurable σ-a.e. projection valued matrix function.

PROOF The if part is trivial. So assume K is an invariant subspace of $L^2(\sigma I)$ where I is the $n \times n$ identity matrix. Let $e_1, \ldots, e_n$ be the standard orthonormal basis in $\mathbb{C}^n$. Let ψ_i denote the orthogonal projection of the constant function e_i onto K. We apply the Gram–Schmidt procedure to the vectors $\psi_1(\lambda), \ldots, \psi_n(\lambda)$ to obtain locally an orthonormal set $\varphi_1(\lambda), \ldots, \varphi_{n(\lambda)}(\lambda)$. The number $n(\lambda)$ of elements in the set varies with λ. It is clear from the way the φ_i are constructed that they are measurable functions. Let $\varphi_i(\lambda)$ have $\varphi_{ij}(\lambda)$ as its components relative to the standard orthonormal basis in $\mathbb{C}^n$. Define a projection P by

$$(Pf)(\lambda) = P(\lambda) f(\lambda) = \sum_k (f(\lambda), \varphi_k(\lambda)) \, \varphi_k(\lambda)$$

Relative to the standard orthonormal basis in $\mathbb{C}^n$ $P(\lambda)$ has the matrix representation $p_{ij}(\lambda)$ with $p_{ij}(\lambda) = \sum_k \varphi_{ki}(\lambda) \overline{\varphi_{kj}(\lambda)}$ and is therefore measurable. We will show that $K = PL^2(\sigma I)$. It is clear that $K \subset PL^2(\sigma I)$. Suppose $g \in PL^2(\sigma I)$ is orthogonal to K, then necessarily $\int (g(\lambda), \varphi_j(\lambda)) \lambda^n \, d\sigma = 0$ for all n. This means that g is pointwise orthogonal σ-a.e. to all φ_j hence is the zero function by the definition of P.

Clearly a subspace K is invariant if and only if its orthogonal complement $K^\perp$ is invariant. If $P^\perp$ is the projection valued function corresponding to $K^\perp$ then we have $P^\perp = I - P$.

The next theorem is a characterization of all B-homomorphisms of $L^2(\sigma I)$.

Theorem 6-11 Let $X: L^2(\sigma I) \to L^2(\sigma I)$ be a B-homomorphism. Then there exists a measurable σ-a.e. bounded matrix function Ξ such that

$$(XF)(\lambda) = \Xi(\lambda) F(\lambda) \qquad \text{for all} \qquad F \in L^2(\sigma I) \qquad (6\text{-}37)$$

Conversely any operator X defined by (6-37) is a B-homomorphism.

PROOF If Ξ is a measurable matrix function satisfying for some $M > 0$ $\|\Xi(\lambda)\| \leq M$ σ-a.e. then for X defined by (6-37) we have

$$\|Xf\|^2 = \int \|\Xi(\lambda) f(\lambda)\|^2 \, d\sigma \leq M^2 \int \|f\|^2 \, d\sigma$$

which shows that $\|X\| \leq M$. It is obvious that X is a B-homomorphism.

Conversely assume X is a bounded B-homomorphism. We suppose that I is the identity matrix in $\mathbb{C}^n$ and $e_1, \ldots, e_n$ is an orthonormal basis there. Since $Xe_i \in L^2(\sigma I)$ we choose some representative for it, say, ξ_i which is defined up to a set of σ-measure zero. We define $\Xi(\lambda)$ by $\Xi(\lambda) e_i = \xi_i(\lambda)$ and extend Ξ by linearity to all of $\mathbb{C}^n$. For any Borel function φ and all $x \in \mathbb{C}^n$

we have $f = \varphi x \in L^2(\sigma I)$ and therefore

$$\int |\varphi(\lambda)|^2 \, \|\Xi(\lambda)\,\xi\|^2 \, d\sigma = \|Xf\|^2 \le \|X\|^2 \cdot \|f\|^2 = \|X\|^2 \int |\varphi(\lambda)|^2 \, \|\xi\|^2 \, d\sigma$$

From this we conclude that $\|\Xi(\lambda)\,\xi\| \le \|X\| \cdot \|\xi\|$. Furthermore, from its definition, Ξ is measurable.

If σI is a scalar type measure then we will write $U_{\mathbf{M}}^{\sigma}$ for the isometric embedding of $L^2(\mathbf{M})$ into $L^2(\sigma I)$. Here we assume $\mathbf{M} \,|\, \sigma I$ or equivalently $d\mathbf{M} = H(\lambda)^* \, H(\lambda) \, d\sigma$. If $M(\lambda)$ is the Radon–Nikodym derivative of $\mathbf{M}$ with respect to σ then $M(\lambda) = H(\lambda)^* \, H(\lambda)$ σ-a.e. It will be of interest to have a concrete representation for $(U_{\mathbf{M}}^{\sigma})^*$, the adjoint of the isometric embedding $U_{\mathbf{M}}^{\sigma}$.

Theorem 6-12 Let $\mathbf{M}$ be a matrix measure and $\mathbf{M}\,|\,\sigma I$ with

$$d\mathbf{M} = M(\lambda) \, d\sigma = H(\lambda)^* \, H(\lambda) \, d\sigma \tag{6-38}$$

Let P be the projection valued function corresponding to the invariant subspace $U_{\mathbf{M}}^{\sigma} L^2(\mathbf{M})$ of $L^2(\sigma I)$. Then we have

$$(U_{\mathbf{M}}^{\sigma})^* \, G = H^* P G \tag{6-39}$$

for all $G \in L^2(\sigma I)$ where H^* is the pseudoinverse of H.

PROOF   Let $F \in L^2(\mathbf{M})$ and $G \in L^2(\sigma I)$, then

$$(F, (U_{\mathbf{M}}^{\sigma})^* \, G) = (U_{\mathbf{M}}^{\sigma} F, G) = \int (H(\lambda) \, F(\lambda), G(\lambda)) \, d\sigma$$

$$= \int (H(\lambda) \, F(\lambda), P(\lambda) \, G(\lambda)) \, d\sigma$$

Since $PG \in U_{\mathbf{M}}^{\sigma} L^2(\mathbf{M})$ there exists an element $G_0 \in L^2(\mathbf{M})$ such that $HG_0 = PG$. By the definition of the pseudoinverse we have $H = HH^*H$, therefore

$$PG = HG_0 = HH^*HG_0 = HH^*PG$$

Using this equality we obtain

$$(F, (U_{\mathbf{M}}^{\sigma})^* \, G) = \int (H(\lambda) \, F(\lambda), P(\lambda) \, G(\lambda)) \, d\sigma$$

$$= \int (H(\lambda) \, F(\lambda), H(\lambda) \, H(\lambda)^* \, P(\lambda) \, G(\lambda)) \, d\sigma$$

$$= \int (H(\lambda)^* \, H(\lambda) \, F(\lambda), H(\lambda)^* \, P(\lambda) \, G(\lambda)) \, d\sigma$$

$$= \int (d\mathbf{M} F, H^* P G)$$

which proves (6-39).

Using this theorem we can obtain a representation for the adjoint of any isometric embedding.

Corollary 6-13 Let $\mathbf{M}$ and $\mathbf{N}$ be matrix measures such that $\mathbf{M}|\mathbf{N}$ and let cI be a scalar-type measure divisible by both $\mathbf{M}$ and $\mathbf{N}$. Assume $d\mathbf{M} = H^* H \, d\sigma$ and $d\mathbf{N} = K^* K \, d\sigma$ then

$$(U_{\mathbf{M}}^{\mathbf{N}})^* \, F = H^* Q K F \tag{6-40}$$

for all $F \in L^2(\mathbf{N})$, where Q is the projection valued function corresponding to the invariant subspace $U_{\mathbf{M}}^\sigma L^2(\mathbf{M})$ of $L^2(\sigma I)$.

PROOF We have $U_{\mathbf{M}}^\sigma = U_{\mathbf{N}}^\sigma U_{\mathbf{M}}^{\mathbf{N}}$ and hence $(U_{\mathbf{M}}^\sigma)^* = (U_{\mathbf{M}}^{\mathbf{N}})^* (U_{\mathbf{N}}^\sigma)^*$. Since $U_{\mathbf{M}}^\sigma$ is isometric we have

$$(U_{\mathbf{M}}^{\mathbf{N}})^* = (U_{\mathbf{M}}^\sigma)^* \, U_{\mathbf{N}}^\sigma \tag{6-41}$$

Applying Theorem 6-12 to (6-41) yields (6-40).

The next two results are instances of lifting theorems. They describe complicated $\mathcal{B}$-homomorphisms between two spaces of type $L^2(\mathbf{M})$ in terms of $\mathcal{B}$-homomorphisms of $L^2(\sigma I)$ which have been described in Theorem 6-11.

Lemma 6-14 Let $\mathbf{M}$ be a matrix measure and assume $\mathbf{M}|\sigma I$. Let $X: L^2(\sigma I) \to L^2(\mathbf{M})$ be a $\mathcal{B}$-homomorphism. Then there exists a $\mathcal{B}$-homomorphism $\overline{X}: L^2(\sigma I) \to L^2(\sigma I)$ for which

$$X = (U_{\mathbf{M}}^\sigma)^* \, \overline{X} \tag{6-42}$$

and $\|\overline{X}\| = \|X\|$. This implies the existence of a measurable σ-a.e. bounded matrix function Ξ, with $\|\Xi\|_\infty = \|X\|$ in terms of which we have the representation

$$XF = H^* P \Xi F \qquad \text{for} \qquad F \in L^2(\sigma I) \tag{6-43}$$

P is the projection valued matrix function corresponding to $U_{\mathbf{M}}^\sigma L^2(\mathbf{M})$.

Conversely any map $X: L^2(\sigma I) \to L^2(\mathbf{M})$ defined by (6-42) where $\overline{X}$ is a $\mathcal{B}$-homomorphism is also a $\mathcal{B}$-homomorphism and

$$\|X\| = \|\overline{X}\| = \|\Xi\|_\infty \tag{6-44}$$

PROOF If $\overline{X}: L^2(\sigma I) \to L^2(\sigma I)$ is a $\mathcal{B}$-homomorphism then so is its composition with $(U_{\mathbf{M}}^\sigma)^*$ and obviously (6-44) holds.

Conversely let $X: L^2(\sigma I) \to L^2(\mathbf{M})$ be a $\mathcal{B}$-homomorphism. Define $\overline{X}: L^2(\sigma I) \to L^2(\sigma I)$ by

$$\overline{X}F = U_{\mathbf{M}}^\sigma X F \qquad \text{for} \qquad F \in L^2(\sigma I) \tag{6-45}$$

Clearly $\overline{X}$ as a product of $\mathcal{B}$-homomorphisms is also one and since $U_{\mathbf{M}}^\sigma$ is isometric $\|\overline{X}\| = \|X\|$. By Theorem 6-11 there exists a σ-a.e. bounded measurable matrix function Ξ for which $\overline{X}F = \Xi F$ and hence (6-43) holds by an application of Theorem 6-12.

Theorem 6-15 Let **M** and **N** be matrix measures and $X: L^2(\mathbf{M}) \to L^2(\mathbf{N})$ a B-homomorphism. Let σI be a positive scalar-type measure divisible by both **M** and **N** and let $d\mathbf{M} = H^*H \, d\sigma$ and $d\mathbf{N} = K^*K \, d\sigma$. Let P and Q be the measurable projection valued functions corresponding to $U_\mathbf{M}^\sigma L^2(\mathbf{M})$ and $U_\mathbf{N}^\sigma L^2(\mathbf{N})$, respectively. Then there exists a B-homomorphism $\overline{X}: L^2(\sigma I) \to L^2(\sigma I)$ satisfying $\|\overline{X}\| = \|X\|$ for which

$$XF = (U_\mathbf{N}^\sigma)^* \overline{X} U_\mathbf{M}^\sigma F \qquad \text{for} \qquad F \in L^2(\mathbf{M}) \tag{6-46}$$

Moreover, there exists a measurable σ-a.e. bounded matrix function Ξ satisfying

$$\|\Xi\|_\infty = \|\overline{X}\| = \|X\| \tag{6-47}$$

$$\Xi(\lambda) = \Xi(\lambda) P(\lambda) = Q(\lambda) \Xi(\lambda) \qquad \sigma\text{-a.e.} \tag{6-48}$$

and for which

$$XF = K^* \Xi H F \qquad \text{for all} \qquad F \in L^2(\mathbf{M}) \tag{6-49}$$

Conversely every operator X defined by (6-49) for Ξ measurable and σ-a.e. bounded is a B-homomorphism from $L^2(\mathbf{M})$ into $L^2(\mathbf{N})$.

PROOF If X is given by (6-49) then it is clearly a B-homomorphism and satisfies (6-47). Let us assume therefore that $X: L^2(\mathbf{M}) \to L^2(\mathbf{N})$ is a B-homomorphism. Define $Y: L^2(\sigma I) \to L^2(\mathbf{N})$ by

$$YF = X(U_\mathbf{M}^\sigma)^* F \tag{6-50}$$

Y is a B-homomorphism as a product of such and $Y|\{U_\mathbf{M}^\sigma L^2(\mathbf{M})\}^\perp = 0$ or equivalently stated $YP^\perp L^2(\sigma I) = 0$ which reduces to

$$YP^\perp = 0 \tag{6-51}$$

If we apply now Lemma 6-14 then we obtain

$$Y = (U_\mathbf{N}^\sigma)^* \overline{X} \tag{6-52}$$

for a B-homomorphism $\overline{X}: L^2(\sigma I) \to L^2(\sigma I)$. Now $\overline{X}F = \Xi F$ where Ξ is a measurable σ-a.e. bounded matrix function that satisfies $\|\Xi\|_\infty = \|\overline{X}\| = \|Y\|$. Since by (6-51) $\overline{X} P^\perp L^2(\sigma I) = U_\mathbf{N}^\sigma Y P^\perp L^2(\sigma I) = 0$ we have

$$\Xi P^\perp = 0 \tag{6-53}$$

which is equivalent to

$$\Xi = \Xi P \tag{6-54}$$

Also since $\overline{X} = U_\mathbf{N}^\sigma Y$ we have

$$\Xi L^2(\sigma I) \subset U_\mathbf{N}^\sigma L^2(\mathbf{N}) = Q L^2(\sigma I)$$

which implies

$$Q^\perp U = 0 \tag{6-55}$$

or equivalently

$$\Xi = Q\Xi \tag{6-56}$$

and (6-48) is proved. We note also that (6-48) implies the equality

$$\Xi P^\perp = Q^\perp \Xi \tag{6-57}$$

Representation (6-49) follows now from (6-50), (6-52) and the formulas for J_M^σ and $(U_N^\sigma)^*$.

We note for future reference that $X^*: L^2(\mathbf{N}) \to L^2(\mathbf{M})$ is also a B-homomorphism. In terms of the notation of the previous theorem we have the following corollary.

Corollary 6-16 If $X: L^2(\mathbf{M}) \to L^2(\mathbf{N})$ is the B-homomorphism having the representation (6-49) with (6-48) satisfied then $X^*: L^2(\mathbf{N}) \to L^2(\mathbf{M})$ is a B-homomorphism having the representation

$$X^*G = (U_M^\sigma)^* \bar{X}^* U_N^\sigma G \qquad \text{for} \qquad G \in L^2(\mathbf{N}) \tag{6-58}$$

or more specifically

$$X^*G = H^* P\Xi^* KG \tag{6-59}$$

where

$$\Xi(\lambda)^* = \Xi(\lambda)^* Q(\lambda) = P(\lambda)\Xi(\lambda)^* \tag{6-60}$$

holds σ-a.e.

For the analysis of the deeper properties of intertwining operators we will introduce the several relevant notions of coprimeness. All definitions will be relative to a fixed positive scalar measure σ. A measurable projection valued $n \times n$ matrix P will be called *trivial* with respect to σ, or σ-*trivial*, if $P(\lambda) = I$ σ-a.e. Two measurable, $n \times m$ and $n \times l$, respectively, matrix functions A and B are called σ-*left coprime* if there exists no σ-nontrivial projection function P for which $A = PA$ and $B = PB$. We denote the σ-left coprimeness of A and B by $(A, B)_L^\sigma = I$. Analogously we define σ-*right coprimeness* and denote it by $(A, B)_R^\sigma = I$. There is also a stronger notion of coprimeness. We say A and B are *strongly σ-left coprime*, and write $[A, B]_L^\sigma = I$, if there exists a $\delta > 0$ such that for all ξ, $\|\xi\| = 1$ we have

$$\|A(\lambda)^* \xi\| + \|B(\lambda)^* \xi\| \geq \delta \qquad \sigma\text{-a.e.} \tag{6-61}$$

Again the analogous notion of strong σ-right coprimeness is introduced in the same manner. The above definitions extend easily to the coprimeness of a finite number of matrix functions.

As expected the coprimeness relations are connected with the ideal structure in the algebra of bounded measurable functions.

Theorem 6-17

(c) Let $A_1, \ldots, A_p$ be bounded measurable $n \times m_i$ matrix valued functions. Then there exist bounded measurable $m_i \times n$ matrix valued functions

B_i such that

$$\sum_{i=1}^{p} A_i(\lambda)\, B_i(\lambda) = I \qquad \sigma\text{-a.e.} \tag{6-62}$$

if and only if

$$[A_1, \ldots, A_p]_L^\sigma = I \tag{6-63}$$

(b) Let $A_1, \ldots, A_p$ be measurable $m_i \times n$ matrix functions. Then there exist $n \times m_i$ matrix functions B_i such that

$$\sum_{i=1}^{p} B_i(\lambda)\, A_i(\lambda) = I \qquad \sigma\text{-a.e.} \tag{6-64}$$

if and only if

$$[A_1, \ldots, A_p]_R^\sigma = I \tag{6-65}$$

PROOF Assume there exist B_i such that (6-62) holds. Taking adjoints and applying the resulting equality to a unit vector ξ we have

$$\xi = \sum B_i(\lambda)^* A_i(\lambda)^* \xi$$

and hence

$$1 = \|\xi\| \le \sum \|B_i(\lambda)^* A_i(\lambda)^* \xi\|$$
$$\le \sum \|B_i(\lambda)^*\| \, \|A_i(\lambda)^* \xi\|$$
$$\le B \sum \|A_i(\lambda)^* \xi\|$$

where $B = \max_i \|B_i(\lambda)^* \xi\|$. Equivalently we have

$$\sum \|A_i(\lambda)^* \xi\| \ge B^{-1}$$

that is $[A_1, \ldots, A_p]_L^\sigma = I$.

Conversely assume $A_1, \ldots, A_p$ are strongly σ-left coprime. From (6-63) it follows that

$$\sum \|A_i(\lambda)^* \xi\|^2 \ge \delta^2 \tag{6-66}$$

for some $\delta > 0$ and all unit vectors ξ. Inequality (6-66) can be rewritten as $\sum_i A_i(\lambda)\, A_i(\lambda)^* \ge \delta^2 I$. Thus $\sum_i A_i(\lambda)\, A_i(\lambda)^*$ is measurable and invertible in the algebra of all bounded measurable $n \times n$ matrix functions. Define B_i by $B_i(\lambda) = A_i(\lambda)^* \left(\sum_j A_j(\lambda)\, A_j(\lambda)^*\right)^{-1}$. Then the B_i are bounded and measurable and (6-62) holds. Part (b) follows by a simple duality argument.

The following corollary justifies the distinction between σ-left coprimeness and strong σ-left coprimeness.

Corollary 6-18 If $A_1, \ldots, A_p$ are bounded measurable $n \times m_i$ matrix valued functions then $[A_1, \ldots, A_p]_L^\sigma = I$ implies $(A_1, \ldots, A_p)_L^\sigma = I$.

PROOF Assume $[A_1, \ldots, A_p]_L^\sigma = I$. Then there exist B_i such that $\sum_i A_i B_i = I$. From this it follows that $A_1, \ldots, A_p$ cannot have a common σ-nontrivial projection valued left factor. Thus the σ-left coprimeness of $A_1, \ldots, A_p$ follows.

The various coprimeness relations provide the language in which to phrase the next result.

Theorem 6-19 Let $X: L^2(\mathbf{M}) \to L^2(\mathbf{N})$ be a B-homomorphism having the representation (6-49) with relation (6-48) satisfied. Then
(a) X has dense range if and only if

$$(\Xi, Q^\perp)_L^\sigma = I \tag{6-67}$$

(b) X is one-to-one if and only if

$$(\Xi, P^\perp)_R^\sigma = I \tag{6-68}$$

(c) X has a bounded right inverse if and only if

$$[\Xi, Q^\perp]_L^\sigma = I \tag{6-69}$$

(d) X has a bounded left inverse if and only if

$$[\Xi, P^\perp]_R = I \tag{6-70}$$

PROOF
(a) The range of X is dense in $L^2(\mathbf{N})$ if and only if the range of $\overline{X}$ is dense in $U_N^\sigma L^2(\mathbf{N}) = Q L^2(\sigma I)$. This occurs if and only if the span of the two linear manifolds $\{\Xi H F \mid F \in L^2(\mathbf{M})\}$ and $Q^\perp L^2(\sigma I)$ is all of $L^2(\sigma I)$. Now $\{\Xi H F \mid F \in L^2(\mathbf{M})\} = \Xi P L^2(\sigma I)$ and since $\Xi P^\perp = Q^\perp \Xi$ it follows that $\Xi P^\perp L^2(\sigma I) \subset Q^\perp L^2(\sigma I)$. Hence X has dense range if and only if the span of $\Xi L^2(\sigma I)$ and $Q^\perp L^2(\sigma I)$ is $L^2(\sigma I)$. Since the span of two invariant subspaces is an invariant subspace we apply Theorem 6-10 on the characterization of invariant subspaces to obtain the result that

$$\Xi L^2(\sigma I) \vee Q^\perp L^2(\sigma I) = L^2(\sigma I) \tag{6-71}$$

if and only if (6-67) holds.
(b) This follows from (a) by a duality argument. X is one-to-one if and only if $X^*: L^2(\mathbf{N}) \to L^2(\mathbf{M})$ has dense range. Now X^* is given by (6-59) with relation (6-60) holding. By applying part (a) X^* has dense range if and only if

$$(\Xi^*, P^\perp)_L^\sigma = I \tag{6-72}$$

which is equivalent to (6-68).
(c) Assume (6-69) holds. By Theorem 6-17 there exist matrix valued functions Θ and R such that

$$\Xi(\lambda)\,\Theta(\lambda) + Q^\perp(\lambda)\,R(\lambda) = I \qquad \sigma\text{-a.e.} \tag{6-73}$$

Define maps $Y: L^2(\mathbf{N}) \to L^2(\mathbf{N})$ and $\bar{Y}: L^2(\sigma I) \to L^2(\sigma I)$ by

$$\bar{Y}F = \Theta F \qquad \text{for} \qquad F \in L^2(\sigma I) \tag{6-74}$$

and

$$YF = (U_{\mathbf{M}}^\sigma)^* \bar{Y}U_{\mathbf{N}}^\sigma F \qquad \text{for} \qquad F \in L^2(\mathbf{N}) \tag{6-75}$$

Obviously Y and $\bar{Y}$ are bounded linear operators. We claim $XY = I$. Let $F \in L^2(\mathbf{N})$ then

$$XYF = (U_{\mathbf{N}}^\sigma)^* \bar{X}U_{\mathbf{M}}^\sigma(U_{\mathbf{M}}^\sigma)^* \bar{Y}U_{\mathbf{N}}^\sigma F$$

Since $U_{\mathbf{M}}^\sigma$ is an isometry $U_{\mathbf{M}}^\sigma(U_{\mathbf{M}}^\sigma)^*$ is the projection on the range of $U_{\mathbf{M}}^\sigma$ which is just the multiplication by the projection valued function P. So

$$XYF = K^* Q\Xi P\Theta KF$$

and using the equality $\Xi P = Q\Xi$ as well as $Q^2 = Q$ yields

$$XYF = K^*\Xi\Theta KF$$

From (6-73) we have $\Xi\Theta = I - Q^\perp R$ and since $QQ^\perp = 0$ we have

$$XYF = K^* QKF = (U_{\mathbf{N}}^\sigma)^* U_{\mathbf{N}}^\sigma F = F$$

To prove the necessity of the condition (6-69) for the existence of a bounded right inverse for X it suffices, by duality considerations, to prove the necessity of the condition (6-70) for the existence of a bounded left inverse for X. Thus assume (6-70) is not satisfied. We will show the existence of a sequence of functions F_n in $L^2(\mathbf{M})$ such that $\lim \|F_n\| = 1$ and $\lim \|XF_n\| = 0$. This would imply the nonexistence of a bounded left inverse for X. Since (6-70) is not satisfied then for all $n > 0$ there exists a unit vector ξ_n for which

$$\|\Xi(\lambda)\xi_n\| + \|P^\perp(\lambda)\xi_n\| < \frac{1}{n} \tag{6-76}$$

for all λ in a set Λ_n of positive σ-measure. Let χ_{Λ_n} be the characteristic function of the set Λ_n then

$$\psi_n(\lambda) = [\sigma(\Lambda_n)]^{-1/2} \chi_{\Lambda_n}\xi_n$$

is a function in $L^2(\sigma I)$ of norm one. We decompose Ψ_n relative to the direct sum $L^2(\sigma I) = PL^2(\sigma I) \oplus P^\perp L^2(\sigma I)$ to obtain $\Psi_n = \Phi_n + \Gamma_n$ with

$$\Phi_n = [\sigma(\Lambda_n)]^{-1/2} P\chi_{\Lambda_n}\xi_n$$

and

$$\Gamma_n = [\sigma(\Lambda_n)]^{-1/2} P^\perp\chi_{\Lambda_n}\xi_n$$

Since $\Phi_n \in PL^2(\sigma I) = U_{\mathbf{M}}^\sigma L^2(\mathbf{M})$ we have $\Phi_n = U_{\mathbf{M}}^\sigma F_n$ for some $F_n \in L^2(\mathbf{M})$ with $\|F_n\| = \|\Phi_n\|$. We note also that

$$\|\Gamma_n\| = [\sigma(\Lambda_n)]^{-1/2} \|P^\perp\chi_{\Lambda_n}\xi_n\| < \frac{1}{n}$$

and therefore

$$\lim \|F_n\|^2 = \lim \left[\|\psi_n\|^2 - \|\Gamma_n\|^2 \right] = 1$$

We will show now that $\lim \|XF_n\| = 0$.

$$XF_n = (U_N^\sigma)^* \overline{X} U_M^\sigma F_n = (U_N^\sigma)^* \overline{X} \Phi_n$$

and

$$\overline{X}\Phi_n = \overline{X}(\Psi_n - \Gamma_n) = [\sigma(\Lambda_n)]^{-1/2} \Xi \chi_{\Lambda_n} \xi_n - [\sigma(\Lambda_n)]^{-1/2} \Xi P^\perp \chi_{\Lambda_n} \xi_n$$

We now give the following estimate

$$\|XF_n\| = \|(U_N^\sigma)^* \overline{X}\Phi_n\| \le \|\overline{X}\Phi_n\|$$

$$\le [\sigma(\Lambda_n)]^{-1/2} \|\chi_{\Lambda_n} \Xi \xi_n\| + [\sigma(\Lambda_n)]^{-1/2} \|\Xi \chi_{\Lambda_n} P^\perp \xi_n\|$$

$$\le [\sigma(\Lambda_n)]^{-1/2} \left\{ \int_{\Lambda_n} \|\Xi(\lambda)\,\xi_n\|^2 \, d\sigma \right\}^{1/2}$$

$$+ [\sigma(\Lambda_n)]^{-1/2} \|\Xi\|_\infty \left\{ \int_{\Lambda_n} \|P^\perp(\lambda)\,\xi_n\|^2 \, d\sigma \right\}^{1/2} \le (1 + \|\Xi\|_\infty) \frac{1}{n}$$

which completes the proof of (c).

(d) Follows from (c) by duality.

One final remark is in order. Throughout this chapter we have dealt with self-adjoint operators, but essentially all results can be easily translated to the case of one or a pair of unitary operators. This is best done through the use of the Cayley transform.

Thus let U be a unitary operator in a Hilbert space H. Since $\mathrm{Ker}(I - U) = \mathrm{Ker}(I - U^*)$ by Lemma 2-5 it follows that this is a reducing subspace of U. Thus without loss of generality we will assume $1 \notin \sigma_p(U)$. Define A by

$$A = i(I + U)(I - U)^{-1}$$

then A is a, not necessarily bounded, self-adjoint operator and can be represented as

$$A = \int \lambda E(d\lambda)$$

for some spectral measure E on $\mathbb{R}$. Since the Cayley transform can be inverted, its inverse given by

$$U = (A - iI)(A + iI)^{-1}$$

and as for real λ, the function $\psi(\lambda) = (\lambda - i)/(\lambda + i)$ is continuous and satisfies $|\psi(\lambda)| = 1$, it follows by the functional calculus for self-adjoint operators that

$$U = \int_{-\infty}^{\infty} \frac{\lambda - i}{\lambda + i} E(d\lambda) \tag{6-77}$$

Define now a spectral measure F on the unit circle $\mathbb{T}$ by

$$F(\sigma) = E[\psi^{-1}(\sigma)]$$

for each Borel set σ on the unit circle. Then a change of variable in the integral representation yields

$$U = \int e^{it} F(dt)$$

which is the form of the spectral theorem for unitary operators.

In an analogous way we can obtain the spectral representation theory for unitary operators as well as the characterization of intertwining operators. We will use these results in the sequel without formalizing them as theorems.

There is, however, one difference which should be pointed out. A subspace which is invariant for a self-adjoint operator A is automatically a reducing subspace. This is no longer the case for a unitary operator. Thus if μ is a measure defined on the unit circle then there are subspaces of $L^2(\mu I)$ which are invariant but not reducing. We study these subspaces in Sec. 12. Therefore in this case Theorem 6-10 is a characterization of the reducing subspaces of the unitary operator $M_{\chi,\mu I}$. There is also a difference regarding the definition of cyclicity. A unitary operator U in H is called cyclic and x a cyclic vector if the smallest reducing subspace for U which contains x is all of H. This is equivalent to the density in H of all linear combinations of the vectors $U^n x$, $n \in \mathbb{Z}$. This differs from the standard definition of cyclicity.

7. THE DOUGLAS FACTORIZATION THEOREM AND RELATED RESULTS

This section is devoted to a factorization result of Douglas and various corollaries of it.

Theorem 7-1 Let A and B be bounded operators in a Hilbert space H. Then the following statements are equivalent.
(a) $A = BC$ for some bounded operator C.
(b) $AA^* \leq \lambda^2 BB^*$ for some $\lambda > 0$.
(c) Range $A \subset$ Range B.

PROOF Statement (c) follows trivially from (a). Similarly if (a) holds then $A^* = C^*B^*$ and therefore for each x in H

$$\|A^* x\|^2 = \|C^* B^* x\|^2 \leq \|C^*\|^2 \|B^* x\|^2$$

which is equivalent to (b) with $\lambda = \|C^*\|$. Now assume (b) holds for all vectors x in H, then $\|A^* x\| \leq \lambda \|B^* x\|$. Define an operator D from Range B^* into H by $DB^* x = A^* x$ then $\|DB^* x\| \leq \lambda \|B^* x\|$ and D extends by continuity to the closure of Range B^*. Since $\{$Range $B^*\}^\perp = \text{Ker } B$ then if we define $D|\operatorname{Ker} B = 0$ then D is a well-defined bounded operator for which $DB^* = A^*$. So (a) holds with $C = D^*$. Finally assume (c) holds. Let B_0 be the restriction of B to $\{\operatorname{Ker} B\}^\perp$. Obviously B_0 is an injective operator into $\operatorname{Range} B$ which implies that B_0^{-1} exists as a closed operator from $\operatorname{Range} B$ into $\{\operatorname{Ker} B\}^\perp$. Since $\operatorname{Range} A \subset \operatorname{Range} B$ the operator $C = B_0^{-1}A$ is a closed well-defined operator from H into $\{\operatorname{Ker} B\}^\perp$ and, by the closed graph theorem, is necessarily bounded. From this we get $R = BC$ which completes the proof.

We single out the special case $\lambda = 1$.

Corollary 7-2 Let A and B be bounded operators in a Hilbert space H. Then there exists a contraction Z such that

$$A = BZ \tag{7-1}$$

if and only if

$$AA^* \le BB^* \tag{7-2}$$

The above corollary can be strengthened somewhat if we have equality in (7-2).

Corollary 7-3 Let A and B be bounded operators in a Hilbert space H. Then

$$AA^* = BB^* \tag{7-3}$$

if and only if there exists a partial isometry U with $\overline{\operatorname{Range} B^*}$ as its final space, such that

$$A = BU \tag{7-4}$$

PROOF Assume there exists a partial isometry U such that (7-4) holds. Then UU^* is the orthogonal projection on the final space of U, that is, on $\overline{\operatorname{Range} B^*}$. Thus $UU^*B^* = B^*$ which implies (7-3). Conversely assume (7-3). Then for all x in H we have $\|A^*x\| = \|B^*x\|$. Define V by $VB^*x = A^*x$ and $V|\operatorname{Ker} B = 0$ then V extends by continuity to a partial isometry with $\overline{\operatorname{Range} B^*}$ as initial space. Thus (7-4) follows with $U = V^*$. Obviously the initial space of V is the final space of U.

We note that Theorem 7-1 holds just as well for operators A and B whose domain of definition are different Hilbert spaces. The interesting case is when B acts on a direct sum of Hilbert spaces.

Theorem 7-4 Let $H_0, H_1, \ldots, H_n$ and K be Hilbert spaces and let $A_i \in B(H_i, K)$. Then there exist $Z_i \in B(H_0, H_i)$ such that

$$A_0 = \sum_{i=1}^{n} A_i Z_i \tag{7-5}$$

if and only if for some $\lambda > 0$

$$A_0 A_0^* \leq \lambda^2 \sum_{i=1}^{n} A_i A_i^* \tag{7-6}$$

The operators Z_i can be taken to satisfy

$$\sum Z_i^* Z_i \leq I_{H_0} \tag{7-7}$$

if and only if (7-6) holds with $\lambda \leq 1$.

PROOF We define $B: H_1 \oplus \cdots \oplus H_n \to K$ by $B(x_1, \ldots, x_n) = \sum_{i=1}^{n} A_i x_i$, then its adjoint is given by $B^* y = (A_1^* y, \ldots, A_n^* y)$. Condition (7-6) is equivalent to (b) of Theorem 7-1. The operator $C: H_0 \to H_1 \oplus \cdots \oplus H_n$ is given by $Cx = (Z_1 x, \ldots, Z_n x)$ and the result follows from Theorem 7-1 and its corollary as $C^* C = \sum Z_i^* Z_i$.

A different extension of Theorem 7-1 is obtained by relaxing somewhat condition (c) of the theorem. It suffices to assume the existence of a subset H_0 of the second category in H whose image under A is included in the range of E. This uses a somewhat stronger form of the closed graph theorem as given in [96]. This observation is the key to the following.

Theorem 7-5 Let $K, H_0, H_1, \ldots$ be Hilbert spaces and let $A_i \in B(H_0, K)$ and $B_i \in B(H_i, K)$. Assume for each x in H_0 there exists an index i such that $A_i x \in \text{Range } B_i$, then for some index m $\text{Range } A_m \subset \text{Range } B_m$ and hence $A_m = B_m C$ for some $C \in B(H_0, H_m)$.

PROOF Let $M_i = \{x \in H_0 | A_i x \in \text{Range } B_i\}$. By assumption $K = \bigcup_{i=1}^{\infty} M_i$ and since H_0 is complete we have, by the Baire category theorem [96], that at least one of the M_i, say M_n, is of the second category in H_0. By the remarks preceding the theorem we have $A_m = B_m C$.

8. SHIFTS, ISOMETRIES, AND THE WOLD DECOMPOSITION

Let N be a Hilbert space. By $l^2(-\infty, \infty; N)$ we denote the Hilbert space of all doubly infinite sequences $\{a_n\}_{n=-\infty}^{\infty}$ with $a_n \in N$ and $\sum_{n=-\infty}^{\infty} \|a_n\|^2 < \infty$. Similarly we define $l^2(0, \infty; N)$ as the space of all one-sided sequences $\{a_n\}_{n=0}^{\infty}$ for which $\sum_{n=0}^{\infty} \|a_n\|^2 < \infty$.

In $l^2(-\infty, \infty; N)$ we define an operator U by

$$U\{x_n\} = \{y_n\} \qquad \text{with} \qquad y_n = x_{n-1} \tag{8-1}$$

We call U the *bilateral right shift* in $l^2(-\infty, \infty; N)$. Its adjoint U^* acts by

$$U^*\{x_n\} = \{y_n\} \qquad \text{with} \qquad y_n = x_{n+1} \tag{8-2}$$

Clearly both U and U^* are isometric hence both are unitary operators U^* is called the *bilateral left shift*.

Similarly we define the *unilateral right shift* S in $l^2(0, \infty; N)$

$$S\{x_n\} = \{y_n\} \tag{8-3}$$

where

$$y_n = \begin{cases} x_{n-1} & n > 0 \\ 0 & n = 0 \end{cases}$$

Again it is obvious that S is an isometry. However, its adjoint S^* given by

$$S^*\{x_n\} = \{y_n\} \tag{8-4}$$

with

$$y_n = x_{n+1} \qquad n \geq 0$$

is not isometric as it has a nontrivial null space. S^* is referred to as the *left shift* or sometimes as the *backward shift*.

The shift operators have been introduced in a concrete way. However, it is easy to abstract the properties which characterize shifts up to unitary equivalence. Let us begin with the unilateral shift.

Consider the subspace L of $l^2(0, \infty; N)$ consisting of all sequences $\{x_n\}$ for which $x_n = 0$ for $n > 0$. The subspace L has the following properties

(a) $L \perp S^n L$ for $n > 0$ and
(b) $l^2(0, \infty; N) = \oplus_{n=0}^{\infty} S^n L$.

Let now V be a general isometry in a Hilbert space H. A subspace L will be called a *wandering subspace* for an isometry V if $L \perp V^n L$ for $n > 0$. Thus we can form the orthogonal direct sum of the subspaces $V^n L$ to obtain $\oplus_{n=0}^{\infty} V^n L$. If we have $H = \oplus_{n=0}^{\infty} V^n L$ then V is clearly unitarily equivalent to the right shift in $l^2(0, \infty; L)$. The *multiplicity* of the right shift is defined as the dimension of L where L is a spanning wandering subspace. L is uniquely determined by V and we have

$$L = \{\text{Range } V\}^{\perp} = \text{Ker } V^* \tag{8-5}$$

We note that two unilateral shifts V and V' are unitarily equivalent if and only if they have the same multiplicity. Equal multiplicity follows from unitary equivalence by (8-5). Conversely if V and V_1 are of the same multiplicity, let L and L_1 be their corresponding spanning wandering subspaces. Let $\{e_\alpha | \alpha \in A\}$ and $\{e'_\alpha | \alpha \in A\}$ be orthonormal bases for L and L_1, respectively, then $\{V^n e_\alpha | n \geq 0, \alpha \in A\}$ and $\{V_1^n e'_\alpha | n \geq 0, \alpha \in A\}$ are orthonormal bases for H and H_1, respectively. Define a map $\psi: H \to H_1$ by $\psi(V^n e_\alpha) = V_1^n e'_\alpha$ for all $n \geq 0$, $\alpha \in A$ and extend ψ by linearity. Obviously ψ is unitary and $\psi V = V_1 \psi$. Thus the unitary equivalence of V and V_1 is proved.

Let S be the right shift in $l^2(0, \infty; N)$. Let $x = \{\xi_n\}_{n=0}^{\infty}$ then $x \in \text{Range } S^n$ if and only if $\xi_i = 0$ for $i = 0, \ldots, n-1$. Thus $\bigcap_{n=0}^{\infty} \text{Range } S^n = \{0\}$. This yields another characterization of unilateral shifts.

Lemma 8-1 Let V be an isometry in a Hilbert space H then V is unitarily equivalent to a unilateral right shift if and only if

$$\bigcap_{n=0}^{\infty} V^n H = \{0\} \tag{8-6}$$

Proof It remains to prove the if part. Let $L = \{\text{Range } V\}^{\perp} = H \ominus VH = \text{Ker } V^*$. L is a wandering subspace for V for $V^n L \subset V^n H \subset VH \perp L$. From $L = H \ominus VH$ it follows that $H = L \oplus VH$ and since V is isometric $VH = VL \oplus V^2 H$. Thus by an induction argument we have $H = L \oplus VL \oplus \cdots \oplus V^{n+1} H$. So

$$\left\{ \bigoplus_{i=0}^{n} V^i L \right\}^{\perp} = V^{n+1} H \tag{8-7}$$

and hence

$$\left\{ \bigoplus_{n=0}^{\infty} V^n L \right\}^{\perp} = \bigcap_{n=0}^{\infty} V^n H \tag{8-8}$$

From the last equality it follows that (8-6) implies V is a unilateral shift.

This lemma contains the essence of the next result generally known as the Wold decomposition.

Theorem 8-2 Let V be an isometry in a Hilbert space, then there exists a unique decomposition of H into a direct sum of reducing subspaces of V, $H = H_0 \oplus H_1$ such that $V|H_0$ is unitary and $V|H_1$ is a unilateral shift.

Proof Let $H_0 = \bigcap_{n=0}^{\infty} V^n H$ then $H_0 = \bigcap_{n=m}^{\infty} V^n H$ as $V^n H$ is a monotonically decreasing sequence of subspaces. The invariance of H_0 is obvious. Now $V^* H_0 = V^* \bigcap_{n=1}^{\infty} V^n H = \bigcap_{n=0}^{\infty} V^n H = H_0$. Thus H_0 is reducing. Define L by $L = H \ominus VH$ then by (8-8) $H_1 = H \ominus H_0 = \{\bigcap_{n=0}^{\infty} V^n H\}^{\perp} = \bigoplus_{n=0}^{\infty} V^n L$. Since H_1 is the orthogonal complement of a reducing subspace of V it is also reducing. Now V and V^* are clearly isometric when restricted to H_0 and $V|H_1$ is a unilateral shift.

As the unilateral shift could be defined abstractly so can the bilateral shift. If U is the bilateral right shift in $l^2(-\infty, \infty; N)$ we note that if we consider $l^2(0, \infty; N)$ as naturally embedded in $l^2(-\infty, \infty; N)$ the following properties hold

$$Ul^2(0, \infty; N) \subset l^2(0, \infty; N)$$

$$\bigcap_{n=-\infty}^{\infty} U^n l^2(0, \infty; N) = \{0\}$$

and

$$\overline{\bigcup_{n=-\infty}^{\infty} U^n l^2(0, \infty; N)} = l^2(-\infty, \infty; N)$$

Taking these properties as our model we define an *outgoing subspace D* for a unitary operator U acting in a Hilbert space H if it satisfies

$$UD \subset D$$

$$\bigcap_{n=-\infty}^{\infty} U^n D = \{0\} \tag{8-9}$$

$$\overline{\bigcup_{n=-\infty}^{\infty} U^n D} = H$$

The definition of outgoing subspaces as well as the following theorem which gives an intrinsic characterization of bilateral shifts are due to Lax and Phillips [82].

Theorem 8-3 Let D be an outgoing subspace for a unitary operator U then U is unitarily equivalent to the bilateral shift in $l^2(-\infty, \infty; N)$ for some Hilbert space N.

PROOF Let us define N by $N = D \ominus VD$. Since D is invariant under U, $U|D$ is isometric, hence applying the Wold decomposition we have $D = \{\oplus_{n=0}^{\infty} U^n N\} \oplus H_0$ with H_0 reducing and $U|H_0$ unitary. Since $\bigcap_{n=-\infty}^{\infty} U^n D = \{0\}$ we have necessarily $H_0 = \{0\}$ and the last condition of (8-9) implies $H = \oplus_{n=-\infty}^{\infty} U^n N$. Clearly $D = \oplus_{n=0}^{\infty} U^n N$. Thus H is isomorphic to $l^2(-\infty, \infty; N)$ and U to the bilateral right shift.

9. CONTRACTIONS, DILATIONS, AND MODELS

The Wold decomposition for isometries in a Hilbert space can be extended to all contractions. We will say that a contraction T in a Hilbert space H is *completely nonunitary* if there exists no nontrivial reducing subspace of T in which T acts unitarily.

After decomposing a contraction T into its unitary and completely nonunitary parts we will introduce the notion of isometric and unitary dilations and see how the study of a large class of contractions can be facilitated by identifying them as parts of special isometric or unitary operators.

We start by introducing some notation. Given a contraction T then both $(I - T^*T)$ and $(I - TT^*)$ are positive operators and hence have, by Theorem 5-7, unique positive square roots. Let us define

$$D_T = (I - T^*T)^{1/2} \tag{9-1}$$

and

$$D_{T^*} = (I - TT^*)^{1/2} \tag{9-2}$$

Since $TD_T^2 = T(I - T^*T) = (I - TT^*)T = D_{T^*}^2 . T$ it follows by induction that for every complex polynomial p we have

$$Tp(D_T^2) = p(D_{T^*}^2)\,T \tag{9-3}$$

Choosing a sequence of polynomials approximating uniformly the square root function on $[0, 1]$ we have in the limit

$$TD_T = D_{T^*}T \tag{9-4}$$

and by taking adjoints also

$$T^*D_{T^*} = D_T T^* \tag{9-5}$$

Also we note that for all $x \in H$

$$\|x\|^2 = \|Tx\|^2 + \|D_Tx\|^2 = \|T^*x\|^2 + \|D_{T^*}x\|^2 \tag{9-6}$$

D_T and D_{T^*} are called the *defect operators* of T and they give a measure of the distance of T and T^* from being isometric. Thus T is isometric if and only if $D_T = 0$ and similarly for T^*. We let

$$\mathfrak{D} = \overline{\mathrm{Range}\,D_T} \tag{9-7}$$

and

$$\mathfrak{D}_* = \overline{\mathrm{Range}\,D_{T^*}} \tag{9-8}$$

and call them the *defect spaces* of T and their respective dimensions the *defect numbers* of T.

Theorem 9-1 Let T be a contraction in a Hilbert space H, then there exists a unique decomposition of H into a direct sum of $H = H_0 \oplus H_1$ of reducing subspaces of T such that $T|H_0$ is unitary and $T|H_1$ is completely nonunitary.

PROOF We define H_0 by $H_0 = \{x|\ \|T^nx\| = \|T^{*n}x\| = \|x\|, n \geq 0\}$. To see that H_0 is actually a subspace we note that $\|T^nx\| = \|x\|$ if and only if $(I - T^{*n}T^n)x = 0$ and $\|T^{*n}x\| = \|x\|$ if and only if $(I - T^nT^{*n})x = 0$. Thus H_0 is the intersection of the kernels of all operators of the form $(I - T^{*n}T^n)$ and $(I - T^nT^{*n})$ and hence is a subspace of H. Next we show that H_0 is invariant under T. Let $x \in H_0$ then $\|T^n(Tx)\| = \|T^{n+1}x\| = \|x\| = \|Tx\|$. Since T is a contraction we have $\|x\| = \|Tx\|$ if and only if $x = T^*Tx$. Using this we have for $x \in H_0$ $\|T^{*n}Tx\| = \|T^{*n-1}T^*Tx\| = \|T^{*n-1}x\| = \|x\| = \|Tx\|$ and the invariance of T with respect to T is proved. As the definition of H_0 was symmetric with respect to T and T^* it follows that H_0 is invariant also under T^* and is therefore a reducing subspace for T. From the definition of H_0 it is clear that $T|H_0$ is unitary. Let $H_1 = H_0^\perp$ then necessarily $T|H_1$ is completely nonunitary for if $L \subset H_1$ is a reducing subspace of T in which T acts unitarily then for $x \in L$ we have $\|x\| = \|T^nx\| = \|T^{*n}x\|$. Thus $L \subset H_0 \cap H_1 = \{0\}$, which completes the proof.

Contrary to the case of isometries where the structure was determined by the unitary part, completely described by the spectral theorem, and the unilateral shift the structure of the general contraction in a Hilbert space is as complicated as its completely nonunitary part which is generally difficult to describe. There is, however, a special subclass of contractions which is closely related to the shift operators and which points out the importance of shift operators as models for other operators.

Theorem 9-2 Let T be a contraction in a Hilbert space H. T^* is unitarily equivalent to the restriction of a left shift to one of its invariant subspaces if and only if $T^{*n} \to 0$ strongly.

PROOF Let $M \subset l^2(0, \infty; N)$ be an invariant subspace of S^* the left shift then clearly $(S^*|M)^n = S^{*n}|M \to 0$ strongly and so must every operator unitarily equivalent to it.

Conversely let $T^{*n} \to 0$ strongly. We define a new norm in H in such a way as to have $\|x\|^2 = \sum_{n=0}^{\infty} \|T^{*n}x\|_1^2$ for all $x \in H$. Since $\|x\|^2 - \|T^*x\|^2 = \|x\|_1^2$ we must have

$$\|x\|_1^2 = \|D_{T^*}x\|^2 \tag{9-9}$$

Let $\mathfrak{D}_*$ be defined by (9-8) and S the right shift in $l^2(0, \infty; \mathfrak{D}_*)$ then if we define a map $W: H \to l^2(0, \infty; \mathfrak{D}_*)$ by $Wx = (D_{T^*}x, D_{T^*}T^*x, D_{T^*}T^{*2}x, \ldots)$ it follows that

$$\|Wx\|^2 = \sum_{n=0}^{\infty} \|D_{T^*}T^{*n}x\|^2 = \sum_{n=0}^{\infty} \{\|T^{*n}x\|^2 - \|T^{*n+1}x\|^2\}$$

$$= \lim_{n \to \infty} \{\|x\|^2 - \|T^{*n+1}x\|^2\} = \|x\|^2$$

Obviously W is isometric, Range W is S^*-invariant and $WT^* = S^*W$ which proves that T^* is unitarily equivalent to $S^*|\,\text{Range}\,W$.

Let P be the orthogonal projection of $l^2(0, \infty; \mathfrak{D}_*)$ onto Range W then for all $x, y \in \text{Range}\,W$ we have

$$(S^{*n}x, y) = (x, S^n y) = (Px, S_y^n) = (x, PS^n y)$$

and hence

$$(S^*|\,\text{Range}\,W)^{*n} = PS^n|\,\text{Range}\,W \tag{9-10}$$

which implies that T of Theorem 9-2 is unitarily equivalent to the operator $PS|\,\text{Range}\,W$ and more generally T^n is unitarily equivalent to $PS^n|\,\text{Range}\,W$.

This brings us to make the following definitions. Let M be a subspace of a Hilbert space H, P the orthogonal projection of H onto M. Let T be a bounded operator in M and A a bounded operator in H. We say that A is a *dilation* of T if

$$T = PA|M \tag{9-11}$$

and a *strong dilation* of T if

$$T^n = PA^n|M \qquad \text{for all} \qquad n \geq 0 \tag{9-12}$$

We also refer to T as the *compression* of A. If the operator A has special properties as for example if it is isometric or unitary we will speak of (strong) isometric or unitary dilations. Given any dilation A of T we define $H_0 = \bigvee_{n=0}^{\infty} A^n M$. H_0 is clearly invariant under A and obviously $A|H_0$ is also a dilation of T and a strong dilation if A is a strong dilation. If we are interested in uniqueness of strong dilations then the condition

$$H = \bigvee_{n=0}^{\infty} A^n M \tag{9-13}$$

is a natural one to impose.

Next we proceed to study the question of existence and uniqueness of strong isometric and unitary dilations of contractions in a Hilbert space.

Theorem 9-3 Let T be a contraction in a Hilbert space H. Then there exists a strong isometric dilation V in a Hilbert space $K \supset H$ satisfying

$$K = \bigvee_{n=0}^{\infty} V^n H \tag{9-14}$$

Any two isometric dilations satisfying condition (9-14) are unitarily equivalent.

PROOF Consider the Hilbert space $l^2(0, \infty; H)$ and the embedding $\rho: H \to l^2(0, \infty; H)$ given by $\rho(x) = (x, 0, 0, \ldots)$. Let P be the projection of $l^2(0, \infty; H)$ onto $\rho(H)$. We define a map $W: l^2(0, \infty; H) \to l^2(0, \infty; H)$ by

$$W(x_0, x_1, \ldots) = (Tx_0, D_T x_0, x_1, x_2, \ldots) \tag{9-15}$$

By virtue of (9-6) W is an isometry and is clearly a strong dilation of T. Since (9-14) is not necessarily satisfied we let $K = \bigvee_{n=0}^{\infty} W^n H$ and let $V = W|K$ which proves the existence.

Let now V and V_1 be strong isometric dilations satisfying condition (9-14). Since vectors of the form $\{\sum_{j=0}^{n} V^j x_j | n \geq 0, x_j \in H\}$, $\{\sum_{j=0}^{n} V_1^j x_j | n \geq 0, x_j \in H\}$ span K and K_1, respectively, we define a map $\varphi: K \to K_1$ by

$$\varphi\left(\sum_{j=0}^{n} V^j x_j\right) = \sum_{j=0}^{n} V_1^j x_j \tag{9-16}$$

Since V and V_1 are isometric we have, assuming without loss of generality that $i \geq j$,

$$(V^i x_i, V^j x_j) = (V^{j*} V^i x_i, x_j) = (V^{i-j} x_i, x_j)$$
$$= (V^{i-j} x_i, P x_j) = (P V^{i-j} x_i, x_j) = (T^{i-j} x_i, x_j)$$

Thus it follows that for all i and j $(V^i x_i, V^j x_j) = (V_1^i x_i, V_1^j x_j)$ and hence

$$\left\| \sum_{i=0}^{n} V^i x_i \right\|^2 = \sum_{i=0}^{n} \sum_{j=0}^{n} (V^i x_i, V^j x_j)$$

$$= \sum_{i=0}^{n} \sum_{j=0}^{n} (V_1^i x_i, V_1^j x_j) = \left\| \sum_{i=0}^{n} V_1^i x_i \right\|^2$$

So φ defined by (9-16) is isometric on a dense subset in K and hence has an isometric extension to all of K. Its range includes a dense subset of K_1 and hence φ, having closed range, is necessarily onto and therefore unitary. The relation $\varphi V = V_1 \varphi$ follows trivially from the definition of φ and this proves the unitary equivalence of V and V_1.

For a given contraction T the isometric dilation V that satisfies (9-14), unique up to unitary equivalence, will be called the *minimal isometric dilation*. In that case we will also say that V^* is the *minimal coisometric dilation* of T^*.

We note that for each $x \in H$ we have $PV^{n+1}x = T^{n+1}x = TT^n x = PVPV^n x$ or $PV(I - P)V^n x = 0$ for all $x \in H$. If we assume that V is the minimal isometric dilation of T then the set of vectors of the form $V^n x$, $x \in H$ spans K and hence we have $PV(I - P) = 0$ or $PV = PVP$. Taking adjoints we obtain

$$V^*P = PV^*P \tag{9-17}$$

which is equivalent to H being invariant under V^*. Also given $x, y \in H$ we have $(V^*x, y) = (x, Vy) = (x, PVy) = (x, Ty) = (T^*x, y)$ which implies

$$T^* = V^* | M \tag{9-18}$$

Thus we have proved the following.

Theorem 9-4 Let T be a contraction in a Hilbert space H and let V be its minimal isometric dilation acting in the dilation space K. Then H is invariant under V^* and (9-18) holds.

It is convenient to have a concrete representation for the minimal isometric dilation of a contraction T. This can be done and is summed up by the following.

Theorem 9-5 Let T be a contraction in a Hilbert space H. Then the operator V defined by the operator matrix

$$V = \begin{pmatrix} T & & & & \\ D_T & 0 & & & \\ & I_{\mathfrak{D}} & 0 & & \\ & & \ddots & \ddots & \\ & & & I_{\mathfrak{D}} & \\ & & & & \ddots \end{pmatrix} \tag{9-19}$$

Since

acting in the Hilbert space

$$K = H \oplus \mathfrak{D} \oplus \mathfrak{D} \oplus \cdots \tag{9-20}$$

is the minimal isometric dilation of T. V is unitarily equivalent to a unilateral right shift if and only if $T^{*n} \to 0$ strongly.

PROOF Let $(x_0, x_1, \ldots) \in K$ then $V(x_0, x_1, \ldots) = (Tx_0, D_T x_0, x_1, \ldots)$ and by (9-6) V is isometric. It is clearly well defined as for each $x_0 \in H$ we have $Dx_0 \in \mathfrak{D}$. Thus we have only to show that $K = \bigvee_{n=0}^{\infty} V^n H$. Since obviously $V^n H \subset K$ the inclusion $\bigvee_{n=0}^{\infty} V^n H \subset K$ clearly holds. Now for each $x \in H$ $V(x_0, 0, \ldots) - (Tx_0, 0, \ldots) = (0, Dx_0, 0, \ldots)$ which implies $H \bigvee VH = H \oplus \mathfrak{D} \oplus \{0\} \oplus \cdots$ and as $V^n[\{0\} \oplus \mathfrak{D} \oplus \{0\} \oplus \{0\} \oplus \cdots] = \underbrace{\{0\} \oplus \cdots \oplus \{0\}}_{n+1} \oplus \mathfrak{D} \oplus \{0\} \oplus \cdots$ the result follows.

Finally if V is a unilateral right shift then by (9-18) $T^* = V^*|H$ and hence $T^{*n} \to 0$ strongly. Conversely assume $T^{*n} \to 0$ strongly then by Theorem 9-2 T has an isometric dilation S which is a unilateral right shift. It follows that the minimal isometric dilation is, up to unitary equivalence, S restricted to a reducing subspace of S. However, a unilateral shift reduced to a reducing subspace is also a unilateral shift. This together with the uniqueness of the minimal isometric dilation proves the theorem.

The importance of Theorems 9-2 and 9-5 will become clear once functional models for shifts are available. In that case a natural functional model is obtained for any contraction whose isometric dilation is a unilateral shift. These models are extremely useful for spectral analysis.

Given a contraction T we may apply the previous construction to T^* instead of T to obtain the operator matrix

$$\begin{pmatrix} T^* & & & \\ D_{T^*} & 0 & & \\ & I_{\mathfrak{D}_*} & 0 & \\ & & I_{\mathfrak{D}_*} & \cdot \\ & & & \cdot & \cdot \\ & & & & \cdot & \cdot \end{pmatrix} \tag{9-21}$$

acting in

$$K' = H \oplus \mathfrak{D}_* \oplus \mathfrak{D}_* \oplus \cdots \tag{9-22}$$

as the minimal isometric dilation of T^*.

We can put now the two pieces together to obtain a matrix representation for the minimal unitary dilation of a contraction T. The resultant matrix (9-23) is known as the *Schäfer matrix*.

Theorem 9-6 Let T be a contraction in a Hilbert space H. Then there is a minimal strong unitary dilation of T and two minimal unitary dilations of T are unitarily equivalent.

PROOF Let $K = \cdots \oplus \mathfrak{D}_* \oplus \mathfrak{D}_* \oplus H \oplus \mathfrak{D} \oplus \mathfrak{D} \oplus \cdots$ where $\mathfrak{D}$ and $\mathfrak{D}_*$ are the defect spaces of T defined by (9-7) and (9-8), respectively. Define U to be the lower triangular operator matrix

$$U = \begin{pmatrix} \cdot & & & & & & \\ & \cdot & & & & & \\ & & I_{\mathfrak{D}_*} & & & & \\ & & & I_{\mathfrak{D}_*} & & & \\ & & & & D_{T^*} & T & \\ & & & & -T^* & D_T & \\ & & & & & & I_{\mathfrak{D}} \\ & & & & & & & I_{\mathfrak{D}} \\ & & & & & & & & \cdot \\ & & & & & & & & & \cdot \end{pmatrix} \qquad (9\text{-}23)$$

acting in $K = \cdots \oplus \mathfrak{D}_* \oplus \mathfrak{D}_* \oplus H \oplus \mathfrak{D} \oplus \mathfrak{D} \oplus \cdots$. Thus if $x = (\ldots, x_{-1}, x_0, x_1, \ldots) \in K$ with the x_0 coordinate in H we have

$$Ux = (\ldots, x_{-2}, Tx_0 + D_{T^*}x_{-1}, D_Tx_0 - T^*x_{-1}, x_1, x_2, \ldots) \quad (9\text{-}24)$$

To show that U is well defined in K it suffices to show that $D_Tx_0 - T^*x_{-1} \in \mathfrak{D}$. However, $x_{-1} \in \mathfrak{D}_*$ and this follows by virtue of the relation $T^*D_{T^*} = D_TT^*$. To show that U is isometric we observe that

$$\| Tx_0 + D_{T^*}x_{-1} \|^2 + \| D_Tx_0 - T^*x_{-1} \|^2$$

$$= \| Tx_0 \|^2 + \| D_{T^*}x_{-1} \|^2 + 2\,\text{Re}(Tx_0, D_{T^*}x_{-1}) + \| D_Tx_0 \|^2$$

$$\quad + \| T^*x_{-1} \|^2 - 2\,\text{Re}(D_Tx_0, T^*x_{-1})$$

$$= \{ \| Tx_0 \|^2 + \| D_Tx_0 \|^2 \} + \{ \| T^*x_{-1} \|^2 + \| D_{T^*}x_{-1} \|^2 \}$$

$$= \| x_0 \|^2 + \| x_{-1} \|^2$$

Here we used relation (9-5) to get $(Tx_0, D_{T^*}x_{-1}) = (x_0, T^*D_{T^*}x_{-1}) = (x_0, D_TT^*x_{-1}) = (D_Tx_0, T^*x_{-1})$ as well as (9-6). Since U^* has a similar matrix representation an analogous computation shows that U^* is isometric and hence the unitarity of U follows. That U is a strong dilation of T can be observed from the lower triangularity of the matrix U or equivalently from (9-24).

The proof of minimality is along the lines of the proof of Theorem 9-3 whereas the unitary equivalence of two minimal unitary dilations of T is similar to the proof of the unitary equivalence of two minimal isometric dilations of T. Details are omitted.

It should be observed that the matrix representation (9-19) of the minimal isometric dilation of T is just the lower right-hand corner of the Schäfer matrix

(9-23), and hence the restriction of the minimal unitary dilation U of T to the invariant subspace $\bigvee_{n=0}^{\infty} U^n H$ is the minimal isometric dilation V of T.

We introduce now a notation that simplifies the description of the geometry of the dilation space. We define subspaces $\mathfrak{L}$ and $\mathfrak{L}_*$ of the dilation space K by

$$\mathfrak{L} = \overline{(U - T)H} \qquad (9\text{-}25)$$

and

$$\mathfrak{L}_* = \overline{(U^* - T^*)H} \qquad (9\text{-}26)$$

Clearly if a vector x in K is written as an infinite sequence $(\ldots, x_{-2}, x_{-1}, x_0, x_1, x_2, \ldots)$ with $x_0 \in m$, $x_i \in \mathfrak{D}$ for $i > 0$ and $x_i \in \mathfrak{D}_*$ for $i < 0$ then $\mathfrak{L}$ is the space of all vectors for which $x_i = 0$ for $i \neq 1$ and $\mathfrak{L}_*$ the space of all vectors for which $x_i = 0$ for $i \neq -1$. Moreover for $n > 0$, $x \in U^n \mathfrak{L}$ if and only if $x_i = 0$ for $i \neq n$ and $x \in U^{*n} \mathfrak{L}_*$ if and only if $x_i = 0$ for $i \neq -n$. The dilation space K has now the direct sum representation

$$K = \cdots \oplus U^* \mathfrak{L}_* \oplus \mathfrak{L}_* \oplus H \oplus \mathfrak{L} \oplus U \mathfrak{L} \oplus \cdots \qquad (9\text{-}27)$$

Obviously the infinite direct sums of subspaces $\oplus_{n=-\infty}^{\infty} U^n \mathfrak{L}$ and $\oplus_{n=-\infty}^{\infty} U^{*n} \mathfrak{L}$ are subspaces of the dilation space K and a natural question is under what circumstances one or the other actually equals K.

Theorem 9-7 Let T be a contraction in a Hilbert space H and let U be its minimal unitary dilation acting in K. Then
(a) $K = \oplus_{n=-\infty}^{\infty} U^n \mathfrak{L}$ if and only if T^n tends to zero strongly
(b) $K = \oplus_{n=-\infty}^{\infty} U^{*n} \mathfrak{L}_*$ if and only if T^{*n} tends to zero strongly
(c) $K = (\oplus_{n=-\infty}^{\infty} U^n \mathfrak{L}) \vee (\oplus_{n=-\infty}^{\infty} U^{*n} \mathfrak{L}_*)$ if and only if T is completely non-unitary.

PROOF Assume T^n tends to zero strongly. Since $\oplus_{n=-\infty}^{\infty} U^n \mathfrak{L}$ is obviously a reducing subspace for U and since $\bigvee_{n=-\infty}^{\infty} U^n H = K$ by minimality it is sufficient to show that $H \subset \oplus_{n=-\infty}^{\infty} U^n \mathfrak{L}$. So for $x \in H$ we write $x - U^{-n} T^n x$ as a telescopic series

$$x - U^{-n}T^n x = \sum_{j=0}^{n-1} (U^{-j}T^j x \, N \, U^{-j-1}T^{j+1}x)$$

$$= \sum_{j=0}^{n-1} U^{-j}(I - U^{-1}T)\, T^j x = \sum_{j=0}^{n-1} U^{-j-1}(U - T)\, T^j x$$

which implies that $x - U^{-n}T^n x \in \oplus_{j=1}^{n} U^{-j}\mathfrak{L}$. But since $T^n x$ tends to zero we have by passing to the limit that $x \in \oplus_{j=1}^{\infty} U^{-j}\mathfrak{L} \subset \oplus_{j=-\infty}^{\infty} U^j\mathfrak{L}$. Since $V^* = U^{-1}|\oplus_{j=1}^{\infty} U^{-j}\mathfrak{L}$ is the minimal isometric dilation of T^* the converse is contained in Theorem 9-2. Thus part (a) is proved and part (b) follows by symmetry.

Finally to prove (c) let N be defined by

$$N = \left\{ \left(\bigoplus_{n=-\infty}^{\infty} U^n \mathfrak{L} \right) \bigvee \left(\bigoplus_{n=-\infty}^{\infty} U^{*n} \mathfrak{L}_* \right) \right\}^{\perp}$$

We will show that $N = \{0\}$ if and only if T is completely nonunitary. To this end we need to identify the vectors in N. By the representation (9-27) of the dilation space K it is clear that $N \subset H$. If $x \in N$ then for all $y \in H$ and $n > 0$ we have

$$0 = (x, U^{-n}(U - T)y) = (U^{n-1}x, y) - (U^n x, Ty) = (T^{n-1}x, y) - (T^n x, Ty)$$

In particular the choice $y = T^{n-1}x$ yields the equality $\|T^{n-1}x\| = \|T^n x\|$. In an analogous fashion we show $\|T^{*n-1}x\| = \|T^{*n}x\|$ for all $n > 0$. Thus N is equal to the subspace $H_0 = \{x | \|x\| = \|T^n x\| = \|T^{*n}x\|, n \geq 0\}$. By Theorem 9-1 T is completely nonunitary if and only if $H_0 = \{0\}$ and this completes the proof of the theorem.

We conclude this section with a proof of Naimark's theorem on unitary representations of groups. In Sec. 4 we saw already the close connection between positive definite functions and integral representations. For example $\{c_n\}_{n=-\infty}^{\infty}$ is a positive definite sequence if and only if $c_n = \int e^{int} \, d\mu$ for a positive measure on $\mathbb{T}$. If we introduce in the Hilbert space $L^2(\mu)$ the unitary operator U defined by $Uf = Xf$ and if φ is the function in $L^2(\mu)$ defined by $\varphi(e^{it}) = 1$ then $c_n = (U^n \varphi, \varphi)$ for all $n \in \mathbb{Z}$. We can consider U^n to be a unitary representation of the group $\mathbb{Z}$. The formula $c_n = (U^n \varphi, \varphi)$ can be considered also as a dilation result. If c_n is considered as a contraction on the Hilbert space of complex numbers we can consider $\mathbb{C}$ as the subspace H of $L^2(\mu)$ of all constant functions. If P is the orthogonal projection of $L^2(\mu)$ onto H then $c_n = PU^n|H$. This circle of ideas can be generalized considerably.

Given a group G and a complex Hilbert space H then a $B(H)$-valued function T is said to be *positive definite* if

$$\sum_i \sum_j (T(g_i^{-1} g_j) h_j, h_i) \geq 0 \tag{9-28}$$

for every finite set of vectors $h_i \in H$ and any $g_i \in G$. A *unitary representation* of G is a homomorphism U of G into the set of unitary operators on H which satisfies $U(e) = I$. The connection between positive definite functions on groups and unitary representations is the content of the following theorem of Naimark.

Theorem 9-8 Let $U(g)$ be a unitary representation of the group G in the Hilbert space K and let H be a subspace of K with P the orthogonal projection of K on H. Then

$$T(g) = PU(g)|H \tag{9-29}$$

is a positive definite function which is weakly continuous if $U(g)$ is.

Conversely if $T(g)$ is a positive definite $B(H)$-valued function on G for which $T(e) = I$ then there exists a Hilbert space $K \supset H$ and a unitary representation $U(g)$ of G such that (9-29) holds together with

$$K = \bigvee_{g \in G} U(g) H \qquad (9\text{-}30)$$

If G is a topological group and $T(g)$ weakly continuous then so is the unitary representation.

PROOF Assume $U(g)$ is a unitary representation of the group G. Then with $T(g)$ defined by (9-29) we have for $h_i \in H$

$$\sum_{i=1}^{n} \sum_{j=1}^{n} (T(g_i^{-1} g_j) h_i, h_j) = \sum_{i=1}^{n} \sum_{j=1}^{n} (P U(g_i)^* U(g_j) h_j, h_i)$$

$$= \sum_{i=1}^{n} \sum_{j=1}^{n} (U(g_i) h_i, U(g_j) h_j) = \left\| \sum_{i=1}^{n} U(g_i) h_i \right\|^2 \geq 0$$

Continuity of T follows from that of U by (9-29).

To prove the converse let T be a positive definite $B(H)$-valued function on G. Let L be the set of all finitely nonzero H-valued functions defined on G. L becomes a complex linear space by the definitions of addition and multiplication by scalars. In L we introduce the inner product

$$(f, f') = \sum_{h \in G} \sum_{g \in G} (T(g^{-1} h) f(h), f'(g)) \qquad (9\text{-}31)$$

where the sum, having only a finite number of nonzero terms, makes sense.

By the assumption of positive definiteness the above inner product is a nonnegative definite Hermitian form. It may happen, however, that $(f, f) = 0$ without $f = 0$. We denote by N the set of all elements f in L for which $(f, f) = 0$. From the Schwarz inequality it follows that if $f \in N$ then $(f, f') = 0$ for all $f' \in L$, in particular N is a subspace. To make the inner product definite we factor out the set of null elements. Let therefore $K' = L/N$ be the set of all equivalence classes modulo the null elements and K the completion of K' in the induced norm topology. So K is a Hilbert space. H can be considered as embedded in K by the following considerations. Define for a given $x \in H$ the function $f_x \in L$ by $f_x(e) = x$ and $f_x(g) = 0$ for $g \neq e$. Since

$$(f_x, f_y) = \sum_{h \in G} \sum_{g \in G} (T(g^{-1} h) f_x(h), f_y(g)) = (f_x(e), f_y(e)) = (x, y)$$

H is isometrically embedded in L and hence also in L/N and eventually in K.

Define now a translation semigroup in L as follows. Given $f \in L$ let

$$U(g_0) f(h) = f(g_0^{-1} h) = f_{g_0}(h)$$

Given $f, f' \in L$ then

$$(U(g_0) f, U(g_0) f') = (f_{g_0}, f'_{g_0}) = \sum_{h \in G} \sum_{g \in G} (T(g^{-1}h) f_{g_0}(h), f'_{g_0}(g))$$

$$= \sum \sum (T(g^{-1}h) f(g_0^{-1}h), f'(g_0^{-1}g))$$

$$= \sum \sum (T(g^{-1}g_0 g_0^{-1}h), f'(g_0^{-1}g))$$

Now as h and g vary over all elements of G so do $g_0^{-1}h$ and $g_0^{-1}g$. Hence

$$(U(g_0) f, U(g_0) f') = (f, f')$$

which shows that $U(g_0)$ is isometric for each $g_0 \in G$. Since $U(g_0)$ is invertible, in fact $U(g_0)^{-1} = U(g_0^{-1})$, $U(g)$ is actually a unitary representation of G in L for which N is a reducing subspace. Passing to K' and K we still have a unitary representation which we denote by the same letter.

Let $x, y \in H$ then

$$(U(g_0) x, y) = \sum_{g \in G} \sum_{h \in G} (T(g^{-1}h) U(g_0) x(h), y(g))$$

$$= \sum_{g \in G} \sum_{h \in G} (T(g^{-1}h) x(g_0^{-1}h), y(g))$$

Now $y(g) = e$ for $g = e$ and zero otherwise and $x(g_0^{-1}h) = x$ for $h = g_0$ and zero otherwise. Therefore it follows that

$$(U(g_0) x, y) = (T(g_0) x, y) \tag{9-32}$$

which is equivalent to (9-29). Since L is clearly the set of all finite linear combinations of translates of elements of H then $L = \bigvee_{g \in G} U(g) h$ and (9-30) follows from that by going to the quotient space K' first and then by completion.

Assuming $T(g)$ is weakly continuous then each function of the form $(T(g) x, y)$ is continuous. Equality (9-29) shows that $(U(g_0) x, y)$ is continuous for x and y in L and since $\|U(g_0)\| = 1$ continuity follows for all x, y in K. This completes the proof.

Naimark's theorem yields another approach to the existence of minimal unitary dilations of contractions. Let T be a contraction in the Hilbert space H. Define a $B(H)$-valued function $T(n)$ on $\mathbb{Z}$ by

$$T(n) = \begin{cases} T^n & n \geq 0 \\ T^{*|n|} & n < 0 \end{cases}$$

then Halperin [65] has observed that $T(n)$ is a positive function on $\mathbb{Z}$. The existence of a strong unitary dilation of T follows by a straightforward application of Naimark's theorem. A similar approach works for the unitary dilation of contractive semigroups.

10. SEMIGROUPS OF OPERATORS

Semigroups of operators arise in the solution of differential equations with operator coefficients in the same way as the exponential of a matrix is used to solve
first order constant linear systems of differential equations. Thus semigroups are
operator solutions of the functional equation of the exponential function. Specifically, a one parameter semigroup of operators in a Banach space X is a family
$\{T(t)|t \geq 0\}$ of bounded operators on X which satisfies

(a) $T(0) = I$ and
(b) $T(t + s) = T(t)\,T(s)$ for all $t, s \geq 0$

Our first object in the study of semigroups is to find an exponential representation for the semigroup. This is possible if some continuity assumptions are
made. It turns out that the assumption of continuity of the semigroup in the
operator norm is much too restrictive and would exclude most of the applications.
Strong continuity is sufficient for the development of the exponential representation. We say that the semigroup $\{T(t)\}$ is strongly continuous if for each $x \in X$
the X-valued function $T(t)\,x$ is norm continuous on $[0, \infty)$.

Given the strongly continuous semigroup $\{T(t)\}$ we define the *infinitesimal
generator* of the semigroup to be the operator

$$Ax = \lim_{t \to 0^+} \frac{T(t)\,x - x}{t} \tag{10-1}$$

where D_A the domain of definition of A is the set of vectors for which the limit
(10-1) exists.

Lemma 10-1

(a) The set D_A is a linear manifold and A is linear on D_A.
(b) If $x \in D_A$ then $T(t)\,x$ has a strong derivative given by

$$\frac{d}{dt} T(t)\,x = AT(t)\,x = T(t)\,Ax \tag{10-2}$$

(c) For x in D_A

$$T(t)\,x - x = \int_0^t T(\tau)\,Ax\,d\tau = \int_0^t AT(\tau)\,x\,d\tau$$

PROOF (a) is obvious from the definition. Let $x \in D_A$, $t \geq 0$ and $\tau > 0$ then

$$\frac{T(t + \tau) - T(t)\,x}{\tau} = T(t)\,\frac{T(\tau)\,x - x}{\tau} = \frac{T(\tau)\,T(t)\,x - T(t)\,x}{\tau}$$

$$\lim_{\tau \to 0} T(t)\,\frac{T(\tau)\,x - x}{\tau} = T(t)\,Ax$$

it follows that

$$\lim_{\tau \to 0} \frac{T(\tau)\, T(t)\, x - T(t)\, x}{\tau} = T(t)\, Ax$$

Thus $T(t)\, x \in D_A$ and by the definition of the infinitesimal generator (10-2) follows. Part (c) follows by integrating (10-2).

Part (c) of the lemma can be strengthened as follows.

Lemma 10-2 For all x in X we have

$$T(t)\, x - x = A \int_0^t T(\tau)\, x \, d\tau$$

PROOF

$$A \int_0^t T(\tau)\, x \, d\tau = \lim_{h \to 0} \frac{T(h) - I}{h} \int_0^t T(\tau)\, x \, d\tau$$

$$= \lim_{h \to 0} \frac{1}{h} \int_0^t \left[T(h + \tau)\, x - T(\tau)\, x \right] d\tau$$

$$= \lim_{h \to 0} \frac{1}{h} \int_t^{t+h} T(\tau)\, x \, d\tau - \lim_{h \to 0} \frac{1}{h} \int_0^h T(\tau)\, x \, d\tau = T(t)\, x - x$$

by the strong continuity of the semigroup.

Corollary 10-3 The linear manifold D_A is dense in H and A is a closed operator.

PROOF For each $x \in X$ the vector $(1/t) \int_0^t T(\tau)\, x \, d\tau$ is in D_A and $x = \lim_{t \to 0} (1/t) \int_0^t T(\tau)\, x \, d\tau$. To show that A is closed let $x_n \in D_A$, $x_n \to x$ and $Ax_n \to y$. By integrating (10-2) we have $T(t)\, x_n - x_n = \int_0^t T(\tau)\, Ax_n \, d\tau$ and by assumption $\lim_{t \to 0} \left(T(t)\, x_n - x_n \right) = T(t)\, x - x$ and $\lim \int_0^t T(\tau)\, Ax_n \, d\tau = \int_0^t T(\tau)\, y \, d\tau$. So $\left(T(t)\, x - x \right)/t = (1/t) \int_0^t T(\tau)\, y \, d\tau$, and hence $x \in D_A$ and $Ax = y$ which proves the closedness of A.

From now on we will assume that the semigroup is a strongly continuous semigroup of contractions, that is, that $\| T(t) \| \leq 1$ for $t \geq 0$. This assumption gives us some information concerning the spectrum of the infinitesimal generator.

Theorem 10-4 Let $\{ T(t) \}$ be a strongly continuous semigroup of contraction then $\rho(A)$, the resolvent set of A, includes the open right-half plane. Moreover for $\lambda > 0$ we have

$$R(\lambda, A) = \int_0^\infty e^{-\lambda t} T(t) \, dt \tag{10-3}$$

PROOF Consider the semigroup $\{e^{-\lambda t}T(t)\}$ having the infinitesimal generator $A - \lambda I$. Applying Lemma 10-2 we have

$$x - e^{-\lambda t}T(t)x = (\lambda I - A)\int_0^t e^{-\lambda \tau}T(\tau)x\,d\tau \qquad \text{for all} \qquad x \in X \qquad (10\text{-}4)$$

and

$$x - e^{-\lambda t}T(t)x = \int_0^t e^{-\lambda \tau}T(\tau)(\lambda I - A)x\,d\tau \qquad \text{for} \qquad x \in D_A \qquad (10\text{-}5)$$

Now the operator $R(\lambda)$ defined by the norm convergent integral

$$R(\lambda) = \int_0^\infty e^{-\lambda \tau}T(\tau)\,d\tau$$

is clearly a well-defined bounded operator. Passing to the limit in (10-4) and (10-5) we obtain

$$x = (\lambda I - A)R(\lambda)x \qquad \text{for all} \qquad x \in X$$

and

$$x = R(\lambda)(\lambda I - A)x \qquad \text{for} \qquad x \in D_A$$

Thus $\lambda > 0$ is in $\rho(A)$ and $R(\lambda) = R(\lambda, A)$.

A strongly continuous semigroup determines uniquely its infinitesimal generator. The converse is also true and is given by the following lemma.

Lemma 10-5 A strongly continuous semigroup is uniquely determined by its infinitesimal generator.

PROOF Assume $\{T_1(t)\}$ and $\{T_2(t)\}$ are two strongly continuous semigroups having the same infinitesimal generator A. Then for all $x \in D_A$ we have also $T_i(t)x \in D_A$, $i = 1, 2$. On differentiating $T_2(t - \tau)T_1(\tau)x$ with $x \in D_A$ we obtain

$$\frac{d}{d\tau}T_2(t - \tau)T_1(\tau)x = T_2(t - \tau)AT_1(\tau)x - AT_2(t - \tau)T_1(\tau)x$$

$$= T_2(t - \tau)AT_1(\tau)x - T_2(t - \tau)AT_1(\tau)x = 0$$

Thus $T_2(t - \tau)T_1(t)x$ is a constant vector. Evaluating it at $\tau = 0$ and $\tau = t$ we have $T_2(t)x = T_1(t)x$ for all $t > 0$ and $x \in D_A$. By the strong continuity of both semigroups this equality holds for all $x \in X$, that is, the semigroups are equal.

Our next object is the characterization of infinitesimal generators of contractive semigroups, that is of strongly continuous semigroups satisfying $\|T(t)\| \leq 1$.

Moreover we want a procedure which will enable us to reconstruct the semi-group, given its infinitesimal generator. This is the content of the Hille–Yosida theorem.

Theorem 10-6 (Hille–Yosida) A closed densely defined operator A is the infinitesimal generator of a contractive semigroup if and only if

$$\|R(\lambda, A)\| \le \lambda^{-1} \qquad \text{for all} \qquad \lambda > 0 \qquad (10\text{-}6)$$

PROOF Assume $\{T(t)\}$ is a contractive semigroup. By (10-3) we have $R(\lambda, A) = \int_0^\infty e^{-\lambda t} T(t) \, dt$ and hence for $\lambda > 0$

$$R(\lambda, A)\| = \left\| \int_0^\infty e^{-\lambda t} T(t) \, dt \right\| \le \int_0^\infty e^{-\lambda t} \|T(t)\| \, dt \le \int_0^\infty e^{-\lambda t} \, dt = \lambda^{-1}$$

To prove the converse we construct a family of approximating semigroups having bounded infinitesimal generators. Let $A_\lambda = \lambda^2 R(\lambda, A) - \lambda I$ and define $T_\lambda(t)$ by

$$T_\lambda(t) = e^{t A_\lambda} \qquad (10\text{-}7)$$

$T_\lambda(t)$ is well defined as e^{tz} is an entire function and so $T_\lambda(t) = \sum_{n=0}^\infty t^n (A_\lambda^n/n!)$. From (10-6) we have for $x \in D_A$

$$\|\lambda R(\lambda, A) x - x\| = \|R(\lambda, A) A x\| \le \lambda^{-1} \|A x\|$$

and hence

$$\lim_{\lambda \to \infty} \lambda R(\lambda, A) x = x \qquad \text{for} \qquad x \in D_A \qquad (10\text{-}8)$$

Since D_A is dense and the set of operators $\lambda R(\lambda, A)$ is uniformly bounded then (10-8) holds for all $x \in H$. In particular since

$$\lambda R(\lambda, A) A x = \lambda [R(\lambda, A)(A - \lambda I) x + R(\lambda, A) \lambda x]$$
$$= \lambda^2 R(\lambda, A) x - \lambda x = A_\lambda x$$

we have for all $x \in D_A$ that

$$\lim_{\lambda \to \infty} A_\lambda x = A x \qquad (10\text{-}9)$$

The approximating semigroups $\{T_\lambda(t)\}$ are also contractive semigroups as is seen from the following estimate

$$\|T_\lambda(t)\| = \left\| e^{-\lambda t} e^{\lambda^2 R(\lambda, A) t} \right\| = \left\| e^{-\lambda t} \sum_{n=0}^\infty \lambda^{2n} \frac{R(\lambda, A)^n t^n}{n!} \right\|$$

$$\le e^{-\lambda t} \sum_{n=0}^\infty \frac{t^n \lambda^n \|\lambda R(\lambda, A)\|^n}{n!} \le e^{-\lambda t} \sum_{n=0}^\infty \frac{t^n \lambda^n}{n!} = 1$$

Since $T_\lambda(t)$ and $T_\mu(t)$ commute for all $\lambda, \mu \geq 0$ we have

$$T_\lambda(t) - T_\mu(t) = \int_0^t \left[\frac{d}{d\tau} T_\lambda(t) T_\mu(t - \tau) \right] d\tau$$

$$= \int_0^t T_\lambda(\tau) T_\mu(t - \tau)(A_\lambda - A_\mu) d\tau$$

from which the estimate

$$\| T_\lambda(t) x - T_\mu(t) x \| \leq t \| A_\lambda x - A_\mu x \|$$

follows. Thus from (10-9) it follows that $\lim_{\lambda \to \infty} T_\lambda(t) x$ exists for all $x \in D_A$ uniformly on compact subsets of $[0, \infty)$. Again the uniform boundedness of the $T_\lambda(t)$ implies the existence of the limit for all $x \in H$. Define now $T(t)$ by

$$T(t) x = \lim_{\lambda \to \infty} T_\lambda(t) x \qquad (10\text{-}10)$$

for $t \geq 0$. We show first that $\{T(t)\}$ is a strongly continuous semigroup.

$$T(t + s) x = \lim_{\lambda \to \infty} T_\lambda(t + s) x = \lim_{\lambda \to \infty} T_\lambda(t) T_\lambda(s) x = T(t) T(s) x$$

So $T(t)$ is a semigroup. By the inequality $\| T(t) x \| = \lim_{\lambda \to \infty} \| T_\lambda(t) x \| \leq \| x \|$ the semigroup is contractive.

Finally as

$$\| x - T(t) x \| \leq \| x - T_\lambda(t) x \| + \| T_\lambda(t) x - T(t) x \|$$

and since T_λ converges to $T(t) x$ uniformly on compact subsets, the strong continuity of $T(t)$ follows from that of $T_\lambda(t)$. To conclude we will show that A is the infinitesimal generator of the semigroup $\{T(t)\}$. Let B be the infinitesimal generator of $\{T(t)\}$. For $x \in D_A$ we have

$$T_\lambda(h) x - x = \int_0^h T_\lambda(\tau) A_\lambda x \, d\tau$$

and hence also

$$\frac{T_\lambda(h) x - x}{h} = \frac{1}{h} \int_0^h T_\lambda(\tau) A_\lambda x \, d\tau$$

Let $\lambda \to \infty$ and we obtain

$$\frac{T(h) x - x}{h} = \frac{1}{h} \int_0^h T(\tau) Ax \, d\tau$$

But the last equality implies $x \in D_B$ and $Bx = Ax$ which is equivalent to $B \supset A$. Now for $\lambda > 0$, $\lambda I - A$ is invertible and has no proper extension so necessarily $B = A$, which completes the proof.

Applying Theorem 3-10 which characterizes the maximal accretive operators we can restate the Hille–Yosida theorem as follows.

Theorem 10-7 A closed densely defined operator is the infinitesimal generator of a strongly continuous semigroup of contractions if and only if it is maximal accretive.

Given a strongly continuous semigroup $\{T(t)\}$, now specifically assumed to act in a Hilbert space H, we define the adjoint semigroup to be the family of adjoints $\{T(t)^* \,|\, t \geq 0\}$. It is clear that $\{T(t)^*\}$ is also a semigroup. Since $\lim_{t \to 0} (T(t)^* x, y) = \lim_{t \to 0} (x, T(t) y) = (x, y)$ for all $x, y \in H$ it follows that

$$0 \leq \|T(t)^* x - x\|^2 = \|T(t)^* x\|^2 - (T(t)^* x, x) - (x, T(t)^* x) + (x, x)$$
$$\xrightarrow[t \to 0]{} \|T(t)^* x\|^2 - \|x\|^2 \leq 0$$

Thus necessarily $\lim_{t \to 0} T(t)^* x = x$ which shows that the adjoint semigroup is also strongly continuous. The expected relation between the infinitesimal generators of adjoint semigroups holds and is given by the following theorem.

Theorem 10-8 The infinitesimal generator of the adjoint semigroup is the adjoint of the infinitesimal generator of the original semigroup.

PROOF Let A and B be the infinitesimal generators of the semigroups $\{T(t)\}$ and $\{T(t)^*\}$, respectively, and let D_A and D_B be their respective domains of definition. Choose $x \in D_A$ and $y \in D_B$, then

$$(Ax, y) = \lim_{t \to 0} \left(\frac{T(t) x - x}{t}, y \right) = \lim_{t \to 0} \left(x, \frac{T(t)^* y - y}{t} \right) = (x, By)$$

So $B \subset A^*$. Conversely if $x \in D_A$ and $y \in D_{A^*}$

$$(T(t) x - x, y) = \int_0^t (AT(\tau) x, y) \, d\tau = \int_0^t (x, T(t)^* A^* y) \, d\tau$$

which implies $T(t)^* y - y = \int_0^t T(\tau)^* A^* y \, d\tau$. Dividing by $t > 0$ and letting t tend to zero we have $By = A^* y$ or $B \supset A^*$. So $B = A^*$ and the proof is complete.

If A is the infinitesimal generator of a contractive semigroup then, by Theorem 3-9, its Cayley transform T defined by

$$T = (A + I)(A - I)^{-1} \tag{10-11}$$

is a contraction. We call this contraction the *infinitesimal cogenerator* of the semigroup. Since the Cayley transform is invertible there is a bijective correspondence between a semigroup and its cogenerators. The cogenerator reflects in a more faithful way the properties of the semigroup. In the following theorems that characterize the infinitesimal generators of isometric and unitary semigroups are some instances of this relation. We precede these theorems by showing that a semigroup and its cogenerator have the same invariant subspaces.

Theorem 10-9 Let $\{T(t)\}$ be a contractive semigroup and T its infinitesimal cogenerator. A subspace M is invariant under the semigroup if and only if it is invariant under T.

PROOF Assume M is a subspace invariant under the semigroup. Since $T = I + 2(A - I)^{-1}$ and by Theorem 10-4 $(I - A)^{-1} x = \int_0^\infty e^{-t} T(t) x \, dt$ it is clear that M is also invariant under T.

Conversely assume M is invariant under T. For $\lambda > 0$ let A_λ be defined by

$$A_\lambda = \lambda^2 R(\lambda, A) - \lambda I \tag{10-12}$$

then for M to be invariant under A_λ it suffices to show that it is invariant under $R(\lambda, A)$. Now

$$R(\lambda, A) = [\lambda - (T + I)(T - I)^{-1}]^{-1} = (T - I)[\lambda(T - I) - (T + I)]^-$$

$$= (T - I)(\lambda + 1)^{-1} \left[\left(\frac{\lambda - 1}{\lambda + 1}\right) T - I\right]^{-1}$$

For $\lambda > 0$ we have $(\lambda - 1)/(\lambda + 1) < 1$ and hence $\{[(\lambda - 1)/(\lambda + 1)] T - I\}^{-1}$ can be expanded into a uniformly convergent geometric series. This shows that M is invariant under A_λ and hence also under the semigroup $\{T_\lambda(t)\} = \{e^{t A_\lambda}\}$ used in the proof of the Hille–Yosida theorem. Finally since $T(t) x = \lim_{\lambda \to \infty} T_\lambda(t) x$ the result follows.

Theorem 10-10 The following statements are equivalent.
(a) $T(t)$ is a strongly continuous semigroup of isometries.
(b) The infinitesimal generator A satisfies $A \subset -A^*$, or equivalently $(iA) \subset (iA)^*$.
(c) $\text{Re}(Ax, x) = 0$ for all $x \in D_A$.
(d) The infinitesimal cogenerator T of the semigroup is isometric and $1 \notin \sigma_p(T)$.

PROOF Assume $\{T(t)\}$ is isometric. Thus $\|T(t) x\|^2 = \|x\|^2$. Differentiating the equality, assuming $x \in D_A$ we have

$$(AT(t) x, T(t) x) + (T(t), AT(t) x) = 0$$

In particular $(Ax, x) + (x, Ax) = 2 \, \text{Re}(Ax, x) = 0$ which also implies $A \subset -A^*$. Thus (b) and (c) follow. The equivalence of (b) and (c) is obvious. Assume now (c) holds, then

$$\|(A + I) x\|^2 = \|x\|^2 + \|Ax\|^2 = \|(A - I) x\|^2$$

Since Range$(A - I) = H$ we have for all $y \in H$

$$\|Ty\|^2 = \|(A + I)(A - I)^{-1} y\|^2 = \|y\|^2$$

or T is isometric. So (c) implies (d). If T is isometric and $1 \notin \sigma_p(T)$ then by Theorem 3-9 $A = (T + I)(T - I)^{-1}$ is maximal accretive. Now $A + I = 2T(T - I)^{-1}$ and $A - I = 2(T - I)^{-1}$ so for all x in $D_A = \text{Range}(T - I)$ we have

$$\|(A + I)x\| = \|2T(T - I)^{-1}x\| = \|2(T - I)^{-1}x\| = \|(A - I)x\|$$

But the equality $\|(A + I)x\| = \|(A - I)x\|$ is equivalent to $\text{Re}(Ax, x) = 0$ and hence (d) implies (b) and (c). Finally, since $(d/dt)\|T(t)x\|^2 = 2\,\text{Re}(AT(t)x, T(t)x) = 0$ for all $x \in D_A$, we have $\|T(t)x\|$ is a constant and hence necessarily $\|T(t)x\| = \|x\|$ for all $t \geq 0$ and $x \in D_A$. By continuity this holds for all x and the implication of (a) by (d) is proved.

If $T(t)$ is a semigroup of unitary operators then it can be extended to a unitary group, $U(t)$, $t \in R$, by letting

$$U(t) = \begin{cases} T(t) & t \geq 0 \\ T(-t)^* = T(-t)^{-1} & t < 0 \end{cases} \tag{10-13}$$

Theorem 10-11 (Stone) The following statements are equivalent.
a) $U(t)$ is a strongly continuous group of unitary operators
b) The infinitesimal generator of $U(t)$ is skew self-adjoint
c) The cogenerator U of the group is unitary and $1 \notin \sigma_p(U)$.

PROOF Assume $U(t)$ is a group of unitary operators then so is $U_1(t) = U(t)^*$. The infinitesimal generator of the adjoint group $U_1(t)$ is A^*. Applying the previous theorem we have $A \subset -A^*$ and $A^* \subset -A^{**}$. However, since A is closed and densely defined we have, by Corollary 3-2, that $A^{**} = A$ and hence $A^* = -A$ and the equivalence of (a) and (b) follows. If $U(t)$ is unitary both $U = (A + I)(A - I)^{-1}$ and $U_1 = (A^* + I)(A^* - I)^{-1}$ are isometries. However, on the dense set D_{A^*} we have

$$(A^* - I)^{-1}(A^* + I) \subset (A^* + I)(A^* - I)^{-1} = U_1$$

So $U^* \subset U_1$ and by continuity they must be equal. Consequently if U is unitary and $1 \notin \sigma_p(U)$ then U and U^* are the infinitesimal cogenerators of the unitary semigroups $T(t)$ and $T(t)^*$. Define $U(t)$ by (10-13) and it is easily checked that $U(t)$ is a unitary group.

If A is the infinitesimal generator of the strongly continuous group of unitary operators $\{U(t)\}$ then $B = -iA$ is a self-adjoint operator. Thus there exists, by the spectral theorem, a spectral measure E on R such that for all $x \in D_B$

$$Bx = \int_{-\infty}^{\infty} \lambda E(d\lambda) x$$

This implies that the group $\{U(t)\}$ has the integral representation

$$U(t)\,x = \int_{-\infty}^{\infty} e^{i\lambda t} E(d\lambda)\,x \tag{10-14}$$

and conversely for each spectral measure the group $U(t)$ defined by (10-13) is a strongly continuous unitary group.

To give important examples of isometric and unitary semigroups we consider the Hilbert space $L^2(-\infty, \infty)$ of all Lebesgue square integrable functions on $\mathbb{R}$ with the norm given by

$$\|f\|^2 = \int_{-\infty}^{\infty} |f(x)|^2\,dx \tag{10-15}$$

In $L^2(-\infty, \infty)$ we define the group $\{U(t)\}$ by

$$(U(t)\,f)(x) = f(x - t) \tag{10-16}$$

Obviously $\{U(t)\}$ is a unitary group with $U(t)^{-1} = U(t)^* = U(-t)$. To see that $\{U(t)\}$ is strongly continuous we note that for any continuous function f of compact support the L^2 norm of $f - U(t)\,f$ can be made arbitrarily small as f is actually uniformly continuous. Since these functions are dense in $L^2(-\infty, \infty)$ and the norms of $U(t)$ uniformly bounded the strong continuity follows. We call $\{U(t)\,|\,t \in R\}$ the *translation group* with $\{U(t)\,|\,t \geq 0\}$ the *right translation semigroup* and $\{U(t)^*\,|\,t \geq 0\} = \{U(-t)\,|\,t \geq 0\}$ the *left translation semigroup*.

The subspace $L^2(0, \infty)$ considered as a subspace of $L^2(-\infty, \infty)$ is clearly invariant under the right translation semigroup. We define $\{V(t)\,|\,t \geq 0\}$ to be the restriction of the right translation semigroup to $L^2(0, \infty)$. Thus we have

$$(V(t)\,t)(x) = \begin{cases} f(x - t) & x \geq t \\ 0 & x < t \end{cases} \tag{10-17}$$

and $\{V(t)\}$ is a strongly continuous semigroup of isometries, the semigroup of right translations in $L^2(0, \infty)$. The adjoint semigroup is given by

$$(V(t)^* f)(x) = f(x + t) \tag{10-18}$$

which defines the semigroup of left translations. We have

$$\|V(t)^* f\|^2 = \int_0^{\infty} |f(x + t)|^2\,dx = \int_t^{\infty} |f(x)|^2\,dx$$

So $\lim_{t \to 0} V(t)^* f = 0$ for all $f \in L^2(0, \infty)$.

We determine next the infinitesimal generators of the various semigroups. To begin let A be the infinitesimal generator of the group $\{U(t)\}$. Let $f \in D_A$ then $(f(x - t) - f(x))/t$ converges, in $L^2(-\infty, \infty)$, to a function $g \in L^2(-\infty, \infty)$.

Integrating against the characteristic function of the interval (a, b) we have

$$\lim_{t=0} \int_a^b \frac{f(x - t) - f(x)}{t} dx = \int_a^b g(x) dx$$

By a change of variable we have

$$\int_a^b \frac{f(x - t) - f(x)}{t} dx = \frac{1}{t} \int_{a-t}^a f(x) dx - \frac{1}{t} \int_{b-t}^b f(x) dx$$

The function f is locally in L^1 and therefore almost everywhere

$$\lim_{t \to 0} \frac{1}{t} \int_{a-t}^a f(x) dx = f(a)$$

and equally for the other integral. So for almost all b we have

$$f(a) - f(b) = \int_a^a g(x) dx \tag{10-19}$$

We redefine f, on a set of measure zero, so that (10-19) holds for all b. But (10-19) means that f is absolutely continuous and almost everywhere $f'(x) = -g(x)$. Summarizing, if A is the infinitesimal generator of the right translation semigroup in $L^2(-\infty, \infty)$ then

$$D_A = \{f \in L^2(-\infty, \infty)| \ f \text{ absolutely continuous and } f' \in L^2(-\infty, \infty)\} \tag{10-20}$$

and

$$Af = -f' \tag{10-21}$$

In exactly the same way if B is the infinitesimal generator of the left translation semigroup in $L^2(-\infty, \infty)$ then $D_B = D_A$ and

$$Bf = f' \tag{10-22}$$

Let U be the infinitesimal cogenerator of the right translation semigroup. Then U^* is the cogenerator of the left translation semigroup. To obtain explicit formulas for U and U^* we note that $U = (A + I)(A - I)^{-1} = I + 2(A - I)^{-1}$.

Let $Uf = g$ then $f + 2(A - I)^{-1} f = g$. If $(A - I)^{-1} f = h$ then $f = (A - I)h = -h' - h$. So h is the solution of the differential equation

$$h' + h = -f \tag{10-23}$$

that lies in $L^2(-\infty, \infty)$. The general solution of (10-23) is $e^{-t}\{c - \int_0^t e^\tau f(t) d\tau\}$ and for that to be in $L^2(-\infty, \infty)$ we have to have $c = \int_0^\infty e^\tau f(\tau) d\tau = -\int_{-\infty}^0 e^\tau f(t) dt$. This yields

$$(Uf)(t) = f(t) - 2e^{-t} \int_{-\infty}^t e^\tau f(\tau) d\tau \tag{10-24}$$

By similar computation for U^* we obtain

$$(U^*f)(t) = f(t) - 2e^t \int_t^\infty e^{-\tau} f(\tau)\, d\tau \qquad (10\text{-}25)$$

It can be seen directly from (10-24) and (10-25) that $L^2(0, \infty)$ is invariant under U and $L^2(-\infty, 0)$ is invariant under U^*. This is of course evident as a consequence of Theorem 10-9.

The semigroups of left and right translation in $L^2(-\infty, \infty)$ are clearly unitarily equivalent. This follows easily from the spectral analysis that follows but can be seen directly by observing that if we define a map $J: L^2(-\infty, \infty) \to L^2(-\infty, \infty)$ by

$$(Jf)(x) = -f(-x) \qquad (10\text{-}26)$$

then J is unitary, $J^* = J^{-1} = J$ and

$$U(t)J = JU(-t) = JU(t)^* \qquad (10\text{-}27)$$

is satisfied.

It is easy to check that relation (10-27) holds also for the cogenerators, that is we have

$$UJ = JU^* \qquad (10\text{-}28)$$

By Stone's theorem both A and B are skew self-adjoint, and from Theorem 10-8 we have $B = A^* = -A$, so $A = -B = iH$ for some self-adjoint operator H. As a consequence the spectra of A and B lie on the imaginary axis. To determine the spectrum of A and B we note first that the point spectrum of A is empty. Indeed the only solutions of the equation $Af = \lambda f$ are of the form $f(x) = ce^{-\lambda x}$ and none of these is in $L^2(-\infty, \infty)$. To characterize the spectrum the simplest approach is to determine the spectral representation of the unitary group $\{U(t)\}$. This is done through the introduction of the Fourier transformation in $L^2(-\infty, \infty)$.

We define the Fourier transform initially in $L^1(-\infty, \infty)$, the space of all Lebesgue integrable functions on the real line, the norm being given by $\|f\|_1 = \int_{-\infty}^\infty |f(x)|\, dx$. We observe that $L^1(-\infty, \infty)$ is a Banach space, and moreover a commutative Banach algebra under convolution multiplication

$$(f * g)(x) = \int_{-\infty}^\infty f(x - t)g(t)\, dt \qquad \text{for} \qquad f, g \in L^1(-\infty, \infty) \qquad (10\text{-}29)$$

The *Fourier transform* $\mathscr{F}$ is defined by $\mathscr{F}f = \hat{f}$ with

$$\hat{f}(\lambda) = \frac{1}{\sqrt{2\pi}} \int_{-\infty}^\infty f(x) e^{i\lambda x}\, dx \qquad (10\text{-}30)$$

If $f, g \in L^1(-\infty, \infty)$ and $h = f * g$ then we have $\hat{h}(\lambda) = \hat{f}(\lambda)\hat{g}(\lambda)$ or equivalently

$$\mathscr{F}(f * g) = \mathscr{F}(f) \cdot \mathscr{F}(g) \qquad (10\text{-}31)$$

Under suitable smoothness and growth conditions the Fourier transform can be inverted.

To obtain an inversion formula for the Fourier transform we use the idea of a summability kernel, introduced in Sec. 4, with the only difference that the domain of definition is taken to be the real line. Thus a family of functions $\{k_\lambda(x)\}$ defined on $\mathbb{R}$ is a summability kernel if $k_\lambda(x) \geq 0$, $\int_{-\infty}^{\infty} k(x)\,dx = 1$ and $\lim_{\lambda \to \infty} \int_{|x| \geq \delta} k_\lambda(x)\,dx = 0$ for all $\delta > 0$. If $k \in L^1(-\infty, \infty)$ and $k(x) \geq 0$ then $k_\lambda(x) = \lambda k(\lambda x)$ is easily checked to be a summability kernel. The important property of summability kernels is that for all $f \in L^1(-\infty, \infty)$ we have $\lim_{\lambda \to \infty} \| f - k_\lambda * f \|_1 = 0$. Let us pick now

$$k(x) = \frac{1}{2\pi}\left(\frac{\sin x/2}{x/2}\right)^2 = \frac{1}{2\pi}\int_{-1}^{1}(1 - |\xi|)\,e^{-i\xi x}\,d\xi$$

which is the Fejer kernel. That $\int_{-\infty}^{\infty} k(x)\,dx = 1$ can be proved by contour integration and the residue theorem, whereas the representation of $k(x)$ as a Fourier transform follows by integration by parts of $(1/2\pi)\int_{-1}^{1}(1 - |\xi|)\,e^{-i\xi x}\,d\xi = (1/\pi)\int_{0}^{1}(1 - \xi)\cos \xi x\,d\xi$. Since $k_\lambda(x) = (1/2\pi)\int_{-\lambda}^{\lambda}(1 - |\xi|/\lambda)\,e^{-i\xi x}\,d\xi$ we have $(k_\lambda * f)(x) = (1/2\pi)\int_{-\lambda}^{\lambda}(1 - |\xi|/\lambda)\,\hat{f}(\xi)\,e^{-i\xi x}\,d\xi$ and hence

$$f(x) = \lim_{\lambda \to \infty}\frac{1}{2\pi}\int_{-\lambda}^{\lambda}\left(1 - \frac{|\xi|}{\lambda}\right)f(\xi)\,e^{-i\xi x}\,d\xi \tag{10-32}$$

the limit existing in the $L^1(-\infty, \infty)$ norm. In particular we have the uniqueness of the Fourier transform of $L^1(-\infty, \infty)$ functions. For assume $\hat{f}(\xi) = 0$ then (10-32) implies $f = 0$.

We will extend now the definition of the Fourier transform to $L^2(-\infty, \infty)$. Let f be continuously differentiable and of compact support. This guarantees that $\int_{-\infty}^{\infty} |\hat{f}(\lambda)|^2\,d\lambda < \infty$. Define $h(x) = \overline{f(-x)}$ then $\hat{h}(\lambda) = \overline{\hat{f}(\lambda)}$. Let $g = f * h$ then

$$g(x) = \int_{-\infty}^{\infty} f(x - t)\,\overline{f(-t)}\,dt = \int_{-\infty}^{\infty} f(x + \tau)\,\overline{f(\tau)}\,d\tau \tag{10-33}$$

and in particular $g(0) = \int_{-\infty}^{\infty} |f(\tau)|^2\,d\tau$. Passing to the Fourier transform of the convolution $g = f * h$ we have $\hat{g}(\lambda) = |\hat{f}(\lambda)|^2$. From (10-33) it follows, putting $x = 0$, that $\lim_{\lambda \to \infty} \int_{-\lambda}^{\lambda}(1 - |\xi|/\lambda)|\hat{f}(\lambda)|^2\,d\lambda = g(0) = \int_{-\infty}^{\infty} |f(x)|^2\,dx$. As the integrand on the left is nonnegative the limit exists and is equal to $\int_{-\infty}^{\infty} |\hat{f}(\lambda)|^2\,d\lambda$ which proves that

$$\int_{-\infty}^{\infty} |\hat{f}(\lambda)|^2\,d\lambda = \int_{-\infty}^{\infty} |f(x)|^2\,dx \tag{10-34}$$

for all f continuous and of compact support. This is a set that is obviously dense in $L^2(-\infty, \infty)$. For these f we define $\mathscr{F}$ by (10-30) then $\mathscr{F}$ is the isometry defined on a dense subset of $L^2(-\infty, \infty)$ into $L^2(-\infty, \infty)$ and hence has a unique extension by continuity to an isometry defined on all of $L^2(-\infty, \infty)$. Since every twice differentiable function of compact support is a Fourier transform of a continuous integrable function it follows that the range of $\mathscr{F}$ is dense in $L^2(-\infty, \infty)$, and being an isometry, actually all of $L^2(-\infty, \infty)$. Summarizing we have obtained the following.

Theorem 10-12 The operator $\mathscr{F}$ defined on $L^1(-\infty, \infty) \cap L^2(-\infty, \infty)$ by

$$(\mathscr{F}f)(\lambda) = \hat{f}(\lambda) = \frac{1}{\sqrt{2\pi}} \int_{-\infty}^{\infty} f(x) e^{i\lambda x} dx \qquad (10\text{-}35)$$

has a unique extension to a unitary operator of $L^2(-\infty, \infty)$ onto itself. Thus for $f \in L^2(-\infty, \infty)$ we have $\hat{f} \in L^2(-\infty, \infty)$ and

$$\int_{-\infty}^{\infty} |\hat{f}(\lambda)|^2 d\lambda = \int_{-\infty}^{\infty} |f(x)|^2 dx \qquad (10\text{-}36)$$

We call the extended operator $\mathscr{F}$ the *Fourier–Plancherel transform* and (10-36) the *Plancherel identity*.

We record now for later use some of the basic properties of the Fourier–Plancherel transform.

Theorem 10-13 Let $f \in L^2(-\infty, \infty)$ and $\hat{f} = \mathscr{F}f$.
(a) If $t^n f(t) \in L^2(-\infty, \infty)$ then

$$\frac{1}{\sqrt{2\pi}} \int_{-\infty}^{\infty} t^n f(t) e^{i\omega t} dt = \frac{(-i)^n}{\sqrt{2\pi}} \frac{d^n}{d\omega^n} \int_{-\infty}^{\infty} f(t) e^{i\omega t} dt = (-i)^n \hat{f}^{(n)}(\omega)$$
$$(10\text{-}37)$$

(b) If f has n derivatives and $f^{(n)} \in L^2(-\infty, \infty)$ then $f^{(j)} \in L^2(-\infty, \infty)$ for $0 \le j < n$ and

$$\frac{1}{\sqrt{2\pi}} \int_{-\infty}^{\infty} f^{(n)}(t) e^{i\omega t} dt = \frac{(-i\omega)^n}{\sqrt{2\pi}} \int_{-\infty}^{\infty} f(t) e^{i\omega t} dt = (-i\omega)^n \hat{f}(\omega) \quad (10\text{-}38)$$

If f is in $L^2(0, \infty)$ we need the extra assumption $f^{(j)}(0) = 0$ for $0 \le j \le n-1$ to make (10-38) hold.
(c) If $f(t) \in L^2(0, \infty)$ and $e^{\alpha t} f(t) \in L^2(0, \infty)$ then

$$\frac{1}{\sqrt{2\pi}} \int_{0}^{\infty} e^{\alpha t} f(t) e^{i\omega t} dt = \hat{f}(\omega - i\alpha) \qquad (10\text{-}39)$$

We return now to the study of the infinitesimal generator A of the right translation semigroup. We introduce the self-adjoint operator $H = -iA$ then on $D_H = D_A$ given by (10-20) H is defined by

$$Hf = if' \qquad (10\text{-}40)$$

As f and f' are in $L^2(-\infty, \infty)$ by integration by parts we have

$$\frac{1}{\sqrt{2\pi}} \int_{-\infty}^{\infty} e^{i\lambda t} f'(t)\, dt = -i\lambda \frac{1}{\sqrt{2\pi}} \int_{-\infty}^{\infty} e^{i\lambda t} f(t)\, dt \qquad (10\text{-}41)$$

which can be rewritten as

$$\mathscr{F}(Hf) = H'\mathscr{F}(f) \qquad (10\text{-}42)$$

where H' is the self-adjoint operator in $L^2(-\infty, \infty)$ defined by

$$D_{H'} = \{ f \mid f \text{ and } \chi f \in L^2(-\infty, \infty) \} \qquad (10\text{-}43)$$

where χ is the identity function $\chi(\lambda) = \lambda$ and

$$(H'f)(\lambda) = \lambda f(\lambda) \qquad (10\text{-}44)$$

Thus the operators H and H' are unitarily equivalent. Actually by the Fourier–Plancherel transform we have obtained a spectral representation for the operators of differentiation. The spectral measure is given by

$$E(\sigma) f(\lambda) = \chi_\sigma(\lambda) f(\lambda) \qquad (10\text{-}45)$$

where χ_σ is the characteristic function of the Borel set σ. The group $U(t)$ has the representation

$$\bigl(\mathscr{F} U(t) f\bigr)(\lambda) = e^{i\lambda t} (\mathscr{F}f)(\lambda) \qquad (10\text{-}46)$$

which holds for all $t \in \mathbb{R}$. From (10-44) we see that $\sigma(H') = \sigma(H) = \mathbb{R}$ and hence $\sigma(A)$ is the whole imaginary axis.

The situation changes drastically when we consider the left translation semigroup in $L^2(0, \infty)$. In this case the infinitesimal generator B has domain

$$D_B = \{ f \in L^2(0, \infty) \mid f \text{ absolutely continuous and } f' \in L^2(0, \infty) \} \qquad (10\text{-}47)$$

and

$$Bf = f' \qquad (10\text{-}48)$$

If we look for solutions of $Bf = \lambda f$ we have to solve the differential equation $f' = \lambda f$ in $L^2(0, \infty)$. The functions which are constant multiples of $e^{\lambda t}$ for $\operatorname{Re}\lambda < 0$ are in $L^2(0, \infty)$ and solve the equation so $\sigma_p(B) = \{\lambda \mid \operatorname{Re}\lambda < 0\}$. Since we have a contractive semigroup each λ with positive real part is in the resolvent set of B and $\sigma(B)$ being closed is necessarily equal to the closed left half plane.

Next we determine the infinitesimal generator of the right translation semigroup in $L^2(0, \infty)$. We can proceed directly or compute the adjoint of the previously determined infinitesimal generator of the left translation semigroup. We choose the first course and follow the line we took before. Let A be the infinitesimal generator then for $f \in D_A$ we have

$$(Af)(x) = \lim_{t \to 0} \frac{f(x - t) - f(x)}{t} = g(x)$$

the limit existing in $L^2(0, \infty)$. Integrating against the characteristic function of $[0, t]$ we have $\int_0^t \{[f(x - \tau) - f(x)]/\tau\} \, dx \to \int_0^t g(x) \, dx$. But

$$\int_0^t \frac{f(x - \tau) - f(x)}{\tau} \, dx = \frac{1}{\tau} \int_\tau^t f(x - \tau) \, dx - \frac{1}{\tau} \int_0^t f(x) \, dx$$

$$= \frac{1}{\tau} \int_0^{t-\tau} f(x) \, dx - \frac{1}{\tau} \int_0^t f(x) \, dx = -\frac{1}{\tau} \int_{t-\tau}^t f(x) \, dx$$

For almost all t we have $\lim_{\tau \to 0} (1/\tau) \int_{t-\tau}^t f(x) \, dx = f(t)$. Redefining f on a set of measure zero we obtain the representation $f(t) = \int_0^t g(x) \, dx$ for $g \in L^2(0, \infty)$. Thus the domain of definition of A is

$$D_A = \{f \in L^2(0, \infty) \,|\, f \text{ absolutely continuous}, \quad f(0) = 0 \quad f' \in L^2(0, \infty)\} \quad (10\text{-}49)$$

and

$$Af = -f' \qquad \text{for} \qquad f \in D_A \qquad (10\text{-}50)$$

Since A is the adjoint of the previously determined B we have immediately that $\sigma(A) = \{\lambda \,|\, \text{Re} \, \lambda \leq 0\}$. However, λ is an eigenvalue of A if and only if f is a constant multiple of $e^{-\lambda t}$ for $\text{Re} \, \lambda > 0$ and none of these is in D_A so the point spectrum is empty.

To compute the infinitesimal cogenerator V of the right translation semigroup in $L^2(0, \infty)$ we can proceed directly in a way analogous to the determination of the cogenerator U of the right translation semigroup in $L^2(-\infty, \infty)$. However, it is simpler to derive a formula for V out of that of U. Indeed we have $L^2(0, \infty)$ naturally embedded in $L^2(-\infty, \infty)$ where if $f \in L^2(0, \infty)$ we assume $f(x) = 0$ for $x < 0$. Then we have $V(t) = U(t) | L^2(0, \infty)$ for $t \geq 0$. This implies that $V = U | L^2(0, \infty)$ and hence from (10-24) we obtain

$$(Vf)(t) = f(t) - 2e^{-t} \int_0^t e^\tau f(\tau) \, d\tau \qquad (10\text{-}51)$$

By Theorem 10-10 V is isometric and V^* the cogenerator of the left translation semigroup coisometric. If P is the orthogonal projection of $L^2(-\infty, \infty)$ onto $L^2(0, \infty)$ then from $V = U | L^2(0, \infty)$ it follows that for all f in $L^2(0, \infty)$ we have $V^*f = PU^*f$. Using (10-25) we obtain the explicit formula

$$(V^*f)(t) = f(t) - 2e^t \int_t^\infty e^{-\tau} f(\tau) \, d\tau \qquad (10\text{-}52)$$

for all $f \in L^2(0, \infty)$. Naturally we are considering only nonnegative values for t.

The kernel of V^* is the set of all $f \in L^2(0, \infty)$ for which $f(t) = 2e^t \int_t^\infty e^{-\tau} f(t) \, d\tau$. Let us put $g(t) = e^{-t} f(t)$ then $f \in \text{Ker} \, V^*$ if and only if $g(t) = 2 \int_t^\infty g(\tau) \, d\tau$. So g is absolutely continuous and $g'(t) = -2g(t)$ which implies $g(t) = ce^{-2t}$ or $f(t) = ce^{-t}$. So $\text{Ker} \, V^*$ is one dimensional. By induction it is easy to prove that $f \in \text{Ker} \, V^{*n}$ if and only if $f(t) = p_{n-1}(t) e^{-t}$ for some polynomial p_{n-1} of degree $n - 1$ at most. Since $\{\cap_{n=0}^\infty V^n H\}^\perp = \text{span} \, \{\text{Ker} \, V^{*n} \,|\, n \geq 0\}$ and as the set of functions $p(t) e^{-t}$, with p any polynomial, is dense in $L^2(0, \infty)$ it follows that V is a completely non-unitary isometry, equivalent by Lemma 8-1 to a unilateral shift of multiplicity one.

It is easy to make contact now with the special orthonormal basis of $L^2(0, \infty)$ consisting of the Laguerre functions.

Let $f_0(t) = \sqrt{2}\,e^{-t}$, then f_0 is an orthonormal basis for Ker V^* and hence, V being unitarily equivalent to a simple right shift, $\{V^n f_0\}_{n=0}^{\infty}$ is an orthonormal basis for $L^2(0, \infty)$. Define $f_{n+1} = V f_n$ then $f_0(0) = \sqrt{2}$ and from (10-51) it follows that $f_{n+1}(0) = (V f_n)(0) = f_n(0)$. So $f_n(0) = \sqrt{2}$ for all $n > 0$. By differentiating (10-51)

$$f'_{n+1}(t) = f'_n(t) - 2f_n(t) + 2e^{-t}\int_0^t e^{\tau} f_n(\tau)\,d\tau$$

or

$$f'_{n+1} + f_{n+1} = f'_n - f_n \tag{10-53}$$

which together with the obvious relation $f'_0 = -f_0$ implies by summing up the equalities from 0 to $n + 1$ that f_{n+1} is the solution of the recursive set of differential equations

$$f'_{n+1} + f_{n+1} = -2\sum_{i=0}^{n} f_i \qquad f_{n+1}(0) = \sqrt{2} \tag{10-54}$$

Now in $L^2(0, \infty)$ we define a linear transformation W by $(Wf)(t) = (1/\sqrt{2})\,f(t/2)$ then W is an isometry which is clearly invertible with $(W^{-1}g)(t) = \sqrt{2}\,g(2t)$. So W is a unitary map in $L^2(0, \infty)$ and hence maps an orthonormal basis onto another orthonormal basis. If we let φ_n be defined by $\varphi_n = W f_n$ then $\{\varphi_n\}$ is an orthonormal basis in $L^2(0, \infty)$. Moreover transforming (10-54) it follows that the φ_n are defined by the following set of recursive differential equations

$$2\varphi'_{n+1}(t) + \varphi_{n+1}(t) = -2\sum_{i=0}^{n} \varphi_i(t) \qquad \varphi_{n+1}(0) = 1 \tag{10-55}$$

The functions $\{\varphi_n\}$ are known as the *Laguerre functions*.
In summary we have proved the following theorem.

Theorem 10-14 Let $\{V(t)\,|\,t \geq 0\}$ be the right translation semigroup in $L^2(0, \infty)$. Its infinitesimal generator A is given by the operator A defined by (10-50) with domain D_A given by (10-49). The cogenerator V is given by (10-51) is an isometry and with respect to the orthonormal basis $\{f_n\}$ of $l^2(0, \infty)$ consisting of the solutions to the recursive set of differential equations (10-54) it is a simple unilateral shift.

The left translation semigroup $\{V(t)^*\}$ has the infinitesimal generator B given by (10-48) with domain D_B given by (10-47). The cogenerator V^* is coisometric and given by (10-52).

Corollary 10-15 The Laguerre functions $\{\varphi_n\}$ defined by the recursive system of differential equations (10-55) are an orthonormal basis for $L^2(0, \infty)$.

Actually we can use the orthonormal basis $\{f_n\}$ of $L^2(0, \infty)$ to produce an orthonormal basis for $L^2(-\infty, \infty)$. Define $g_n = Jf_n$ for the unitary map J of (10-26). Since J maps $L^2(0, \infty)$ onto $L^2(-\infty, 0)$ it is clear that $\{g_n\}_{-\infty}^{\infty}$ is an orthonormal basis for $L^2(-\infty, 0)$. Moreover since $Uf_n = f_{n+1}$ relation (10-28) implies

$$U^* g_n = g_{n+1} \tag{10-56}$$

To link the two sequences we consider both $L^2(-\infty, 0)$ and $L^2(0, \infty)$ as subspaces of $L^2(-\infty, \infty)$. It is important to note that the restriction of an $L^2(0, \infty)$ to the negative half axis is zero. By a direct computation we have $U^* f_0 = g_C$. We state the results concerning the translation group in $L^2(-\infty, \infty)$ as a theorem.

Theorem 10-16 Let $\{U(t) | t \geq 0\}$ be the right translation semigroup in $L^2(-\infty, \infty)$. Its infinitesimal generator A is given by (10-21) with domain given by (10-20). The cogenerator U is unitary and given by (10-24) The sequence of functions $\{\ldots, g_{-2}, g_{-1}, g_0, f_0, f_1, \ldots\}$ with the $\{f_n\}$ defined by (10-54) and the $\{g_n\}$ by $g_n(t) = -f_n(-t)$ or alternately by the recursive system of differential equations

$$g'_{n+1} - g_{n+1} = \sum_{i=0}^{n} g_i \qquad g_{n+1}(0) = -\sqrt{2} \tag{10-57}$$

for the half line $(-\infty, 0]$, is an orthonormal basis for $L^2(-\infty, \infty)$. With respect to this basis U is a bilateral right shift. The cogenerator U^* of $\{U(t) | t \leq 0\} = \{U(t)^* | t \geq 0\}$ is given by (10-25) and is a bilateral left shift of multiplicity one.

The same analysis of the translation semigroups can be carried out in the vectorial case. Thus instead of scalar valued function we consider, for a given separable Hilbert space N, the space of all N-valued weakly measurable functions satisfying $\|f\|^2 = \int_{-\infty}^{\infty} \|f(x)\|_N^2 \, dx < \infty$. We denote this space by $L^2(-\infty, \infty; N)$ and define $L^2(0, \infty; N)$ analogously. We just note that in this case the respective cogenerators are shifts of multiplicity equal to $\dim N$.

A semigroup $\{T(t)\}$ is called a *completely nonunitary semigroup* if there exists no nontrivial subspace reducing all $T(t)$ in which $T(t)$ acts unitarily.

As for contractions given a contraction semigroup $\{T(t)\}$ we can separate its unitary and completely nonunitary parts.

Theorem 10-17 Given a contraction semigroup $\{T(t)\}$ in a Hilbert space H then there exists a unique direct sum decomposition $H = H_0 \oplus H_1$ that reduce $T(t)$, such that $\{T_0(t)\} = \{T(t) | H_0\}$ is a unitary semigroup and $\{T_1(t)\} = \{T(t) | H_1\}$ is a completely nonunitary semigroup.

PROOF Let $\{T(t)\}$ be a contractive semigroup and T its infinitesimal cogenerator. Let $T = T_0 \oplus T_1$ be its unique decomposition into its unitary and completely nonunitary part. The existence of this decomposition has been proved in Theorem 9-1. Now 1 does not belong to $\sigma_p(T)$ if and only if it does not belong to $\sigma_p(T_i)$, $i = 0, 1$. Thus T_0 and T_1 are also infinitesimal cogenerators of contraction semigroups $\{T_0(t)\}$ and $\{T_1(t)\}$ acting in H_0 and H_1, respectively. By Theorem 10-10 the semigroup $\{T_0(t)\}$ is unitary. By the same theorem $\{T_1(t)\}$ must be completely nonunitary for if $\{T_1(t)\}$ had as a direct summand a unitary semigroup then the cogenerator T_1 could not be completely nonunitary.

More can be said about the relation of the semigroup and its cogenerator if we assume $T(t)^*$ tends strongly to zero. This turns out to be equivalent to T^{*n} tending strongly to zero. To prove it we need the following result which is extremely important for its own sake. It is the continuous analogue of Theorem 9-2 and points out the importance of the translation semigroups in providing universal models for contraction semigroups.

Theorem 10-18 A contraction semigroup $\{T(t)\}$ is unitarily equivalent to the adjoint of the semigroup of left translations restricted to a left invariant subspace of some $L^2(0, \infty; N)$ space if and only if $T(t)^*$ tends strongly to zero.

PROOF Let $\{V(t)\}$ be the right translation semigroup in $L^2(0, \infty; N)$ where N is a separable Hilbert space. Assume M is a left invariant subspace of $L^2(0, \infty; N)$ and let P be the orthogonal projection of $L^2(0, \infty; N)$ on M. Define $W(t)$ by

$$W(t) = PV(t)|M \qquad (10\text{-}58)$$

then $\{W(t)\}$ is strongly continuous semigroup and

$$W(t)^* = V(t)^*|M \qquad (10\text{-}59)$$

Since $\|W(t)^* f\|^2 = \int_t^\infty \|f(x)\|^2 \, dx$ we have $\lim_{t \to \infty} \|W(t)^* f\| = 0$ for all $f \in M$. Consequently, if $\{T(t)\}$ is unitarily equivalent to $\{W(t)\}$ we must have $\lim_{t \to \infty} \|T(t)^* x\|^2 = 0$ for all $x \in H$.

Conversely assume $\lim_{t \to \infty} \|T(t)^* x\| = 0$ for all $x \in H$. We want to identify a vector $x \in H$ with the vectorial function $T(t)^* x$ and construct a new norm in H such that this identification will be unitary. So we want to have

$$\|x\|^2 = \int_0^\infty \|T(t)^* x\|_1^2 \, dt \qquad (10\text{-}60)$$

and since this has to hold for all vectors in particular we must have

$$\|T(s)^* x\|^2 = \int_0^\infty \|T(t)^* T(s)^* x\|_1^2 \, dt$$

$$= \int_0^\infty \|T(t + s)^* x\|_1^2 \, dt = \int_s^\infty \|T(t) x\|_1^2 \, dt \qquad (10\text{-}61)$$

Differentiating (10-64) with respect to s and letting B denote the infinitesimal generator of $\{T(t)^*\}$ we obtain

$$\left(BT(s)^*\, x,\, T(x)^*\, x\right) + \left(T(s)^*\, x,\, BT(s)^*\, x\right) = -\| \,T(s)^*\, x\|_1^2 \qquad (10\text{-}62)$$

and by evaluating at $s = 0$

$$\|x\|_1^2 = -(Bx, x) - (x, Bx) = -2\,\mathrm{Re}(Bx, x) \qquad (\,0\text{-}63)$$

Now B being the infinitesimal generator of a contractive semigroup is maximal accretive so indeed the norm defined by (10-63) is nonnegative.

Now for every x in D_B we take (10-63) as the definition of the norm. From (10-61) we have

$$\frac{d}{ds}\, \| T(s)^*\, x\|^2 = -\| T(s)^*\, x\|_1^2$$

and integrating over $(0, t)$

$$\int_0^t \| T(s)^*\, x\|_1^2\, ds = \|x\|^2 - \| T(t)^*\, x\|^2 \qquad (10\text{-}64)$$

By our assumption $T(t)^*$ tends strongly to zero so (10-60) follows from (10-64). Define N_0 as the set of all vectors x in D_B for which $\|x\|_1 = 0$, and let N be the completion of D_B in the new norm. N is a Hilbert space and since D_B is dense in H and the map from D_B to $L^2(0, \infty; N)$ defined by $x \to T(t)^*\, x$ is isometric it can be extended to all of H. If M is the image of H under this map then clearly M is left invariant. For if $V(t)^*$ is the left translation semigroup in $L^2(0, \infty; N)$ then

$$V(t)^*\, T(s)^*\, x = T(t + s)^*\, x = T(t)^*\, T(s)^*\, x = T(s)^*\, T(t)^*\, x \qquad (10\text{-}65)$$

Equality (10-65) shows not only the left invariance of M but also that the action of $T(t)^*$ in H is mapped into the action of $V(t)^*$ in M. Thus the unitary equivalence of $\{T(t)^*\}$ and $V(t)^*|M$. This proves the theorem.

Theorem 10-19 Let $\{T(t)\}$ be a strongly continuous semigroup and T its infinitesimal cogenerator. Then the conditions

$$\lim_{t \to \infty} T(t)^*\, x = 0 \qquad \text{for all} \qquad x \in H \qquad (10\text{-}66)$$

and

$$\lim_{n \to \infty} T^{*n}x = 0 \qquad \text{for all} \qquad x \in H \qquad (10\text{-}67)$$

are equivalent.

PROOF Assume (10-66) holds. By the previous theorem we may assume that $\{T(t)^*\}$ is the left translation semigroup restricted to a left invariant subspace M of some $L^2(0, \infty; N)$ space. Its cogenerator V^* is given by (10-52) and by Theorem 10-9. M is also invariant under V^* so $T^* = V^*|M$. Now V^* is unitarily equivalent to a left shift so (10-67) follows.

Conversely assume (10-67) holds. By Theorem 9-2 we may assume that T^* is the left shift restricted to a left invariant subspace of $L^2(0, \infty; N)$, or for that matter that it is the operator V^* given by (10-52) restricted to a left invariant subspace of $L^2(0, \infty; N)$. Therefore, the semigroup $T(t)^*$ is unitarily equivalent to the left translation semigroup restricted to a left translation invariant subspace of $L^2(0, \infty; N)$ and consequently (10-68) holds.

Theorem 10-18 provided an isometric dilation for a strongly continuous semigroup provided $T(t)^* \to 0$ strongly. Actually every strongly continuous contractive semigroup $\{T(t)\}$ has a unitary dilation in the sense that there exists a Hilbert space $K \supset H$ and a strongly continuous unitary semigroup $\{U(t)\}$ in K such that

$$T(t) = PU(t)|H \qquad (10\text{-}68)$$

with P the orthogonal projection of K on H.

Theorem 10-20 Every strongly continuous contractive semigroup $\{T(t)\}$ in a Hilbert space H has a unitary dilation.

PROOF There are several ways to proceed. We can check that the function $T_t(t)$ defined on $\mathbb{R}$ by

$$T_t = \begin{cases} T(t) & t \geq 0 \\ T(-t)^* & t < 0 \end{cases} \qquad (10\text{-}69)$$

is a positive definite function on $\mathbb{R}$ and apply the Naimark dilation theorem.

Otherwise let T be the cogenerator of the semigroup $\{T(t)\}$. We know that T is a contraction and $1 \notin \sigma_p(T)$. Let U be the minimal unitary dilation of T then $1 \notin \sigma_p(U)$ and hence U is the cogenerator of a unitary semigroup $\{U(t)\}$. This semigroup provides the unitary dilation of T.

Next we describe another application of Naimark's theorem. Let F be a positive operator measure on $\mathbb{R}$, that is a $B(H)$-valued function defined on the Borel subsets of $\mathbb{R}$ such that $(F(\cdot)x, x)$ is a positive measure for each $x \in H$. The question that naturally arises is when can F be dilated to a spectral measure on a larger space K.

Theorem 10-21 Let F be a positive operator measure on $\mathbb{R}$ which satisfies $F(\sigma) \leq I$ for each Borel set σ. Then there exists a Hilbert space $K \supset H$ and a spectral measure E in K such that for each Borel set σ

$$F(\sigma) = PE(\sigma)\big|H \qquad (10\text{-}70)$$

PROOF We can use the positive operator measure F to construct a function $T(t)$ on $\mathbb{R}$ by letting

$$T(t) = \int e^{i\lambda t} F(d\lambda)$$

where the integral certainly exists in the weak sense. We check that $T(t)$ is a positive definite function on $\mathbb{R}$. Let $x_i \in H$ and $t_i \in \mathbb{R}$ then

$$\sum_{i=1}^{n} \sum_{j=1}^{n} (T(t_i - t_j) x_i, x_j) = \sum_{i=1}^{n} \sum_{j=1}^{n} \int e^{i\lambda(t_i - t_j)} (F(d\lambda) x_i, x_j)$$

$$= \int \left(F(d\lambda) \sum_{i=1}^{n} e^{i\lambda t_i} x_i, \sum_{j=1}^{n} e^{i\lambda t_j} x_j \right) \geq 0$$

By Naimark's theorem there exists a unitary representation $U(t)$ of $\mathbb{R}$ in $K \supset H$ such that

$$(T(t) x, y) = (U(t) x, y) \qquad \text{for all} \qquad x, y \in H$$

Applying Stone's theorem we have $U(t) = \int e^{i\lambda t} E(d\lambda)$ for some spectral measure E in K. It follows that for all $x, y \in H$

$$\int e^{i\lambda t} (F(d\lambda) x, y) = \int e^{i\lambda t} (E(d\lambda) x, y)$$

By the uniqueness of the Fourier–Stieltjes transform the measures $(F(\cdot) x, y)$ and $(E(\cdot) x, y)$ are equal, that is for each Borel subset σ of $\mathbb{R}$ we have

$$(F(\sigma) x, y) = (E(\sigma) x, y) \tag{10-71}$$

Condition (10-71) is equivalent to (10-70).

With the results obtained so far we are in a position to give an abstract characterization of the group of translations in $L^2(-\infty, \infty)$, characterization analogous to the one obtained in Theorem 8-3 for bilateral shifts in $l^2(-\infty, \infty)$.

Let $\{U(t)\}$ be any strongly continuous group of unitary operators. A subspace D is an *outgoing subspace* for the group $\{U(t)\}$ if it satisfies $U(t) D \subset D$ for $t > 0$, $\cap_{t \in \mathbb{R}} U(t) D = \{0\}$ and $\overline{\cup_{t \in \mathbb{R}} U(t) D} = H$.

Lemma 10-22 D is an outgoing subspace for the unitary group $\{U(t)\}$ f and only if it is an outgoing subspace for the unitary cogenerator U of the group.

PROOF By Theorem 10-9 D is invariant under the semigroup $\{U(t) | t \geq 0\}$ if and only if D is invariant under the cogenerator U. If a subspace M is invariant under all $U(t)$ then it is invariant under all U^n and $U^n M = M = U(t) M$.

Let now $M = \cap_{t \in \mathbb{R}} U(t) D$ and $M' = \cap_{n \in \mathbb{Z}} U^n D$. Since $U(t) M = M$ for all t and $U^n M' = M'$ for all n we have also $U^n M = M$ and $U(t) M' = M'$

so consequently

$$M = \bigcap_{n \in \mathbb{Z}} U^n M \subset \bigcap_{n \in \mathbb{Z}} U^n D = M' = \bigcap_{t \in \mathbb{R}} U(t) M' \subset \bigcap_{t \in \mathbb{R}} U(t) D = M$$

So $M = M'$ and $\cup_{n \in \mathbb{Z}} U^n D = \{0\}$ if and only if $\cap_{t \in \mathbb{R}} U(t) D = \{0\}$. Finally, let $N = \cup_{t \in \mathbb{R}} U(t) D$ and $N' = \cup_{n \in \mathbb{Z}} U^n D$. N and N' are invariant under all $U(t)$ and U^n hence

$$N = \overline{\bigcup_{n \in \mathbb{Z}} U^n N} \supset \overline{\bigcup_{n \in \mathbb{Z}} U^n D} = N' = \overline{\bigcup_{t \in \mathbb{R}} U(t) N'} \supset \overline{\bigcup_{t \in \mathbb{R}} U(t) D} = N$$

So $N = N'$ and $N = H$ if and only if $N' = H$ which completes the proof.

Theorem 10-23 Let D be an outgoing subspace for a strongly continuous group $\{U_1(t)\}$ of unitary operators in a Hilbert space H. Then $\{U_1(t)\}$ is unitarily equivalent to the group of translations $\{U(t)\}$ in some $L^2(-\infty, \infty; N)$ space.

Proof Let U_1 be the cogenerator of $\{U_1(t) | t \geq 0\}$ then U_1 is unitary and D is an outgoing space with respect to U_1. By Theorem 8-3 we may assume without loss of generality that U_1 is the right shift in $l^2(-\infty, \infty; N)$ for some auxiliary Hilbert space N, and D is $l^2(0, \infty; N)$. If $\{f_n\}$ and $\{g_n\}$ are the sequences of functions defined by (10-54) and (10-57), respectively, and $x = \{\xi_n\}_{n=-\infty}^{\infty}$ is in $l^2(-\infty, \infty; N)$ then we define a map $\varphi: l^2(-\infty, \infty; N) \to L^2(-\infty, \infty; N)$ by $\varphi\{\xi_n\} = \sum_{n=0}^{\infty} \xi_n f_n + \sum_{n=0}^{\infty} \xi_{-n} g_{n-1}$.

It is clear that φ is a unitary map and, as a consequence of Theorem 10-16

$$\varphi U_1 = U \varphi \tag{10-72}$$

where U is given by (10-24). But U is the cogenerator of the right translation semigroup in $L^2(-\infty, \infty; N)$, so $\{U_1(t)\}$ and $\{U(t)\}$ are unitarily equivalent. This completes the proof.

11. THE LIFTING THEOREM

This section is devoted to a proof of the lifting theorem describing the commutant of a given contraction in terms of the commutant of its minimal isometric dilation.

We begin by presenting a theorem that is instrumental in the proof of the main result.

Theorem 11-1 Let T_1 and T_2 be contractions in the Hilbert spaces H_1 and H_2, respectively and let $X \in B(H_2, H_1)$. A necessary and sufficient condition that the operator on $H_1 \oplus H_2$ defined by the operator matrix

$$\begin{pmatrix} T_1 & X \\ 0 & T_2 \end{pmatrix}$$

be a contraction is that there exists a contraction $C \in B(H_2, H_1)$ such that

$$X = D_{T_1^*} C D_{T_2} \tag{11-1}$$

where D_T and D_{T^*} are defined by $(I - T^*T)^{1/2}$ and $(I - TT^*)^{1/2}$, respectively.

PROOF Assume the operator defined by the matrix

$$\begin{pmatrix} T_1 & X \\ 0 & T_2 \end{pmatrix}$$

is a contraction. Then

$$\begin{pmatrix} T_1 & X \\ 0 & T_2 \end{pmatrix} \begin{pmatrix} T_1 & X \\ 0 & T_2 \end{pmatrix}^* = \begin{pmatrix} T_1 & X \\ 0 & T_2 \end{pmatrix} \begin{pmatrix} T_1^* & 0 \\ X^* & T_2^* \end{pmatrix} \leq \begin{pmatrix} I_{H_1} & 0 \\ 0 & I_{H_2} \end{pmatrix} \tag{11-2}$$

But

$$\begin{pmatrix} T_1 & X \\ 0 & T_2 \end{pmatrix} \begin{pmatrix} T_1^* & 0 \\ X^* & T_2^* \end{pmatrix} = \begin{pmatrix} T_1 T_1^* + XX^* & XT_2^* \\ T_2 X^* & T_2 T_2^* \end{pmatrix}$$

$$= \begin{pmatrix} T_1 T_1^* & 0 \\ 0 & 0 \end{pmatrix} + \begin{pmatrix} XX^* & XT_2^* \\ T_2 X^* & T_2 T_2^* \end{pmatrix}$$

which implies that

$$\begin{pmatrix} 0 & X \\ 0 & T_2 \end{pmatrix} \begin{pmatrix} 0 & X \\ 0 & T_2 \end{pmatrix}^* \leq \begin{pmatrix} D_{T_1^*}^2 & 0 \\ 0 & I_{H_2} \end{pmatrix} \tag{11-3}$$

Applying Corollary 7-2 this is equivalent to the existence of a contraction in $H_1 \oplus H_2$ represented by a matrix

$$\begin{pmatrix} K_{11} & K_{12} \\ K_{21} & K_{22} \end{pmatrix}$$

which satisfies

$$\begin{pmatrix} 0 & X \\ 0 & T_2 \end{pmatrix} = \begin{pmatrix} D_{T_1^*} & 0 \\ 0 & I_{H_2} \end{pmatrix} \begin{pmatrix} K_{11} & K_{12} \\ K_{21} & K_{22} \end{pmatrix}$$

From this equality it follows that $K_{21} = 0$, $K_{22} = T_2$ and $X = D_{T_1^*} K_{12}$. Moreover if

$$\begin{pmatrix} K_{11} & K_{12} \\ 0 & T_2 \end{pmatrix}$$

is a contraction, then so is

$$M = \begin{pmatrix} 0 & K_{12} \\ 0 & T_2 \end{pmatrix} = \begin{pmatrix} K_{11} & K_{12} \\ 0 & T_2 \end{pmatrix} \begin{pmatrix} 0 & 0 \\ 0 & I_{H_2} \end{pmatrix}$$

Since M is a contraction we have

$$M^*M = \begin{pmatrix} 0 & 0 \\ K_{12}^* & T_2^* \end{pmatrix}\begin{pmatrix} 0 & K_{12} \\ 0 & T_2 \end{pmatrix} = \begin{pmatrix} 0 & 0 \\ 0 & K_{12}^*K_{12} + T_2^*T_2 \end{pmatrix}$$

$$\leq \begin{pmatrix} I_{H_1} & 0 \\ 0 & I_{H_2} \end{pmatrix}$$

from which $K_{12}^*K_{12} \leq I_{H_2} - T_2^*T_2 = D_{T_2}^2$ follows. Another application of Corollary 7-2 yields the existence of a contraction C such that $K_{12} = CD_{T_2}$ and (11-1) is proved. Conversely if (11-1) holds for a contraction C then we define $K_{12} = CD_{T_2}$ which implies $K_{12}^*K_{12} \leq I_{H_2} - T_2^*T_2$ and hence that M defined as above is a contraction. Since obviously

$$\begin{pmatrix} 0 & X \\ 0 & T_2 \end{pmatrix}\begin{pmatrix} 0 & D_{T_1^*}CD_{T_2} \\ 0 & T_2 \end{pmatrix} = \begin{pmatrix} D_{T_1} & 0 \\ 0 & I_{H_2} \end{pmatrix}\begin{pmatrix} 0 & CD_{T_2} \\ 0 & T_2 \end{pmatrix}$$

relation (11-3) follows and, adding

$$\begin{pmatrix} T_1 T_1^* & 0 \\ 0 & 0 \end{pmatrix}$$

to both sides, also (11-2) and the proof is complete.

The characterization of the commutant of the general contraction in a Hilbert space will be based on that of partial isometries. Let Q be a partial isometry in a Hilbert space K. Let H be the final space of Q and G its orthogonal complement in K. Thus $K = H \oplus G$ and with respect to this direct sum decomposition Q has a matrix representation

$$Q = \begin{pmatrix} T & S \\ 0 & 0 \end{pmatrix} \tag{11-4}$$

By Theorem 2-3

$$QQ^* = \begin{pmatrix} TT^* + SS^* & 0 \\ 0 & 0 \end{pmatrix}$$

is the orthogonal projection on the final space of Q, hence $TT^* + SS^* = I_H$.

Theorem 11-2 Let G and H be Hilbert spaces, T a contraction in H and $S \in B(G, H)$ for which $TT^* + SS^* = I_H$. Let Q be the partial isometry defined by (11-4) acting in $H \oplus G$ and let X be an operator in H that commutes with T. Then there exists an operator

$$Y = \begin{pmatrix} X & A \\ 0 & B \end{pmatrix} \tag{11-5}$$

in $H \oplus G$ such that Y commutes with Q and $\|Y\| = \|X\|$.

PROOF We may assume without loss of generality that $\|X\| = 1$. An operator Y defined by (11-5) commutes with Q, assuming X commutes with T, if and only if

$$TA + SB = XS \tag{11-6}$$

To prove the theorem we will show the existence of operators $A \in B(G, H)$ $B \in B(G, G)$ such that (11-6) holds and Y is a contraction. This, in view of Theorem 11-1 is equivalent to $\|B\| \leq 1$ and $A = D_{X^*}CD_B$ for some contraction C. Since X and T commute we have $XTT^*X^* = TXX^*T^*$. Using the fact that X is a contraction the inequality

$$I_H - TXX^*T^* \geq XX^* - XTT^*X^* = X(I - TT^*)X^* = XSS^*X^*$$

follows. Replacing I_H by $TT^* + SS^*$ we obtain

$$XSS^*X^* \leq TD_{X^*}^2 \cdot T^* + SS^* \tag{11-7}$$

which by an application of Theorem 7-4 guarantees the existence of operators K and B which satisfy

$$TD_{X^*}K + SB = XS \tag{11-8}$$

and

$$B^*B + K^*K = I_G \tag{11-9}$$

From (11-9) clearly B itself is a contraction and $K^*K \leq D_B^2$. Applying Corollary 7-2 we infer the existence of a contraction C such that $K = CD_B$. Letting $A = D_{X^*}CD_B$ completes the proof.

We proceed to the statement and proof of the lifting theorem. Since we can start with the minimal isometric dilations of both T and T^* we can state the result in terms of the minimal coisometric or isometric dilations of T. We begin with the first.

Theorem 11-3 Let T be a contraction in a Hilbert space H and let W be the minimal coisometric dilation of T acting in $K \supset H$. Given an operator X in H that commutes with T there exists an operator Y in K that commutes with W and satisfies $YH \subset H$, $X = Y|H$ and $\|Y\| = \|X\|$.

PROOF By the remarks following Theorem 9-5 we have the matrix representation (9-21) for W in $K = H \oplus \mathfrak{D}_* \oplus \mathfrak{D}_* \oplus \cdots$. Let us define $H_0 = H \oplus \{0\} \oplus \{0\} \oplus \cdots$ and $H_n = H \oplus \mathfrak{D}_* \oplus \cdots \oplus \mathfrak{D}_* \oplus \{0\} \oplus \cdots$. We observe that H_n is invariant under W and define $W_n = W|H_n$. Thus with respect to the direct sum $H \oplus \mathfrak{D}_* \oplus \cdots \oplus \mathfrak{D}_*$ W_n has the matrix representation

$$W_n = \begin{pmatrix} T & D_{T^*} & & & \\ 0 & I_{\mathfrak{D}_*} & & & \\ & 0 & \ddots & & \\ & & \ddots & I_{\mathfrak{D}_*} & \\ & & & 0 \end{pmatrix} \tag{11-10}$$

Using the equality $TT^* + D_{T^*}^2 = I_H$ a simple matrix calculation yields

$$W_n W_n^* = \begin{pmatrix} I_H & & & & \\ & I_{\mathfrak{D}_*} & & & \\ & & \ddots & & \\ & & & I_{\mathfrak{D}_*} & \\ & & & & 0 \end{pmatrix}$$

or $W_n W_n^*$ is the projection on $H \oplus \mathfrak{D}_* \oplus \cdots \oplus \mathfrak{D}_* \oplus \{0\}$ which means that W_n^* is isometric on H_{n-1} embedded naturally in H_n. Since W_n^* is clearly zero on $H_n \ominus H_{n-1}$ we have that W_n is a partial isometry with H_{n-1} as final space. In terms of the direct sum decomposition $H_n = H_{n-1} \oplus \mathfrak{D}_{T^*} W_n$, for $n \geq 2$, has the matrix representation

$$W_n = \begin{pmatrix} W_{n-1} & S_{n-1} \\ 0 & 0 \end{pmatrix}$$

and $W_{n-1} W_{n-1}^* + S_{n-1} S_{n-1}^* = I_{H_{n-1}}$.

We define now inductively a sequence of operators $\{Y_n\}$ with Y_n acting in H_n. Let $Y_0 = X$. Applying Theorem 11-2 there exists an operator Y_1 in H_1 such that H_0 is invariant under Y_1, $X = Y_1 | H_0$ and $\|Y_1\| = \|X\|$. Suppose that for $1 \leq k \leq n$ we defined Y_k acting in H_k such that

$$Y_k W_k = W_k Y_k, \qquad Y_k H_{k-1} \subset H_{k-1}, \qquad Y_{k-1} = Y_k | H_{k-1}$$
$$\text{and} \qquad \|Y_k\| = \|X\| \tag{11-11}$$

Applying Theorem 11-2 once again we have the existence of an operator Y_{n+1} in H_{n+1} satisfying (11-11) for $k = n + 1$. If we embed all spaces H_n naturally in K then we can consider $\{Y_n\}$ as a sequence of operators in K, by letting $Y_n | H_n^\perp = \{0\}$. In the linear manifold $L = \cup_{n=0}^\infty H_n$, which is obviously dense in K, we define an operator Y by $Yx = Y_n x$ if $x \in H_n$. From (11-11) it follows that Y is well defined on L and for each $x \in L$ there exists an n such that $\|Yx\| = \|Y_n x\| \leq \|Y_n\| \|x\| = \|X\| \|x\|$ which implies $\|Y\| \leq \|X\|$. Since obviously $\|Y\| \geq \|Y_n\|$ we have actually the equality $\|Y\| = \|X\|$. Thus Y can be extended to a bounded operator on all of K. Since $Y_k W_k = W_k Y_k$ on H_k we have actually $YW = WY$ on all of L and by continuity this holds also in K. This completes the proof.

As a direct corollary we have this alternative form of the lifting theorem.

Theorem 11-4 Let T be a contraction in a Hilbert space H and let V be the minimal isometric dilation of T acting in $K \supset H$. Given an operator X in H that commutes with T there exists an operator Y in K that commutes with V and satisfies $Y(K \ominus H) \subset K \ominus H$, $\|Y\| = \|X\|$ and

$$X = P_H Y | H \tag{11-12}$$

PROOF Let W be the minimal coisometric dilation of T^* then $V = W^*$ is the minimal isometric dilation of T. Since X^* commutes with T^* there exists an operator Z commuting with W and satisfying $ZH \subset H$, $\|Z\| = \|X^*\|$ and $Z|H = X^*$. Define $Y = Z^*$ then Y commutes with V and $\|Y\| = \|X\|$. The invariance of $K \ominus H$ under Y is equivalent to that of H under Z. Finally for all x, y in H we have $(Xx, y) = (x, X^*y) = (x, Zy) = (Z^*x, y) = (Yx, y) = (P_H Yx, y)$ and (11-12) follows.

The assumption of the minimality of the strong isometric dilation is not essential and it will be convenient to have a formulation of Theorem 11-3 which is independent of it. Naturally the same holds for Theorem 11-4.

Theorem 11-5 Let T be a contraction in a Hilbert space H and let W be a coisometric dilation of T acting in a Hilbert space $K \supset H$. For every operator X that commutes with T there exists an operator Y in K that commutes with W and satisfies $YH \subset H$, $X = Y|H$ and $\|Y\| = \|X\|$.

PROOF Let $K_1 \subset K$ be the smallest reducing subspace of W that contains H. Then $W_1 = W|K_1$ is, up to unitary equivalence, the minimal coisometric dilation of T. Let Y_1 be the operator in K_1 that satisfies $Y_1 W_1 = W_1 Y_1$, $Y_1 H \subset H$ and $Y_1|H = X$, which exists by virtue of Theorem 11-3. Let K_2 be the orthogonal complement of K_1 in K. Relative to the direct sum decomposition $K = K_1 \oplus K_2$ define an operator Y by the 2×2 operator matrix

$$Y = \begin{pmatrix} Y_1 & 0 \\ 0 & 0 \end{pmatrix}$$

then clearly Y has the required properties.

The above two theorems generalize easily to the context of intertwining operators for two contractions T_1 and T_2 that act in the Hilbert space H_1 and H_2, respectively.

Theorem 11-6 Let, for $i = 1, 2$, T_i be contractions in the Hilbert spaces H_i and let W_i be some coisometric dilations acting in the Hilbert space K_i. Let $X: H_1 \to H_2$ be an operator that intertwines T_1 and T_2, that is, satisfies $T_2 X = X T_1$. Then there exists an operator $Y: K_1 \to K_2$ such that $Y W_1 = W_2 Y$, $Y H_1 \subset H_2$, $X = Y|H_1$ and $\|Y\| = \|X\|$.

PROOF Define operators $\hat{T}$ and $\hat{X}$ on the direct sum $H_1 \oplus H_2$ by the 2×2 operator matrices

$$\hat{T} = \begin{pmatrix} T_1 & 0 \\ 0 & T_2 \end{pmatrix} \quad \text{and} \quad \hat{X} = \begin{pmatrix} 0 & 0 \\ X & 0 \end{pmatrix} \tag{11-13}$$

then $XT_1 = T_2X$ is equivalent to $\hat{X}\hat{T} = \hat{T}\hat{X}$. Since obviously

$$W = \begin{pmatrix} W_1 & 0 \\ 0 & W_2 \end{pmatrix}$$

is a coisometric dilation of $\hat{T}$ we can apply Theorem 11-5 to get the existence of an operator

$$\hat{Y} = \begin{pmatrix} Z_1 & Z_2 \\ Y & Z_3 \end{pmatrix}$$

on $K_1 \oplus K_2$ such that $\hat{Y}\hat{W} = \hat{W}\hat{Y}$, $\hat{Y}|H = \hat{X}$ and $\|\hat{Y}\| = \|\hat{X}\|$. It follows immediately that Y has the required properties.

Theorem 11-7 Let, for $i = 1, 2$, T_i be contractions in the Hilbert spaces H_i and let V_i be some isometric dilations acting in the Hilbert spaces $K_i \supset H_i$. Let $X: H_1 \to H_2$ be an operator that satisfies $XT_1 = T_2X$. Then there exists an operator $Y: K_1 \to K_2$ such that $YV_1 = V_2Y$, $Y(K_1 \ominus H_1) \subset K_2 \ominus H_2$, $\|Y\| = \|X\|$ and

$$X = P_{H_2}Y|H_1 \tag{11-14}$$

We note that in Theorem 11-5 and 11-7 the assumption of minimality of the isometric or coisometric dilation has not been made.

12. ELEMENTS OF H^2 THEORY

In the previous two sections we obtained abstract models for certain contractions and a characterization of operators intertwining two contractions. More information, especially of spectral nature, could be obtained if the Hilbert spaces under consideration would have more structure. Thus our aim will be to transform the Hilbert spaces, by way of the Fourier transformation, into function spaces. The relevant spaces turn out to be H^2 spaces and this section is devoted to a short survey of the necessary background.

Let $\mathbb{T}$ be the unit circle and D the open unit disc. We denote by $L^p(\mathbb{T})$, or simply by L^p, the Banach space of all functions whose pth power is integrable with respect to the normalized Lebesque measure with the usual conventions. Thus we have for $1 \leq p \leq p' \leq \infty$ that $L^1 \supset L^p \supset L^{p'} \supset L^\infty$. L^∞ is defined as the space of all essentially bounded functions. Each $f \in L^1$ has well-defined Fourier coefficients given by

$$a_n = \frac{1}{2\pi} \int f(e^{it}) e^{-int} \, dt \tag{12-1}$$

We define for $1 \leq p \leq \infty$ the Hardy space H^p to be the closed subspace of L^p consisting of all functions for which $a_n = 0$ when $n < 0$. Since the map $f \to a_n$

is continuous for each space L^p then H^p is a closed subspace and hence also a Banach space. Of course H^2 inherits from L^2 the Hilbert space structure.

For $p = 2$ we have the orthogonal decomposition of L^2 given by $L^2 = H_2 \oplus \bar{H}_0^2$ where $\bar{H}_0^2 = \{f \in L^2 | (1/2\pi) \int f(e^{it}) e^{-int} dt = 0, n \geq 0\}$. For later use we also define $\bar{H}^2$ by $\bar{H}^2 = \{f \in L^2 | (1/2\pi) \int f(e^{it}) e^{-int} dt = 0, n > 0\}$. Clearly $f \in H^2$ if and only if $\tilde{f}$, defined by $\tilde{f}(e^{it}) = \overline{f(e^{it})}$, is in $\bar{H}^2$ which explains the notation.

It is worthwhile to point out right at the beginning the connection of the H^2 spaces with the analysis of shift operators. By the completeness of L^2 it follows that $\sum_{n=-\infty}^{\infty} a_n e^{int}$ converges if and only if $\sum |a_n|^2 < \infty$.

We define the Fourier transform $\mathscr{F}$ to be the map $\mathscr{F} : l^2(-\infty, \infty) \to L^2$ defined by $\mathscr{F}(\{a_n\}_{n=-\infty}^{\infty}) = \sum_{n=-\infty}^{\infty} a_n e^{int}$. It is easily checked that it is a unitary map. Moreover if U is the bilateral right shift in $l^2(\infty, \infty)$ then defining the operator $\bar{U}$, in L^2 by $(U_1 f)(e^{it}) = e^{it} f(e^{it})$ we have

$$\mathscr{F} U = U_1 \mathscr{F} \tag{12-2}$$

Also it is clear that $\mathscr{F} l^2(0, \infty) = H^2$ and the operators $U | l^2(0, \infty)$ and $U_1 | H^2$ are also unitarily equivalent, the unitary equivalence provided by the Fourier transform restricted to $l^2(0, \infty)$.

We will now focus our attention on the structure of invariant subspaces. Thus let μ be a positive Borel measure on the unit circle and let $L^2(\mu)$ be the space of all square integrable functions with respect to μ with the norm $\|f\|^2 = \int |f(e^{it})|^2 d\mu$. In $L^2(\mu)$ we single out the operator U_μ defined by $(U_\mu f)(e^{it}) = e^{it} f(e^{it})$ or alternately $U_\mu f = \chi f$ with χ the identity function on $\mathbb{T}$. Clearly U_μ is a unitary map in $L^2(\mu)$ and the restriction of U_μ to any of its invariant subspaces is isometric. A straightforward application of the Wold decomposition (Theorem 8-2) yields the following.

Theorem 12-1 Let μ be a positive measure on $\mathbb{T}$ and let M be an invariant subspace. Then there exists a unique direct sum decomposition $M = M_0 \oplus M_1$ which reduces the isometry $V = U_\mu | M$ and such that $V | M_0$ is unitary and $\cap_{n=0}^{\infty} V^n M_1 = \{0\}$.

Since each of the two subspaces is invariant under U_μ it is of interest to characterize the two types of invariant subspaces.

Lemma 12-2 Let μ be a Borel measure on $\mathbb{T}$ that satisfies $\int \chi^n d\mu = 0$ for all $n \in \mathbb{Z}$ then $\mu = 0$.

PROOF The measure μ represents a continuous linear functional on $C(\mathbb{T})$ which vanishes on the dense subset of all trigonometric polynomials so is necessarily zero.

Corollary 12-3 Let μ be a real Borel measure on $\mathbb{T}$ for which $\int \chi^n d\mu = 0$ for $n > 0$ then μ is a constant multiple of the normalized Lebesgue measure σ.

Proof Since μ is real we actually have $\int \chi^n \, d\mu = 0$ for all $n \neq 0$. This is also true for the Lebesgue measure σ and hence for some α we have $\int \chi^n (d\mu - \alpha \, d\sigma) = 0$ for all n. By the previous lemma $\mu = \alpha\sigma$.

Theorem 12-4 Let μ be positive Borel measure on $\mathbb{T}$. A subset M of $L^2(\mu)$ is a reducing subspace for U_μ if and only if $M = \chi_E L^2(\mu)$, where χ_E is the characteristic function of a Borel subset E of $\mathbb{T}$.

Proof If $M = \chi_E L^2(\mu)$ then M is a linear manifold. It is closed as the range of the orthogonal projection P_M defined by $P_M f = \chi_E f$, and moreover it clearly reduces U_μ.

Conversely let M be a reducing subspace of U_μ and let P be the orthogonal projection on M. Since M reduces U_μ we have $U_\mu P = P U_\mu$. By the results of Sec. 6, easily adapted to the case of measures on $\mathbb{T}$, P is represented by a multiplication operator by a function π in $L^\infty(\mu)$. Since $\pi^2 = \pi$ we have $\pi(e^{it})^2 = \pi(e^{it})$ almost everywhere with respect to μ. So $\pi = \chi_E$ for some Borel subset E.

The above theorem settles the case of reducing subspaces. At the other extreme we have those invariant subspaces of U_μ which do not contain a reducing direct summand. This is equivalent to the restriction of U_μ being completely nonunitary.

Theorem 12-5 Let μ be a positive measure on $\mathbb{T}$. A subspace M of $L^2(\mu)$ is invariant and $V_\mu = U|M$ completely nonunitary if and only if $M = \varphi H^2$ for some Borel measurable function such that $|\varphi|^2 \, d\mu = d\sigma$.

Proof Let φ be a Borel measurable function for which $|\varphi|^2 \, d\mu = d\sigma$. Define a map $\Phi: H^2 \to L^2(\mu)$ by $\Phi f = \varphi f$ then

$$\|\Phi f\|^2 = \int |\varphi f|^2 \, d\mu = \int |f|^2 |\varphi|^2 \, d\mu = \int |f|^2 \, d\sigma = \|f\|^2$$

So Φ is an isometry and $M = \varphi H^2$ a closed subspace. If U_σ is defined in H^2 by $U_\sigma f = \chi f$ then clearly $U_\mu \Phi = \Phi U_\sigma$. V is completely nonunitary if and only if $\cap_{n=0}^\infty V^n M = \cap_{n=0}^\infty U^n M = \{0\}$. But $\cap_{n=0}^\infty U^n M = \cap_{n=0}^\infty U^n \Phi H^2 = \cap_{n=0}^\infty \Phi U_\sigma^n H^2 = \Phi \cap_{n=0}^\infty U_\sigma^n H^2 = \{0\}$.

Conversely let M be an invariant subspace of $L^2(\mu)$ from which $\cap_{n=0}^\infty U_\mu^n M = \{0\}$. By the Wold decomposition $M = \oplus_{n=0}^\infty U^n L$ where $L = M \ominus U_\mu M$. Given any unit vector φ in L we have $\varphi \perp U_\mu^n \varphi$ for $n > 0$ or $\int \varphi \bar{\varphi} \chi^{-n} \, d\mu = 0$. By Corollary 12-3 we have $|\varphi|^2 \, d\mu = d\sigma$. We claim L is one-dimensional. If φ' is another unit vector in L which is orthogonal to φ then $(U^n \varphi, U^m \varphi') = 0$ for all $n, m \geq 0$. This means $\int \varphi \bar{\varphi}' \chi^k \, d\mu = 0$ for all $k \in \mathbb{Z}$ and hence that $\varphi \bar{\varphi}' \, d\mu = 0$ or $\varphi \bar{\varphi}' = 0$ almost everywhere with respect to μ. But $|\varphi|^2 \, d\mu = |\varphi'|^2 \, d\mu = d\sigma$ which is a contradiction. We conclude that L is one-dimensional and $M = \oplus_{n=0}^\infty U_\mu^n L = \varphi H^2$.

A special case of the previous result is Beurling's characterization of the invariant subspaces of H^2 which proved to be one of the turning points in operator theory. To this end we define a function q in H^∞ to be *inner* if $|q(e^{it})| = 1$ almost everywhere on $\mathbb{T}$. A function $f \in H^2$ is called *outer* if the linear combinations of the functions $f\chi^n$, $n \geq 0$ are dense in H^2.

Theorem 12-6 (Beurling) Let S be defined in H^2 by $Sf = \chi f$, then a subspace M of H^2 is invariant under S if and only if $M = qH^2$ for some inner function q. The inner function q is determined up to a constant of absolute value one.

PROOF If $M = qH^2$ for some inner function q then it is clearly a closed invariant subspace. Conversely if M is invariant under S and since $\cap_{n=0}^{\infty} S^n H^2 = \{0\}$ we have by the previous theorem that $M = qH^2$ for some measurable q such that $|q|^2 \, d\sigma = d\sigma$ or $|q|^2 = 1$ almost everywhere with respect to σ. But as the function 1 is in H^2 we have q in H^2 or q is inner. Finally let q' be another inner function such that $M = q'H^2$. Since $qH^2 = q'H^2$ it follows that $\bar{q}'q$ and $\bar{q}q'$ are in H^2. So $\int \bar{q}q'\chi^n \, d\sigma = 0$ for all $n \neq 0$ and hence $\bar{q}q'$ is a constant λ of modulus one. Thus $q' = \lambda q$.

Corollary 12-7 A subspace M of H^2 is invariant under multiplication by χ if and only if it is invariant under multiplication by all H^∞ functions.

PROOF If M is invariant under multiplication by χ it is, by Beurling's theorem, of the form qH^2 for some inner function. Now since H^2 is invariant under multiplication by H^∞ functions so is $M = qH^2$. The converse is obvious.

Corollary 12-8 If f is in H^2 and $f = 0$ on a set of positive measure the f is the zero function.

PROOF Let M be the invariant subspace spanned by the functions $f\chi^n$. By Beurling's theorem if M is nontrivial we have $M = qH^2$ for some inner function q. Obviously all functions in M are zero whenever f is. Since $q \in M$ we get a contradiction. Thus necessarily f is the zero function.

The same ideas yield the important factorization of H^2 functions into inner and outer factors.

Theorem 12-9 Every nonzero function $f \in H^2$ has a factorization $f = qg$ where q is inner or constant and g is outer. This factorization is unique up to constant factors of absolute value one.

PROOF If f is outer take $g = f$ and $q = 1$. Otherwise let M_f be the subspace spanned by the functions $f\chi^n$ for $n \geq 0$. M is an invariant subspace so, by Beurling's theorem, $M_f = qH^2$ for some inner function q. As $f \in M_f$ we have $f = qg$ for some $g \in H^2$. Since multiplication by an inner function q is an isometry in H^2 then $M_f = qM_g$, M_g being the invariant subspace spanned by the functions $g\chi^n$. This implies that $M_g = H^2$ or that g is outer. If $f = q_1 g_1$ is another factorization of f into inner and outer factors we have $M_f = q_1 H^2$ so q and q_1 are equal up to a constant of modulus one.

The structure of invariant subspaces has been established only for the H^2 case. The same result holds also for all H^p spaces, $1 \le p \le \infty$. Thus a subspace M of H^p is invariant if and only if $M = qH^p$ for some inner function q which is uniquely determined up to a constant of absolute value one. The case $p = \infty$ holds true for those subspaces which are w^*-closed, a characterization which will prove useful to us later. For the proofs of the quoted results we refer to [67].

Theorem 12-10 (F. and M. Riesz) Let μ be a finite Borel measure on $\mathbb{T}$ satisfying

$$\int \chi^n \, d\mu = 0 \qquad \text{for} \qquad n > 0 \qquad (12\text{-}3)$$

Then μ is absolutely continuous with respect to Lebesque measure and hence $d\mu = f \, d\sigma$ for some f in H^1.

PROOF Let ν be the total variation of μ. Then ν is a finite positive measure on $\mathbb{T}$ and $d\mu = h \, d\nu$ for some measurable h satisfying $|h(e^{it})| = 1$ ν-almost everywhere. We also have that μ and ν are equivalent measures.

Let now M be the subspace of $L^2(\nu)$ spanned by the functions $\{\bar{h}\chi^n \mid n \ge 0\}$. If $U_\nu f = \chi f$ for $f \in L^2(\nu)$ then U_ν is unitary and M clearly an invariant subspace. Since $|h(e^{it})| = 1$ ν-almost everywhere the set of functions $\{\bar{h}\chi^n \mid n \in \mathbb{Z}\}$ spans all of $L^2(\nu)$, which implies that $\cup_{n \in \mathbb{Z}} U_\nu^n M = L^2(\nu)$. Next we note that $\cap_{n > 0} U_\nu^n M = \{0\}$ for this is equivalent to showing that the span of all the subspaces $U_\nu^n M$ is M and that is obvious. Applying Theorem 8-2 it follows that U_ν is unitarily equivalent to a unilateral shift of multiplicity one. In particular U_ν and U_σ are unitarily equivalent. This implies the mutual absolute continuity of ν and σ and hence of μ and σ. Thus $d\mu = f \, d\sigma$ for some f in $L^1(\mathbb{T})$ and the assumption (12-3) implies that actually $f \in H^1$.

So far we have considered H^2, and more generally H^1, functions as defined on the unit circle. However, with any function f in H^1 we can associate a uniquely determined analytic function in the unit disc. If f is in H^1 then f has a Fourier series of analytic type, that is, of the form $\sum_{n=0}^{\infty} a_n e^{int}$. By the Riemann–Lebesgue lemma $\lim_{n \to \infty} a_n = 0$ so the sequence $\{a_n\}$ is bounded. This implies that the power series $\sum_{n=0}^{\infty} a_n z^n$ converges absolutely and uniformly on compact subsets of D. Define $\hat{f}$ by $\hat{f}(z) = \sum a_n z^n$ then $\hat{f}$ provides the analytic extension of f into D.

The precise relationship between functions in H^1 and their analytic extensions is our next object. To explore this further we give an alternate definition of the H^p spaces.

We define now $H^p(D)$, for $1 \le p \le \infty$, as the space of all analytic functions in D satisfying

$$\|f\|_p = \lim_{r \to 1} \|f_r\|_p < \infty$$

where f_r is the function on the unit circle defined by $f_r(e^{i\theta}) = f(re^{i\theta})$. The spaces $H^r(D)$ are Banach spaces whereas $H^\infty(D)$ is actually a commutative Banach algebra.

Theorem 12-11 There exists an isometric isomorphism between H^p and $H^p(D)$ given by $f \to \hat{f}$, for all $1 \le p \le \infty$.

PROOF Let $f \in H^p$ and $\hat{f}$ be the analytic extension of f into D. If $P_r(\theta)$ is the Poisson kernel then, as the Fourier coefficients of the convolution of two L^1 functions are the product of the corresponding Fourier coefficients we have $\hat{f}_r = f * P_r$ or $\hat{f}_r(e^{i\theta}) = \sum_{n=0}^\infty a_n r^n e^{in\theta}$. Using the summability properties of the Poisson kernel, in line with the proof of Theorem 4-1, we have $\lim_{r \to 1} \|\hat{f}_r - f\|_p = 0$ for $1 \le p < \infty$ and $\lim \hat{f}_r = f$ in the w^*-topology of H^∞ for $f \in H^\infty$. In particular it follows that $\lim_{r \to 1} \|\hat{f}_r\|_p = \|f\|_p$ so the map $f \to \hat{f}$ is a linear isometry. Conversely suppose $\hat{f} \in H^p(D)$ and $\hat{f}(z) = \sum_{n=0}^\infty a_n z^n$. Then $\|\hat{f}_r\|_p$ is bounded for $0 \le r < 1$ and we may assume, without loss of generality, that $\|\hat{f}_r\|_p \le 1$, for $1 < p \le \infty$ the functions $\hat{f}_r$ lie in the unit ball of a dual Banach space which is w^*-compact by the Banach–Alaoglu theorem [29]. Thus there exists a function f in L^p such that $\hat{f}_r$ converges to f in the w^*-topology of L^p. In particular, as $\chi^n \in L^q$, we have $\lim(\hat{f}_r, \chi^n) = (f, \chi^n)$ for all $n \in \mathbb{Z}$. Now $(\hat{f}_r, \chi^n) = a_n r^n$ so f has the Fourier series $\sum_{n=0}^\infty a_n e^{in\theta}$. By comparing Fourier series we have $\hat{f}_r = f * P_r$ and hence by the first part of the proof $\hat{f}_r \to f$ in L^p norm for $1 < p < \infty$. The case $p = 1$ requires separate treatment. Thus if $\hat{f} \in H^1(D)$ we consider the family of measures $\hat{f}_r d\sigma$ which again we consider to lie in the unit ball of $M(\mathbb{T})$ the space of all finite Borel measures on $\mathbb{T}$. By applying the Banach–Alaoglu theorem once again, there exists a measure $\mu \in M(\mathbb{T})$ such that $f_r d\sigma$ converges to μ in the w^*-topology of $M(\mathbb{T})$. Since all functions χ^n are in $C(\mathbb{T})$ then

$$\lim_{r \to 1} \frac{1}{2\pi} \int e^{int} f_r(e^{it}) \, dt = \int e^{int} \, d\mu$$

and as f_r has a Fourier series of analytic type we have $\int e^{int} \, d\mu = 0$ for $n > 0$. Thus μ is an analytic measure and by the F. and M. Riesz theorem $d\mu = f \, d\sigma$ for some f in H^1. Again $\hat{f}_r = f * P_r$ and hence $\hat{f}_r$ converges to f in L^1 norm. This completes the proof.

As of now we will use both representations of H^p interchangeably and use the same letter to denote both the function on the circle as well as its analytic extension.

In terms of the analytic extension the structure of inner and outer functions can be completely described. A function q is inner if and only if it is of the form

$q(z) = \alpha B(z) S(z)$ where

$$B(z) = z^k \prod_{i=1}^{\infty} \frac{z - \alpha_i}{1 - \bar{\alpha}_i z} \cdot \left(- \frac{\bar{\alpha}_i}{|\alpha_i|} \right) \tag{12-4}$$

is a *Blaschke product* and the condition $\sum_{i=1}^{\infty} 1 - |\alpha_i| < \infty$ is satisfied whereas

$$S(z) = e^{- \int \frac{e^{it} + z}{e^{it} - z} d\mu} \tag{12-5}$$

with μ a positive measure singular with respect to Lebesgue measure is a *singular inner function*. A function f in H^1 is outer if and only if

$$f(z) = e^{- \int \frac{e^{it} + z}{e^{it} - z} k(e^{it}) dt} \tag{12-6}$$

for some real valued k in L^1. In that case we have necessarily $k(e^{it}) = \log |f(e^{it})|$.

We note that since f_r converges to f in H^p a subsequence f_i converges to f almost everywhere. A theorem of Fatou states that actually the nontangential limits of $f(z)$ exist almost everywhere on the unit circle.

An important result concerning H^∞, the proof of which is outside the scope of this book is the corona theorem of Carleson [20, 30, 73]. It is equivalent to the density of the open unit disc in the maximal ideal space of the commutative Banach algebra H^∞. In analytic terms it states that given $a_1, \ldots, a_n \in H^\infty$ there exist $b_1, \ldots, b_n \in H^\infty$ such that $\sum_{i=1}^{n} a_i(z) b_i(z) = 1$ for all z in D if and only if there exists a $\delta > 0$ such that for all z in D

$$\sum_{i=1}^{n} |a_i(z)| \geq \delta$$

If $a_1, \ldots, a_n$ satisfy this condition we say that they are *strongly coprime*. They are called *coprime* if their greatest common inner divisor is 1, and write $a_1 \wedge \cdots \wedge a_n = 1$. Clearly strong coprimeness implies coprimeness.

The study of H^p spaces is not restricted to the unit disc and other domains of definition can be considered. For us the interesting case is that of the upper and lower half planes. Moreover we will consider only the H^2 and H^∞ spaces. In our approach we will stress the close connection between the corresponding spaces in the disc and in a half plane.

Thus let Π_+ be the open upper half plane $\Pi_+ = \{\lambda \,|\, \text{Im } \lambda > 0\}$ and Π_- the lower half plane. We denote by $H^\infty(\Pi_+)$ the space of all functions analytic in Π_+ and satisfying

$$\|f\|_\infty = \sup_{z \in \Pi_+} |f(z)| < \infty \tag{12-7}$$

We let $H^2(\Pi_+)$ be the space of functions analytic in Π_+ and satisfying

$$\|f\|^2 = \sup_{y > 0} \int_{-\infty}^{\infty} |f(x + iy)|^2 \, dx < \infty \tag{12-8}$$

$H^\infty(\Pi_+)$ is clearly a commutative Banach algebra and there is a simple isomorphism between $H^\infty(\Pi_+)$ and H^∞ of the unit disc which is induced by the fractional linear transformation $z = (w - i)/(w + i)$ which maps the upper half plane Π_+ onto the unit disc D. The induced map is given by $\varphi(z) = f(i(1 + z)/(1 - z))$ and its inverse by $f(w) = \varphi((w - i)/(w + i))$, for $f \in H^\infty(\Pi_+)$ and $\varphi \in H^\infty$.

The H^2 spaces of the disc and the upper half plane are also related. For this the Paley–Wiener characterization of $H^2(\Pi_+)$ is instrumental.

Theorem 12-12 (Paley–Wiener) A complex valued function F defined in Π_+ is in $H^2(\Pi_+)$ if and only if

$$F(w) = \frac{1}{\sqrt{2\pi}} \int_0^\infty f(t) e^{itw} \, dt \tag{12-9}$$

for some f in $L^2(0, \infty)$.

PROOF Assume $f \in L^2(0, \infty)$ and let $w = x + iy \in \Pi_+$ then $f_y(x) = F(x + iy) = (1/\sqrt{2\pi}) \int_0^\infty f(t) e^{-yt} e^{ixt} \, dt$. By the Fourier–Plancherel theorem $F_y \in L^2(-\infty, \infty)$ and $\int_{-\infty}^\infty |F_y(x)|^2 \, dx = \int_{-\infty}^\infty |F(x + iy)|^2 = \int_0^\infty |e^{-yt} f(t)|^2 \, dt \leq \int_0^\infty |f(t)|^2 \, dt$. In particular we have $\sup_{0 < y} \int_{-\infty}^\infty |F(x + iy)|^2 \leq \int_0^\infty |f(t)|^2 \, dt < \infty$, and hence to show that $F \in H^2(\Pi_+)$ it suffices to show that it is analytic in Π_+. To begin with if w and w_0 are in the half plane $\operatorname{Im} w > \delta > 0$ then

$$|F(w) - F(w_0)| = \left| \int_0^\infty (e^{itw} - e^{itw_0}) f(t) \, dt \right|$$

$$\leq \left\{ \int_0^\infty |e^{itw} - e^{itw_0}|^2 \, dt \right\}^{1/2} \left\{ \int_0^\infty |f(t)|^2 \, dt \right\}^{1/2}$$

Now in the previously defined half plane the function $|e^{itw} - e^{itw_0}|^2$ is bounded by $4e^{-\delta t}$ which is certainly in $L^1(0, \infty)$. Since for each t we have $\lim_{w \to w_0} |e^{itw} - e^{itw_0}|^2 = 0$ the Lebesgue dominated convergence theorem yields the continuity of F in $\operatorname{Im} w > \delta > 0$. Since $\delta > 0$ was arbitrary F is actually continuous in Π_+. If γ is any closed contour in Π_+ then $\int_\gamma e^{itw} \, dw = 0$ and by the use of Fubini's theorem it follows that $\int_\gamma F(w) \, dw = 0$. So F is analytic by Morera's theorem.

Conversely assume F is in $H^2(\Pi_+)$. Let $F_y(x) = F(x + iy)$ then $F_y \in L^2(-\infty, \infty)$ and hence F_y is the Fourier–Plancherel transform of an $L^2(-\infty, \infty)$ function which we denote by f_y, that is

$$F_y(x) = \frac{1}{\sqrt{2\pi}} \int_{-\infty}^\infty f_y(t) e^{ixt} \, dt \tag{12-10}$$

and by the inversion formula $f_y(t) = 1/(\sqrt{2\pi}) \int_{-\infty}^\infty F_y(x) e^{-itx} \, dx$. We will show that $e^{yt} f_y(t)$ is independent of $y > 0$. We note that

$$\frac{1}{\sqrt{2\pi}} \int_{-\infty}^\infty F_y(x) e^{-itx} \, dx = \frac{1}{\sqrt{2\pi}} \int_{\operatorname{Im} w = y} F(w) e^{-iwt} e^{-yt} \, dw$$

or

$$e^{yt} f_y(t) = \frac{1}{\sqrt{2\pi}} \int_{\text{Im } w = y} F(w) e^{-iwt} dw \qquad (12\text{-}11)$$

The function $F(w) e^{-iwt}$ is analytic in Π_+ and so its integral on any closed contour lying in Π_+ is zero. We integrate it on the positively oriented rectangle whose vertices are at the points $-\xi + iy_1, \xi + iy_1, \xi + iy_2$ and $-\xi + iy_2$, so we have

$$\frac{1}{\sqrt{2\pi}} \int_{-\xi + iy_1}^{\xi + iy_1} F(w) e^{iwt} dw + \int_{\xi + iy_1}^{\xi + iy_2} F(w) e^{-iwt} dw$$

$$+ \int_{\xi + iy_2}^{-\xi + iy_2} F(w) e^{-iwt} dw + \int_{-\xi + iy_2}^{-\xi + iy_1} F(w) e^{-iwt} dw = 0 \quad (12\text{-}12)$$

Let us estimate now the second integral.

$$\left| \int_{\xi + iy_1}^{\xi + iy_2} F(w) e^{-iwt} dw \right|^2 = \left| \int_{y_1}^{y_2} F(\xi + iu) e^{-i(\xi + iu)t} du \right|^2$$

$$\leq \int_{y_1}^{y_2} \left| F(\xi + iu) \right|^2 du \int_{y_1}^{y_2} e^{2ut} du$$

From (12-8) it follows, by Fubini's theorem, that $\int_{-\infty}^{\infty} \int_{y_1}^{y_2} |F(\xi + iu)|^2 \, du \, d\xi$ is finite and hence there exists a sequence of points $\xi_n \to \infty$ such that

$$\lim_{n \to \infty} \int_{y_1}^{y_2} |F(\pm \xi_n + iu)|^2 \, du = 0$$

This in turn implies $\lim\limits_{n \to \infty} \left| \int_{\pm \xi_n + iy_1}^{\pm \xi_n + iy_2} F(w) e^{-iwt} dw \right| = 0$ the limit being independent of t. In the same way we estimate the fourth integral in (12-12). Letting now ξ go to infinity through the sequence ξ_n we have, from (12-12), that

$$\frac{1}{\sqrt{2\pi}} \int_{\text{Im } w = y_2} F(w) e^{-wt} dw = \frac{1}{\sqrt{2\pi}} \int_{\text{Im } w = y_1} F(w) e^{-iwt} dw$$

that is, the integral $(1/\sqrt{2\pi}) \int_{\text{Im } w = y} F(w) e^{-iwt} dw$ is independent of $y > 0$. From (12-11) define a function f by $f(t) = e^{yt} f_y(t)$. Applying the Plancherel identity to $f_y(t) = e^{-yt} f(t)$ we have

$$\int_{-\infty}^{\infty} |e^{-yt} f(t)|^2 \, dt = \int_{-\infty}^{\infty} |f_y(t)|^2 = \int_{-\infty}^{\infty} |F(x + iy)|^2 \, dx \leq \|F\|^2$$

Letting $y \to \infty$ it follows from this that $f(t) = 0$ almost everywhere in $(-\infty, 0)$, while letting $y \to 0$ implies $f \in L^2(0, \infty)$. Substituting $e^{-yt} f(t)$ for $f_y(t)$ in (12-10) we get the required representation (12-9).

As a corollary to the Paley–Wiener theorem we prove the existence of boundary values of $H^2(\Pi_+)$ functions, at least in the $H^2(\Pi_+)$ sense.

Theorem 12-13 If $F \in H^2(\Pi_+)$ then $F = \lim F_y$ exists in $L^2(-\infty, \infty)$ norm and $F(x) = (1/\sqrt{2\pi}) \int_0^\infty e^{ixt} f(t) \, dt$ for some $f \in L^2(0, \infty)$.

PROOF   If $F \in H^2(\Pi_+)$ then $F(w) = (1/\sqrt{2\pi}) \int_0^\infty f(t) e^{iwt} \, dt$ for some $f \in L^2(0, \infty)$. In particular for the restriction to the real axis we have

$$f(x) = \frac{1}{\sqrt{2\pi}} \int_0^\infty f(t) e^{ixt} \, dt$$

Now

$$F_y(x) - F(x) = \frac{1}{\sqrt{2\pi}} \int_0^\infty f(t) e^{ixt} (e^{-yt} - 1) \, dt$$

and by the Plancherel identity (10-36) we have

$$\|F_y - F\|^2 = \int_0^\infty |f(t)|^2 \, |e^{-yt} - 1|^2 \, dt$$

Applying the Lebesgue dominated convergence theorem we have $\lim_{y \to 0} \|F_y - F\| = 0$.

To obtain the pointwise limits almost everywhere we will first seek a concrete isomorphism between H^2 and $H^2(\Pi_+)$. To this end we start with a deeper study of the properties of the set of functions $\{f_n\}_{n=0}^\infty$ that were introduced in Sec. 10 as the uniquely determined solutions of the recursive set of differential equations

$$y_n' + y_n = -2 \sum_{j=0}^{n-1} y_j \qquad y_n(0) = \sqrt{2} \qquad (12\text{-}13)$$

Given a set of functions $\{f_n\}$ then a function of two variables $\psi(t, z)$ is called a *generating function* of $\{f_n\}$ if

$$\psi(t, z) = \sum_{n=0}^\infty f_n(t) z^n \qquad (12\text{-}14)$$

Lemma 12-14 The generating function of the set $\{f_n\}$ defined by (12-13) is

$$\psi(t, z) = \sqrt{2} \, \frac{e^{-t(1+z)/(1-z)}}{1 - z} \qquad (12\text{-}15)$$

The functions f_n defined by (12-13) are explicitly given by

$$f_n(t) = \sqrt{2}\,\frac{e^t}{n!}\,\frac{d^n}{dt^n}(e^{-2t}t^n) \tag{12-16}$$

and

$$f_n(t) = \sqrt{2}\,e^{-t} \sum_{k=0}^{n} \frac{(-2)^k}{k!}\binom{n}{k} t^k \tag{12-17}$$

and the Fourier–Plancherel transform of f_n is given by

$$(\mathscr{F}f_n)(w) = \frac{i}{\sqrt{\pi}}\,\frac{1}{w+i}\left(\frac{w-i}{w+i}\right)^n \tag{12-18}$$

PROOF Let $\varphi(t, z) = \sqrt{2}\,e^{-t(1+z)/(1-z)} = \sum_{n=0}^{\infty} \alpha_n(t)\,z^n$. We will show that the $\{\alpha_n\}$ are solutions of the recursive set of differential equations

$$\begin{cases} \alpha_0' + \alpha_0 = 0 & \alpha_0(0) = \sqrt{2} \\ \alpha_n' + \alpha_n = -2\sum_{j=0}^{n-1}\alpha_j & \alpha_n(0) = 0 \quad\text{for}\quad n>0 \end{cases} \tag{12-19}$$

and that

$$f_n = \sum_{j=0}^{n} \alpha_j \tag{12-20}$$

Indeed, differentiating the series expansion of $\varphi(t, z)$ we have

$$\sum_{n=0}^{\infty} \alpha_n'(t)\,z^n = \sqrt{2}\,\frac{d}{dt}\,e^{-t(1+z)/(1-z)} = -\sqrt{2}\,\frac{1+z}{1-z}\,e^{-t(1+z)/(1-z)} \tag{12-21}$$

the term by term differentiation being permissible as the expansion of $\varphi(t, z)$ is uniformly convergent on finite intervals for each z in D. Now $-(1+z)/(1-z) = -(1+z)\sum_{k=0}^{\infty} z^k = -1 - 2\sum_{k=1}^{\infty} z^k$. So from (12-21) it follows that $-\sum_{n=0}^{\infty} \alpha_n'(t)\,z^n = \sum_{n=0}^{\infty}\{\alpha_n + 2\sum_{j=0}^{n-1}\alpha_j\}\,z^n$. Consequently the α_n are solutions of (12-19) with the correct initial values satisfied as $\varphi(0, z) = 1 + \sum_{n=0}^{\infty}\alpha_n(0)\,z^n$.

Let now $g_n = \sum_{j=0}^{n}\alpha_j$. We will show that the g_n are also solutions of (12-13) and hence coincide with the f_n. First we check that $g_n(0) = \sum_{j=0}^{n}\alpha_j(0) = \alpha_0(0) = \sqrt{2}$, that is the initial conditions are satisfied. Since $\alpha_0 = f_0$ and $\alpha_0 = g_0$ by the definition of $\{g_n\}$ we have $g_0 = f_0$. Assume $g_i = f_i$ for $0 \le i \le n$ and proceed by induction.

$$g_{n+1}' + g_{n+1} = \sum_{j=0}^{n+1}\alpha_j' + \sum_{j=0}^{n+1}\alpha^j = \sum_{j=0}^{n+1}(\alpha_j' + \alpha_j)$$

By the induction hypothesis

$$\sum_{j=0}^{n+1}(\alpha_j' + \alpha_j) = -2\sum_{j=0}^{n+1}\sum_{k=0}^{j-1}\alpha_k = -2\sum_{j=0}^{n} g_j$$

and hence the g_n satisfy (12-13). This proves $g_n = f_n = \sum_{j=0}^{n} \alpha_j$. Now consider

$$\frac{\varphi(t, z)}{1 - z} = \sqrt{2} \frac{e^{-t(1+z)/(1-z)}}{1 - z} = \left\{ \sum_{n=0}^{\infty} \alpha_n(t)\, z^n \right\} \left\{ \sum_{k=0}^{\infty} z^k \right\} = \sum_{n=0}^{\infty} \left\{ \sum_{k=0}^{n} \alpha_k(t) \right\} z^n$$

$$= \sum_{n=0}^{\infty} f_n(t)\, z^n = \psi(t, z)$$

which proves (12-5).

To prove (12-17) we compute the power series expansion of $\psi(t, z)$

$$\psi(t, z) = \sum_{n=0}^{\infty} f_n(t)\, z^n = \sqrt{2} \frac{e^{-t(1+z)/(1-z)}}{1 - z}$$

and hence

$$e^t \psi(t, z) = \sum_{n=0}^{\infty} e^t f_n(t)\, z^n = \sqrt{2} \frac{e^{-2tz/(1-z)}}{1 - z} = \sqrt{2} \sum_{k=0}^{\infty} \frac{(-2)^k t^k}{k!} \frac{z^k}{(1 - z)^{k+1}}$$

By differentiating the power series of $1/(1 - z)$ n times we have

$$\frac{z^k}{(1 - z)^{k+1}} = \sum_{n=k}^{\infty} \binom{n}{k} z^n$$

which yields upon substituting back in the previous equality

$$\sum_{n=0}^{\infty} e^t f_n(t)\, z^n = \sqrt{2} \sum_{k=0}^{\infty} \frac{(-2)^k t^k}{k!} \sum_{n=k}^{\infty} \binom{n}{k} z^n$$

$$= \sqrt{2} \sum_{k=0}^{\infty} \sum_{n=k}^{\infty} \frac{(-2)^k t^k}{k!} \binom{n}{k} z^n$$

$$= \sqrt{2} \sum_{n=0}^{\infty} \left\{ \sum_{k=0}^{n} \frac{(-2)^k}{k!} \binom{n}{k} t^k \right\} z^n$$

Equating the coefficients of equal powers of z proves (12-17).

Apply now the Leibnitz differentiation rule to the nth derivative of $e^{-2t} t^n$.

$$\frac{d^n}{dt^n}(e^{-2t} t^n) = \sum_{k=0}^{n} \binom{n}{k}(-2)^k e^{-2t} n(n-1)\ldots(k+1)\, t^k$$

$$= \sum_{k=0}^{n} \binom{n}{k}(-2)^k e^{-2k} \frac{n!}{k!}\, t^k$$

which is equivalent to (12-16). If we put $l_n(t) = e^t f_n(t)$ then $l_n(t)$ is an nth degree polynomial and the nth Laguerre polynomial $L_n(t)$ is obtained from it by $L_n(t) = (1/\sqrt{2})\, l_n(t/2)$. So for L_n we have

$$L_n(t) = \sum_{k=0}^{n} \frac{(-1)^k}{k!} \binom{n}{k} t^k \qquad (12\text{-}22)$$

as in [12].

To compute the Fourier–Plancherel transforms of the f_n our starting point is (12-16) and we use the computational rules for the Fourier–Plancherel transform outlined in Theorem 10-13.

By direct computation

$$\frac{1}{\sqrt{2\pi}} \int_0^\infty e^{-2t} e^{iwt}\, dt = \frac{i}{\sqrt{2\pi}} \frac{1}{2i + w}$$

from which

$$\frac{1}{\sqrt{2\pi}} \int_0^\infty e^{-2t} t^n e^{iwt}\, dt = \frac{i}{\sqrt{2\pi}} (-i)^n \frac{d^n}{dw^n}\left(\frac{1}{2i + w}\right) = \frac{i^{n+1} n!}{\sqrt{2\pi}\, (2i + w)^{n+1}}$$

Next

$$\frac{1}{\sqrt{2\pi}} \int_0^\infty \left\{ \frac{d^n}{dt^n}(e^{-2t} t_n) \right\} e^{iwt}\, dt = (-iw)^n \sqrt{2\pi}\, \frac{i^{n+1} n!}{(2i + w)^{n+1}}$$

$$= i\, \frac{n!}{\sqrt{2\pi}} \frac{w^n}{(2i + w)^{n+1}}$$

and, finally,

$$\frac{1}{\sqrt{2\pi}} \int_0^\infty \frac{\sqrt{2}}{n!} e^t \left[\frac{d^n}{dt^n}(e^{-2t} t^n) \right] e^{iwt}\, dt = \frac{\sqrt{2}}{n!}\, i\, \frac{n!}{\sqrt{2\pi}} \frac{1}{w + i}\left(\frac{w - i}{w + i}\right)^n$$

$$= \frac{i}{\sqrt{\pi}} \frac{1}{w + i}\left(\frac{w - i}{w + i}\right)^n$$

which proves (12-18).

Theorem 12-15 The map $J: H^2 \to H^2(\Pi_+)$ defined by

$$Jf(w) = \frac{i}{\sqrt{\pi}} \frac{1}{w + i}\, f\left(\frac{w - i}{w + i}\right) \qquad (12\text{-}23)$$

is a unitary map of H^2 onto $H^2(\Pi_+)$ with the inverse map being given by

$$(J^*F)(z) = \frac{2\sqrt{\pi}}{1 - z}\, F\left(i\, \frac{1 + z}{1 - z}\right) \qquad (12\text{-}24)$$

PROOF The set of functions $\{\chi^n\}$ is an orthonormal set in H^2, integration being with respect to the normalized Lebesgue measure. The set of functions $\{f_n\}$ in $L^2(0, \infty)$ defined by (12-13) is an orthonormal basis for $L^2(0, \infty)$, as was shown in Sec. 10. So if we define a map Φ by $\Phi(\chi^n) = f_n$ then Φ has a unique extension to a unitary map of H^2 onto $L^2(0, \infty)$. Compose this map with the Fourier–Plancherel transform $\mathscr{F}$ to obtain, by the Paley–Wiener theorem, a unitary map $J: H^2 \to H^2(\Pi_+)$, defined by $J(\chi^n) = \mathscr{F}(f_n)$.

Let S be the right shift in H^2, that is, $Sf = \chi f$ for $f \in H^2$, then since $f_{n+1} = Vf_n$ with V defined by (10-51), it follows that

$$\Phi(S\chi^n) = \Phi(\chi^{n+1}) = f_{n+1} = Vf_n = V\Phi(\chi^n)$$

and by extension to all of H^2 we have

$$\Phi S = V\Phi \tag{12-25}$$

By applying the Fourier–Plancherel transform we have

$$(\mathscr{F}f_{n+1})(w) = \frac{i}{\sqrt{\pi}}\frac{1}{w+i}\left(\frac{w-i}{w+i}\right)^{n+1} = \frac{w-i}{w+i}(\mathscr{F}f_n)(w)$$

which implies

$$(\mathscr{F}Vf)(w) = \frac{w-i}{w+i}\mathscr{F}(f)(w) \tag{12-26}$$

By composition we also have

$$(JSg)(w) = \frac{w-i}{w+i}(Jg)(w) \tag{12-27}$$

In other words JSJ^* is the multiplication by $(w-i)/(w+i)$ operator in $H^2(\Pi_+)$. Now V is the infinitesimal cogenerator of the right translation semigroup in $L^2(0, \infty)$ so $\mathscr{F}V\mathscr{F}^*$ is the infinitesimal cogenerator of the semigroup $\mathscr{F}V(t)\mathscr{F}^*$ in $H^2(\Pi_+)$. Consequently the infinitesimal generator of $\mathscr{F}V(t)\mathscr{F}^*$ is the operator of multiplication by iw which means that the action of the semigroup $\mathscr{F}V(t)\mathscr{F}^*$ in $H^2(\Pi_+)$ is multiplication by e^{iwt}. This can of course be verified directly as

$$\frac{1}{\sqrt{2\pi}}\int_t^\infty f(\tau - t)e^{iw\tau}\,d\tau = \frac{e^{iwt}}{\sqrt{2\pi}}\int_0^\infty f(\tau)e^{iw\tau}\,d\tau \tag{12-28}$$

for all $f \in L^2(0, \infty)$.

From the definition of J we have, by summing finite powers, that (12-23) is satisfied for any polynomial and by continuity it is true for all $f \in H^2$.

Actually we have proved more than was claimed and we state it as a theorem.

Theorem 12-16 The right translation semigroup $\{V(t)\}$ in $L^2(0, \infty)$ is unitarily equivalent to the semigroup of multiplication by e^{iwt} in $H^2(\Pi_+)$. The Fourier–Plancherel transform provides the unitary equivalence.

As a corollary to Theorem 12-15 we obtain the Poisson formula in the upper half plane and the existence of nontangential boundary values for $H^2(\Pi_+)$ functions.

Theorem 12-17 If $F \in H^2(\Pi_+)$ then for $w = \xi + i\eta$ in the upper half plane

$$F(\xi + i\eta) = \frac{1}{\pi} \int_{-\infty}^{\infty} F(x) \frac{\eta}{(\xi - x)^2 + \eta^2} \, dx \qquad (12\text{-}29)$$

and the nontangential limits of $F(w)$ exist almost everywhere on $\mathbb{R}$. In particular $\lim_{y \to 0} F(x + iy) = F(x)$ almost everywhere.

PROOF Let $F \in H^2(\Pi_+)$ and use Theorem 12-15 to define

$$f(z) = \frac{2\sqrt{\pi}}{1 - z} F\left(i\frac{1 + z}{1 - z}\right)$$

which is in H^2 of the disc. Obviously $(1 - z) f(z)$ is also in H^2 so by the Poisson formula in D we have

$$(1 - z) f(z) = \frac{1}{2\pi} \int (1 - e^{it}) f(e^{it}) \operatorname{Re} \frac{e^{it} + z}{e^{it} - z} \, dt \qquad (12\text{-}30)$$

Define the function F on $\mathbb{R}$ by transforming the boundary values of f, that is

$$F(x) = \frac{i}{\sqrt{\pi}} \cdot \frac{1}{x + i} f\left(\frac{x - i}{x + i}\right)$$

and note that as $e^{it} = (x - i)/(x + i)$ it follows that $ie^{it} \, dt = (2i \, dx)/(x + i)^2$ and hence that $dt = 2 \, dx/(1 + x^2)$. From (12-30) we obtain by a change of variables

$$F(w) = \frac{1}{\pi} \int_{-\infty}^{\infty} F(x) \operatorname{Re} \frac{e^{it} + z}{e^{it} - z} \frac{dx}{1 + x^2}$$

Now

$$\operatorname{Re} \frac{e^{it} + z}{e^{it} - z} = \operatorname{Re} \frac{1}{i} \frac{1 + xw}{x - w} = \operatorname{Im} \frac{1 + xw}{x - w} = \frac{(\operatorname{Im} w)(1 + x^2)}{|x - w|^2}$$

which yields

$$F(w) = \frac{1}{\pi} \int_{-\infty}^{\infty} F(x) \frac{\operatorname{Im} w}{|x - w|^2} \, dx \qquad (12\text{-}31)$$

Formula (12-31) is evidently equivalent to (12-29).

Using Fatou's theorem in the disc we obtain the existence of nontangential boundary values for $F(w)$. We bear in mind that the fractional linear transformation $z = (w - i)/(w + i)$ is a conformal map of Π_+ onto D and hence preserves nontangential arcs.

We have now at hand all that is needed for the characterization of the right translation invariant subspaces of $L^2(0, \infty)$.

A subspace M of $L^2(0, \infty)$ is called *right translation invariant* if $V(t) M \subset M$ for all members $V(t)$ of the right translation semigroup (10-17).

We begin with two lemmas.

Lemma 12-18 A subspace M of $L^2(0, \infty)$ is right translation invariant if and only if its Fourier–Plancherel transform $\mathscr{F} M$ is invariant under multiplication by all functions e^{iwt}, $t \geq 0$.

PROOF Follows from (12-28).

Lemma 12-19 A subspace N of $H^2(\Pi_+)$ is invariant under multiplication by all functions e^{iwt}, $t \geq 0$ if and only if it is invariant under multiplication by all $H^\infty(\Pi_+)$ functions.

PROOF Since for $t \geq 0$ the functions e^{iwt} are in $H^\infty(\Pi_+)$ the if part is trivial.

Conversely assume N is invariant under multiplication by all e^{iwt} for $t \geq 0$. In view of Corollary 12-7 and (12-27) it suffices to prove that N is invariant under multiplication by $(w - i)/(w + i)$ or equivalently by $i/(w + i)$ as $(w - i)/(w + i) = 1 - \{2i/(w + i)\}$. Now

$$\frac{i}{w + i} = \int_0^\infty e^{-t} \cdot e^{iwt}\, dt = \int_0^\infty e^{i(w + i)t}\, dt = \lim_{n \to \infty} \int_0^n e^{i(w + i)t}\, dt = g_n(w)$$

The function $g_n(w)$ can be approximated by exponentials and $g_n(w) \to i/(w + i)$ boundedly pointwise. So invariance of N follows.

Theorem 12-20 (Lax) A subspace M of $L^2(0, \infty)$ is right translation invariant if and only if its Fourier–Plancherel transform has the form $QH^2(\Pi_+)$ for some inner function Q in the upper half plane.

PROOF Let M be a right translation invariant subspace of $L^2(0, \infty)$. By the two preceding lemmas $\mathscr{F} M$ is invariant under multiplication by all $H^\infty(\Pi_+)$ functions. This means that $J^* \mathscr{F} M = \Phi^* M$ is a subspace of H^2 invariant under multiplication by all H^∞ functions. Applying Beurling's theorem we have $\Phi^* M = qH^2$ for some inner function q. Using the explicit representation (12-23) for J we have $\mathscr{F} M = J(qH^2) = QH^2(\Pi_+)$ where

$$Q(w) = q \frac{w - i}{w + i}$$

is obviously an inner function. The other part is trivial.

We conclude this section with a summary of results about vectorial H^p spaces.

Let N be a separable Hilbert space, then we denote by L_N^2 the space of all (equivalent classes) of weakly measurable N-valued functions for which

$$\|F\|^2 = \int \|F(e^{it})\|^2 \, d\sigma < \infty \tag{12-32}$$

Here $\|F(e^{it})\|$ is the pointwise norm of $F(e^{it})$ as a vector in N. L_N^2 is a Hilbert space, the inner product of two functions, $F, G \in L_N^2$ given by

$$(F, G) = \int (F(e^{it}), G(e^{it})) \, d\sigma \tag{12-33}$$

There are two natural ways of expanding elements of L_N^2 into infinite series. Roughly the first corresponds to writing a function F in terms of a fixed orthonormal basis of N, the coordinates being scalar L^2 functions. The other expansion is a Fourier expansion with coefficients in N.

To get the first expansion let $\{e_n | n \geq 0\}$ be a fixed orthonormal basis for N and let $F \in L_N^2$. Define f_n by

$$f_n(e^{it}) = (F(e^{it}), e_n) \tag{12-34}$$

then, by the assumed weak measurability of F, the f_n are measurable functions and as

$$\int |f_n(e^{it})|^2 \, d\sigma = \int |(F(e^{it}), e_n)|^2 \, d\sigma \leq \int \|F(e^{it})\|^2 \, d\sigma = \|F\|^2$$

we have actually $f_n \in L^2$ for all $n \geq 0$. The Parseval equality applied to (12-34) implies that a.e. on the unit circle

$$\|F(e^{it})\|^2 = \sum_{n=0}^{\infty} |f_n(e^{it})|^2 \tag{12-35}$$

and by integration

$$\|F\|^2 = \sum_{n=0}^{\infty} \|f_n\|^2 \tag{12-36}$$

which is one form of the Parseval equality.

In the second representation we write

$$F = \sum_{k=-\infty}^{\infty} \varphi_k e^{ikt} \qquad \text{with} \qquad \varphi_k \in N \tag{12-37}$$

To obtain the φ_k we note that the set $\{e_n e^{ikt} | n \geq 0, k \in \mathbb{Z}\}$ is an orthonormal basis for L_N^2. The orthonormality is obvious and for completeness it suffices to show, by Theorem 1-13, that this orthonormal set is closed. If $G \in L_N^2$ is orthogonal to all $e_n e^{ikt}$ we have

$$\int (G(e^{it}), e_n) e^{-ikt} \, d\sigma = 0 \qquad \text{for all} \qquad n \geq 0, k \in \mathbb{Z} \tag{12-38}$$

Fixing n (12-38) implies $(G(e^{it}), e_n) = 0$ a.e. for each n. So $G(e^{it}) = 0$ a.e. that is, $G = 0$. If $F = \sum_{n=0}^{\infty} f_n e_n$ let $f_n(e^{-it}) = \sum_{k=-\infty}^{\infty} \varphi_{nk} e^{ikt}$ be the scalar Fourier series of f_n. Since $\|f_n\|^2 = \sum_{k=-\infty}^{\infty} |\varphi_{nk}|^2$ we obtain from (12-36) that

$$\|F\|^2 = \sum_{n=0}^{\infty} \|f_n\|^2 = \sum_{n=0}^{\infty} \sum_{k=-\infty}^{\infty} |\varphi_{nk}|^2$$

and hence

$$F(e^{it}) = \sum_{n=0}^{\infty} \sum_{k=-\infty}^{\infty} \varphi_{nk} e^{ikt} e_n = \sum_{k=-\infty}^{\infty} \left\{ \sum_{n=0}^{\infty} \varphi_{nk} e_n \right\} e^{ikt} \qquad (12\text{-}39)$$

and all series are norm convergent. Comparing coefficients with (12-37) we have $\varphi_k = \sum_{n=0}^{\infty} \varphi_{nk} e_n$ and

$$\|F\|^2 = \sum_{k=-\infty}^{\infty} \sum_{n=0}^{\infty} |\varphi_{nk}|^2 = \sum_{k=-\infty}^{\infty} \|\varphi_k\|^2 \qquad (12\text{-}40)$$

holds. As in the scalar case we call the φ_k the Fourier coefficients of F. The space H_N^2 is then defined to be the subspace of L_N^2 of all functions whose negatively indexed Fourier coefficients vanish. Thus $F \in H_N^2$ if and only if $F = \sum_{n=0}^{\infty} f_n e_n$ and $f_n \in H^2$. Functions in H^2 have analytic extensions into D and so do functions in H_N^2 by defining $F(z) = \sum_{n=0}^{\infty} f_n(z) e_n$. The last series converges absolutely (in norm) and uniformly on compact subsets of D hence $F(z)$ represents a vector valued analytic function in D. If $f_n(z) = \sum_{k=0}^{\infty} \varphi_{nk} z^k$ is the Taylor expansion of f_n it follows, with the previous notation, that $F(z) = \sum_{k=0}^{\infty} \varphi_k z^k$. Letting $F_r(z) = F(rz)$ we have $F_r = \sum f_{n,r} e_n$ and $\|F - F_r\|^2 = \sum_{n=0}^{\infty} \|f_n - f_{n,r}\|^2$ which implies that F_r converges to F in the L_N^2 norm. Separability of N coupled with Fatou's theorem shows that actually the nontangential limits of $F(z)$ exist a.e. and are equal to $F(e^{it})$.

In a completely analogous way we can treat operator valued functions. Given two separable Hilbert spaces N and M we say a function $A: \mathbb{T} \to B(N, M)$ is *weakly measurable* if for all $x \in N$ and $y \in M$ the function $(A(\cdot)x, y)$ is measurable. $L_{B(N,M)}^{\infty}$ is the space of all weakly measurable essentially bounded $B(N, M)$ valued functions. The norm given by

$$\|A\|_{\infty} = \text{ess-sup}\{\|A(e^{it})\| \,|\, 0 \le t < 2\pi\} \qquad (12\text{-}41)$$

Each element $A \in L_{B(N,M)}^{\infty}$ has a natural Fourier series associated with it. In fact for a fixed $x \in N$ $A(\cdot)x \in L_M^2$ and hence $A(e^{it})x = \sum_{k=-\infty}^{\infty} A_k(x) e^{ikt}$. Since $A_k(x)$ depends linearly on x it follows, noting that

$$\|A_k(x)\|^2 \le \int \|A(w^{it})x\|^2 \, d\sigma \le \|A\|_{\infty}^2 \cdot \|x\|^2$$

that

$$A_k(x) = A_k x \qquad k \in \mathbb{Z} \qquad (12\text{-}42)$$

for some uniformly bounded set of operators $A_k \in B(N, M)$. Again $H^\infty_{B(N,M)}$ is defined as the subspace of all $L^\infty_{B(N,M)}$ whose negatively indexed Fourier coefficients vanish. Again every $A \in H^\infty_{B(N,M)}$ has an analytic extension into D and is recoverable by strong nontangential limits. In $L^\infty_{B(N,M)}$ we have an important conjugation given by

$$\tilde{A}(e^{it}) = A(e^{-it})^* \tag{12-43}$$

This definition induces a conjugation in $H^\infty_{B(N,M)}$ by the same formula as $A \to \tilde{A}$ naturally preserves analyticity. We saw already in Theorem 6-11 that $L^\infty_{B(N,M)}$ is a representation of the set of all B-homomorphisms of L^2_N, that is every operator commuting with all multiplications by bounded Borel functions is actually multiplication by a $L^\infty_{B(N,M)}$ function. The same is true in the context of H^2_N spaces. Each H^2_N is actually in H^∞-module where for each $\varphi \in H^\infty$ we denote by M_φ the operator of multiplication by φ. The H^∞-homomorphisms are easily determined.

Theorem 12-21 A bounded operator $A\colon H^2_N \to H^2_M$ is an H^∞-homomorphism, that is

$$AM_\varphi = M_\varphi A \tag{12-44}$$

holds for all $\varphi \in H^\infty$, if and only if there exists a unique operator valued analytic function $A \in H^\infty_{B(N,M)}$ such that

$$(AF)(z) = A(z)F(z) \tag{12-45}$$

and

$$\|A\| = \|A\|_\infty = \sup_{|z|<1} \|A(z)\| \tag{12-46}$$

PROOF The direct part is obvious. Conversely assume A is a bounded operator from H^2_N into H^2_M which commutes with all multiplication operators M_φ. For any vector $\xi \in N$ we have $A\xi \in H^2_M$. Using the fact that $AM_\varphi = M_\varphi A$ for all $\varphi \in H^\infty$ we obtain for any vector polynomial $p(z) = \sum_{i=0}^n \xi_i z^i$ that $(A_p)(z) = \sum z^i (A\xi_i)(z)$. Since $(A\xi)(z)$ is linear in ξ there exists an operator valued function $A(z)$ for which $(A\xi)(z) = A(z)\xi$. Thus $(A_p)(z) = A(z)p(z)$ for all vector polynomials. If we restrict ourselves to the unit circle then we define

$$(A(\bar{\chi}^k p))(z) = \bar{z}^k A(z) p(z) = A(z)\bar{z}^k p(z) \tag{12-47}$$

Thus (12-45) holds for all vector trigonometric polynomials, and as these are dense in L^2_N it holds by continuity for all $F \in L^2_N$. It is also clear from (12-47) that the norm of A as an operator from H^2_N into H^2_M is equal to its norm as an operator from L^2_N into L^2_M, therefore (12-46) follows from Theorem 6-11 and Fatou's theorem concerning existence of strong radial limits of A which satisfy the equality

$$\sup_{|z|<1} \|A(z)\| = \text{ess-sup} \|A(e^{it})\|$$

In order to effectively develop the spectral theory we need the vectorial version of Beurling's theorem. Thus a subspace M of H_N^2 is called invariant, or right invariant, if it is an invariant subspace for all M_φ, $\varphi \in H^\infty$.

Theorem 12-22 Let $M \subset H_N^2$ be an invariant subspace then there exists a function $Q \in H_{B(N)}^\infty$ with the following properties

$$\|Q\|_\infty \leq 1 \tag{12-48}$$

$$Q(e^{it}) \text{ is a.e. a partial isometry with a fixed initial space} \tag{12-49}$$

and

$$M = QH_N^2 \tag{12-50}$$

Conversely every subspace M defined in this way is an invariant subspace of H_N^2.

PROOF If M is defined by (12-50) then certainly it is invariant. Moreover, multiplication by Q in H_N^2 is a partial isometry, its initial space given by the set of all $f \in H_N^2$ such that $f(e^{it})$ belongs a.e. to the initial space of Q. It follows that M is closed.

To prove the converse assume M is an invariant subspace. Let $S = M_\chi$ then S is a right unilateral shift in H_N^2. S restricted to the invariant subspace M is then also a unilateral shift. By the Wold decomposition we have $M = \oplus_{n=0}^\infty S^n L$ where $L = M \ominus SM$. Choose an orthonormal basis q_i in L then, because $q_i \in M \ominus SM$, the vectors $q_i(e^{it})$ are pointwise orthogonal a.e. on $\mathbb{T}$. Since $\dim L \leq \dim N$ we can find an orthonormal set e_i in N of the same cardinality. Let K be the subspace spanned by the e_i. Define a map Q by $Qe_i = q_i$ and extend it by linearity and continuity to all of K. Finally, let $Q|K^\perp = 0$. It is clear that $Q(e^{it})$ is a.e. a partial isometry having K as its initial space and it extends to an analytic operator valued function into D as we have $Q(z)\xi = \sum_i \xi_i q_i(z)$ for $\xi = \sum_i \xi_i e_i$. Now every function f in L can be written as $\sum_i \alpha_i q_i(z) = Q(z) \sum_i \alpha_i e_i$ and so every function in M can be written as $\sum \varphi_i(z) q_i(z) = Q(z) \sum \varphi_i(z) e_i$ for H^2 functions φ_i, or as $Q(z)g(z)$ for $g \in H_K^2$. Since $H_N^2 = H_K^2 + H_{K^\perp}^2$ and multiplication by Q annihilates $H_{K^\perp}^2$ we may as well write $M = QH_K^2$. This completes the proof. Of course the function Q is not uniquely determined as our choice of the subspace K of N was arbitrary.

We call the functions described in the previous theorem *rigid* functions. Out of the class of rigid functions we will be interested in a particular subclass which arises out of a special class of invariant subspaces. We say that an invariant subspace M of H_N^2 is an *invariant subspace of full range* if a.e. on the unit circle $\{f(e^{it}) | f \in M\}$ spans N. In this case $\dim L = \dim N$ and we can choose the orthonormal set $\{e_i\}$ to be an orthonormal basis. It is then obvious that in this case the function $Q(e^{it})$ is a.e. unitary. Such functions are called inner functions and they generalize the scalar inner functions. Given an invariant subspace of full range the inner function corresponding to it is only determined up to a choice of an orthonormal basis in N. Thus if Q and Q_1 are inner then $QH_N^2 = Q_1 H_N^2$ if and only if $Q_1 = QU$ where U is a fixed unitary operator in N. If $Q \in H_{B(N)}^\infty$ is inner we will also say that Q is an inner function in N.

There is a natural partial ordering of inner functions induced by the partial ordering of invariant subspaces. We say an inner function Q is *stronger* than P and write $Q < P$ if $QH_N^2 \subset PH_N^2$.

Lemma 12-23 Let P and Q be inner functions in N. Q is stronger than P if and only if

$$Q = PR \tag{12-51}$$

for some inner function R in N.

PROOF $QH_N^2 \subset PH_N^2$ if and only if $P^*QH_N^2 \subset H_N^2$ that is if and only if $R = P^*Q$ is inner. This is equivalent to (12-51).

In the case of a factorization $Q = PR$ of an inner function Q into inner factors we say that P is a *left inner factor* of Q, R a *right inner factor* and Q a *left inner multiple* of R. If I is the identity operator in N and q a scalar inner function then qI is also inner. We call such inner functions *scalar inner functions*. An inner function Q has a scalar inner multiple q if there exists another inner function R such that

$$QR = RQ = qI \tag{12-52}$$

If N is finite dimensional then any inner function in N has a scalar multiple.

Theorem 12-24 Let N be finite dimensional and Q an inner function in N. Then $q = \det Q$ is inner and $qI < Q$.

PROOF Let $\text{adj} Q$ be the classical adjoint of Q. By Cramer's rule we have

$$Q \, \text{adj} Q = qI \tag{12-53}$$

Since the elements of $\text{adj} Q$ are analytic $\text{adj} Q$ is inner. Equality (12-53) shows that q is a scalar multiple of Q which is equivalent to $qI < Q$.

Given two inner functions P and R having scalar multiples p and r, respectively we can consider $M = PH_N^2 \cap RH_N^2$. M is obviously an invariant subspace of H_N^2 and it certainly contains prH_N^2 as

$$prH_N^2 \subset pH_N^2 \cap rH_N^2 \subset PH_N^2 \cap RH_N^2 = M$$

Therefore M is a subspace of full range and has the representation $M = QH_N^2$ for some inner function Q in N. Q is determined up to a constant unitary factor on the right. We sometimes write $Q = P \vee_L R$ and say that Q is the *least common left inner multiple* of P and R. For scalar inner functions p and r we write $p \vee r$ for the least common inner multiple. We define the least common right inner multiple of P and R by $P \vee_R R = (\tilde{P} \vee_L \tilde{R})$.

In an analogous way we can consider the invariant subspace $L = PH_N^2 \vee RH_N^2$. If P and R are inner L has full range so $L = QH_N^2$ for some inner Q. We call Q the *greatest common left inner divisor* of P and R and write $Q = P \wedge_L R$.

If Q is an inner function having a scalar multiple q then $qH_N^2 \subset QH_N^2$. Let J be the set of all functions φ in H^∞ for which $\varphi H_N^2 \subset QH_N^2$. Clearly J is a non-trivial ideal in H^∞ and in fact a w*-closed one. To see the w*-closure of J let φ_α be a net in J that converges to φ in the w*-topology. Since for every $f \in H_N^2$ and $\xi \in N$ the function $(f, Q\xi) \chi^n$ is in L^1 of the circle we have, due to analyticity of $\varphi_\alpha(Q^*f, \xi)$, that for $n > 0$

$$\int \varphi(e^{it}) (Q(e^{it})^* f(e^{it}), \xi) e^{int} \, dt = \lim_\alpha \int \varphi_\alpha(e^{it}) (Q(e^{it})^* f(e^{it}), \xi) e^{int} \, dt = 0$$

This shows that φQ^*f is in H_N^2 and hence φ is in J. Every w*-closed ideal in H^∞ is of the form $J = mH^\infty$ for some inner function m which is uniquely determined up to a constant factor of absolute value one. We call it the *minimal inner function of* Q.

Lemma 12-25 Let Q be an inner function in an n-dimensional space N, and let m be its minimal inner function. If $q = \det Q$ then we have

$$m \,|\, q \quad \text{and} \quad q \,|\, m^n \tag{12-54}$$

PROOF That it divides q follows from the fact that q belongs to the previously defined ideal J. Since m is a scalar multiple of Q we have $mH_N^2 \subset QH_N^2$ or $mI = QR$ for some inner function R. Taking the determinant of the last equality we obtain $m^n = q \cdot \det R$ or $q \,|\, m^n$.

It will turn out to be important to characterize the points of the unit circle where an inner function Q has an analytic continuation to the exterior of the unit disc. For this we need the following version of the Schwarz reflection principle.

Theorem 12-26 Let f and g be analytic in the domains $\Omega_f = \{re^{i\theta} \,|\, 1 - \varepsilon < r < 1, \alpha < \theta < \beta\}$ and $\Omega_g = \{re^{i\theta} \,|\, 1 < r < 1 + \varepsilon, \alpha < \theta < \beta\}$, respectively. Assume that a.e. on $\gamma = \{e^{i\theta} \,|\, \alpha < \theta < \beta\}$ the radial limits of f and g exist and

$$f(e^{it}) = \lim_{r \to 1^-} f(re^{it}) = \lim_{r \to 1^+} g(re^{it}) \tag{12-55}$$

and also that

$$\lim_{r \to 1^-} \int_\gamma |f(e^{it}) - f(re^{it})| \, dt = \lim \int_\gamma |f(e^{it}) - g(re^{it})| \, dt = 0 \tag{12-56}$$

then f and g are analytic continuations of each other across γ.

PROOF Choose, ε_1, α_1, and β_1 so that $0 < \varepsilon_1 < \varepsilon$, $\alpha < \alpha_1 < \beta_1 < \beta$ and the radial limits of f and g exist at $e^{i\alpha_1}$ and $e^{i\beta_1}$. Let Γ be the positively oriented contour along the boundary of the circular strip $\Omega = \{re^{i\theta} \| 1 - \varepsilon < r < 1 + \varepsilon, \alpha_1 < \theta < \beta_1\}$ and define $h(z)$ by

$$h(z) = \begin{cases} f(z) & z \in \Omega_f \\ g(z) & z \in \Omega_g \end{cases}$$

Define now a function $H(z)$ in Ω by

$$H(z) = \frac{1}{2\pi i} \int_\Gamma \frac{h(\zeta)}{\zeta - z} d\zeta \quad \text{for} \quad z \in \Omega$$

H is obviously analytic in Ω and it remains to show that H coincides with f in $\Omega \cap \Omega_f$ and with g in $\Omega \cap \Omega_g$. Let $z \in \Omega \cap \Omega_f$ and for $0 < \delta < \varepsilon_1$ let Γ_1, Γ_2, and Γ_3 be the positively oriented contours along the boundaries of the circular strips

$$\Omega_1 = \{re^{i\theta} | 1 - \varepsilon_1 < r < 1 - \delta, \alpha_1 < \theta < \beta_1\}$$
$$\Omega_2 = \{re^{i\theta} | 1 - \delta < r < 1 + \delta, \alpha_1 < \theta < \beta_1\}$$

and

$$\Omega_3 = \{re^{i\theta} | 1 + \delta < r < 1 + \varepsilon_1, \alpha_1 < \theta < \beta_1\}$$

respectively. Obviously

$$\frac{1}{2\pi i} \int_\Gamma \frac{h(\zeta)}{\zeta - z} d\zeta = \sum_{j=1}^{3} \frac{1}{2\pi i} \int_{\Gamma_j} \frac{h(\zeta)}{\zeta - z} d\zeta \tag{12-57}$$

By Cauchy's theorem if δ is sufficiently small then

$$f(z) = \frac{1}{2\pi i} \int_{\Gamma_1} \frac{h(\zeta)}{\zeta - z} d\zeta$$

whereas

$$\frac{1}{2\pi i} \int_{\Gamma_3} \frac{h(\zeta)}{\zeta - z} d\zeta = 0$$

The middle integral tends to zero with δ by our assumptions on f and g and so H coincides with f in $\Omega \cap \Omega_f$. That H coincides with g on $\Omega \cap \Omega_g$ can be shown similarly.

We observe that L^2-convergence to the boundary values can replace the weaker condition (12-56). Essentially the same result holds for vectorial functions and we will use the theorem freely in that context.

Theorem 12-27 Let $|\lambda| = 1$. An inner function Q has an analytic continuation at λ if and only if there exists a neighborhood V of λ such that for $z \in V \cap D$ $Q(z)$ is invertible and $\|Q(z)^{-1}\|$ is uniformly bounded there.

PROOF Assume Q has an analytic continuation across the unit circle at λ. Since $Q(\lambda)$ is necessarily inner $Q(\lambda)^{-1} = Q(\lambda)^*$. The set of invertible elements is open so in a neighborhood of λ $Q(\lambda)$ is invertible and $Q(z)^{-1}$ uniformly bounded.

To prove the converse assume $Q(z)^{-1}$ is uniformly bounded in $V \cap D$ where V is an open neighborhood of λ. Since Q is inner then for almost all points μ of the unit circle $Q(z)$ has strong radial limits at μ, and the limits $Q(\mu)$ are unitary. Thus for such a point μ and a vector $\eta \in N$ there exists a vector $\xi \in N$ for which $Q(\mu)\xi = \eta$. Now we show that $Q(z)^{-1}\eta$ has also a radial limit at μ. To this end write

$$Q(z)^{-1}\eta = Q(z)^{-1} Q(\mu)\xi = Q(z)^{-1}[Q(z)\xi + (Q(\mu) - Q(z)\xi)]$$

$$= \xi + Q(z)^{-1}(Q(\mu) - Q(z))\xi$$

Now $\|Q(z)^{-1}\|$ is uniformly bounded in $V \cap D$ whereas the radial limit of $Q(z)\xi$ coincides with $Q(\mu)\xi$. Thus we obtain the existence of the radial limit of $Q(z)^{-1}\eta$ and $\lim Q(z)^{-1}\eta = \xi$ as $z \to \mu$ radially. Define now a function $\hat{Q}$ in the exterior of the unit circle by $\hat{Q}(z) = \tilde{Q}(z^{-1})^{-1}$. Obviously $\hat{Q}$ is analytic at all points z where $\tilde{Q}(z^{-1})$ is invertible. Our assumptions guarantee the analyticity and uniform boundedness of $\hat{Q}(z)$ in the intersection of a neighborhood of λ and D_e. The strong radial limits of $\hat{Q}$ exist and $\lim \hat{Q}(R\mu)\xi = Q(\mu)^{*-1}\xi = Q(\mu)\xi$ as $Q(\mu)$ is unitary. We apply now Theorem 12-26 to infer that Q and $\hat{Q}$ are analytic conjunctions of each other.

13. MODELS FOR CONTRACTIONS AND THEIR SPECTRA

We saw already several instances of the desirability of having a functional model for the study of an operator or a semigroup of operators. In this way we obtained spectral representations for self-adjoint and unitary operators and by way of the Fourier–Plancherel transform also a spectral representation of the translation semigroups in $L^2(-\infty, \infty; N)$ and $L^2(0, \infty; N)$. Similarly the bilateral right shifts in $l^2(-\infty, \infty; N)$ and $l^2(0, \infty; N)$ have the multiplication operators M_χ in $L^2_N(\mathbb{T})$ and $M_\chi | H^2_N$ as their models.

By Theorem 9-5, the minimal isometric dilation V of a contraction T is a unilateral right shift if and only if T^{*n} tends strongly to zero. If that is the case we have a functional model for T induced by that of V. Let S be the right shift in H^2_N.

Theorem 13-1 A contraction T is unitarily equivalent to the operator R defined by

$$Rf = P_M Sf \qquad f \in M \tag{13-1}$$

where M is a left invariant subspace of some H^2_N space, P_M being the orthogonal projection on M, if and only if T^{*n} tends to zero strongly.

PROOF The adjoint of R is given by

$$R^* = S^* \mid M \tag{13-2}$$

as $R^*f = P_M \bar{\chi} f$. In terms of the analytic extension of f into D we have for $f(z) = \sum_{n=0}^{\infty} \xi_n z^n$ that $(R^*f)(z) = (f(z) - f(0))/z = \sum_{k=1}^{\infty} \xi_k z^{k-1}$. So $\|R^{*n}f\|^2 = \sum_{k=n}^{\infty} \|\xi_k\|^2$ and $\lim_{n \to \infty} \|R^{*n}f\| = 0$. So if T is unitarily equivalent to R necessarily $\lim_{n \to \infty} \|T^{*n}x\| = 0$ for all $x \in H$.

Conversely assume T^{*n} tends to zero strongly then, by Theorem 9-5, V the minimal isometric dilation of T is a unilateral shift in $K \supset H$, and H is V^* invariant. Representing V as the right shift S in H_N^2 then T is given by (13-2) for some left invariant subspace M of H_N^2.

Applying the Beurling–Lax theorem we obtain the representation $M^\perp = QH_N^2$ for some rigid function Q introduced in Sec. 12. For an N-operator valued rigid function Q we define $H(Q)$ by

$$H(Q) = \{QH_N^2\}^\perp \tag{13-3}$$

and so $H(Q)$ is a left invariant subspace of H_N^2. In $H(Q)$ we define an operator $S(Q)$ by

$$S(Q) f = P_{H(Q)} S f \qquad f \in H(Q) \tag{13-4}$$

then from the previous discussion it follows that

$$S(Q)^* = S^* \mid H(Q) \tag{13-5}$$

So $S(Q)^*$ is a restriction of the left shift operator or in short a *restricted shift*. By abuse of language we refer to $S(Q)$ also as a restricted shift though a compression of a shift would be more accurate. If Q is an inner function this has some justification deriving from the next theorem.

We define a map $J: L_N^2 \to L_N^2$ by

$$(Jf)(e^{it}) = f(e^{-it}) \qquad \text{for} \qquad f \in L_N^2 \tag{13-6}$$

Clearly J is unitary and satisfies $J^2 = I$ and $J = J^* = J^{-1}$. For a given inner function Q we define a map $\tau_Q: L_N^2 \to L_N^2$ by

$$\tau_Q f = \bar{\chi} \tilde{Q}(Jf) \tag{13-7}$$

where as usual $\tilde{Q}(z) = Q(\bar{z})^*$.

Theorem 13-2 For a given inner function Q the operator τ_Q defined by (13-7) is a unitary operator in L_N^2 for which the following relations hold

$$\tau_Q(QH_N^2) = \bar{H}_{0,N}^2 \tag{13-8}$$

$$\tau_Q(\bar{H}_{0,N}^2) = \tilde{Q}H_N^2 \tag{13-9}$$

$$\tau_Q(H(Q)) = H(\tilde{Q}) \tag{13-10}$$

$$\tau_Q^{-1} = \tau_Q^* = \tau_{\tilde{Q}} \tag{13-11}$$

and

$$\tau_Q P_{H(Q)} = P_{H(\tilde{Q})} \tau_Q \tag{13-12}$$

If we restrict τ_Q to $H(Q)$ then the following diagram is commutative

$$
\begin{array}{ccc}
H(Q) & \xrightarrow{\tau_Q} & H(\tilde{Q}) \\
{\scriptstyle S(Q)}\downarrow & & \downarrow{\scriptstyle S(\tilde{Q})^*} \\
H(Q) & \xrightarrow{\tau_Q} & H(\tilde{Q})
\end{array}
\tag{13-13}
$$

and hence $S(Q)$ and $S(\tilde{Q})^*$ are unitarily equivalent.

PROOF Since Q is inner we have for $f \in L_N^2$

$$\|\tau_Q f\|^2 = \int \|e^{-it}\tilde{Q}(e^{it}) f(e^{-it})\|^2 \, d\sigma = \int \|f(e^{-it})\|^2 \, d\sigma = \|f\|^2$$

So τ_Q is isometric and so is $\tau_{\tilde{Q}}$. An easy check yields $\tau_{\tilde{Q}}\tau_Q = I$ so (13-11) is proved. For $f = Qh \in QH_N^2$ we have

$$(\tau_Q f)(e^{it}) = e^{-it}\tilde{Q}(e^{it}) Q(e^{-it}) h(e^{-it}) = e^{-it}h(e^{-it})$$

But $e^{-it}h(e^{-it})$ belongs to $\bar{H}_{0,N}^2$ so $\tau_Q(QH_N^2) \subset Q\bar{H}_{0,N}^2$. Similarly if $h \in \bar{H}_{0,N}^2$ then $(\tau_Q h)(e^{it}) = e^{-it}\tilde{Q}(e^{it}) h(e^{-it})$ which belongs to $\tilde{Q}H_N^2$ and so

$$\tau_Q(\bar{H}_{0,N}^2) \subset \tilde{Q}H_N^2.$$

By the symmetry of the situation relative to the inner functions Q and $\tilde{Q}$ the inclusions are actually equalities. So this implies that (13-10) holds too. The relation (13-12) follows from (13-8)–(13-10) and applying it after U, the bilateral shift in L_N^2, we have

$$\tau_Q P_{H(Q)} Uf = \tau_Q P_{H(Q)} \chi f = P_{H(\tilde{Q})} \tau_Q \chi f$$

$$= P_{H(\tilde{Q})} \chi \tau_Q f = P_{H(\tilde{Q})} U^* \tau_Q f$$

which implies the commutativity of diagram (13-13).

The operator $S(Q)$ is completely determined by the rigid function Q and our aim is to study $S(Q)$ as well as T in terms of Q. The relation between T and $S(Q)$ and the function Q is as of the general completely nonunitary contraction to its characteristic function in the Sz.-Nagy–Foias theory.

Our study of the spectrum of $S(Q)$ starts with the point spectrum.

Theorem 13-3 Let Q be a rigid function.
(a) For $|\lambda| < 1$ $\lambda \in \sigma_p(S(Q)^*)$ if and only if $Q(\lambda)^*$ has a nontrivial null space and

$$\dim \mathrm{Ker}(\bar{\lambda}I - S(Q)^*) = \dim \mathrm{Ker}\, Q(\lambda)^* \tag{13-14}$$

The normalized eigenfunctions of $S(Q)^*$ have the form

$$(1 - |\lambda|^2)^{1/2}\, \xi/(1 - \bar\lambda z) \tag{13-15}$$

where ξ is a unit vector in $\mathrm{Ker}\, Q(\lambda)^*$.

b) If Q is inner, $|\lambda| < 1$, then $\lambda \in \sigma_p(S(Q))$ if and only if $Q(\lambda)$ has a nontrivial null space, and

$$\dim \mathrm{Ker}(\lambda I - S(Q)) = \dim \mathrm{Ker}\, Q(\lambda) \tag{13-16}$$

The normalized eigenfunctions of $S(Q)$ have the form

$$(1 - |\lambda|^2)^{1/2}\, Q(z)\, \xi/(z - \lambda) \tag{13-17}$$

where ξ is a unit vector in $\mathrm{Ker}\, Q(\lambda)$.

PROOF An eigenfunction of $S(Q)^*$ relative to the eigenvalue $\bar\lambda$ satisfies $S(Q)^* f = \bar\lambda f$ or $(f(z) - f(0))/z = \bar\lambda f(z)$, which means that $f(z) = f(0)/(1 - \bar\lambda z)$. Thus $\lambda \in \sigma_p(S(Q)^*)$ if and only if for some $\xi \in N$, $\xi/(1 - \bar\lambda z)$ is orthogonal to QH_N^2. The orthogonality condition is

$$0 = \frac{1}{2\pi}\int \left(Q(e^{it})\, g(e^{it}), \frac{\xi}{1 - \bar\lambda e^{it}} \right) dt = \frac{1}{2\pi}\int (Q(e^{it})\, g(e^{it}), \xi)\, \frac{e^{it}}{e^{it} - \lambda}\, dt$$

$$= \frac{1}{2\pi i}\int (Q(\zeta)\, g(\zeta), \xi)\, \frac{1}{\zeta - \lambda}\, d\zeta = (Q(\lambda)\, g(\lambda), \xi) = (g(\lambda), Q(\lambda)^*\, \xi)$$

Since this holds for all $g \in H_N^2$, $\xi/(1 - \bar\lambda z)$ belongs to $H(Q)$ if and only if $Q(\lambda)^*\, \xi = 0$. To compute the norm of $\xi/(1 - \bar\lambda z)$ we consider the integral

$$\frac{1}{2\pi}\int \left(\frac{\xi}{1 - \bar\lambda e^{it}}, \frac{\xi}{1 - \bar\lambda e^{it}} \right) dt = \frac{\|\xi\|^2}{2\pi}\int \frac{dt}{(1 - \bar\lambda e^{it})(1 - \lambda e^{-it})}$$

$$= \frac{\|\xi\|^2}{2\pi i}\int \frac{ie^{it}\, dt}{(1 - \bar\lambda e^{it})(e^{it} - \lambda)}$$

$$= \frac{\|\xi\|^2}{2\pi i}\int \frac{d\zeta}{(1 - \bar\lambda \zeta)(\zeta - \lambda)} = \frac{\|\xi\|^2}{1 - |\lambda|^2}$$

by a straightforward application of the Cauchy formula. This proves (13-15) and that the map $\xi \to (1 - |\lambda|^2)^{1/2}\, \xi/(1 - \bar\lambda z)$ is a unitary map of $\mathrm{Ker}\, Q(\lambda)^*$ onto $\mathrm{Ker}(\bar\lambda I - S(Q)^*)$. Condition (13-14) is a consequence of that fact.

To prove part *(b)* we use the unitary equivalence of $S(Q)$ and $S(\tilde Q)^*$ proved in Theorem 13-2. $\lambda I - S(Q)$ has a null function if and only if $\bar\lambda I - S(\tilde Q)^*$ has, and this occurs by part *(a)* if and only if $\mathrm{Ker}\, \tilde Q(\bar\lambda)^*$ is nontrivial. But $\tilde Q(\bar\lambda)^* = Q(\lambda)$, so (13-16) follows. The normalized eigenfunctions of $\lambda I - S(Q)$ are the image under $\tau_{\tilde Q}$ of the functions $(1 - |\lambda|^2)^{1/2}\, \xi/(1 - \bar\lambda z)$ and a simple computation yields (13-17).

If N is finite dimensional the analysis can always be made more precise due to the availability of the determinant function. One instance of that is the following.

Lemma 13-4 Let N be finite dimensional and Q a noninner rigid function. Then $\sigma_p(S(Q)^*)$ is equal to the open unit disc D.

PROOF Since Q is not inner $\det Q(e^{it})$ vanishes on a set of positive measure hence vanishes identically. $Q(\lambda)$ is therefore singular for every $\lambda \in D$, and since the space N is finite dimensional $Q(\lambda)$ and $Q(\lambda)^*$ have nontrivial kernels. The result follows from part (a) of Theorem 13-3.

Before proceeding with the analysis of the spectrum of $S(Q)$ we collect some information concerning functions in $H(Q)$.

Lemma 13-5 Let Q be inner. A function $f \in H_N^2$ is in $H(Q)$ if and only if $Q^*f \in \bar{H}_{0,N}^2$.

PROOF The map of L_N^2 onto itself given by $f \to Q^*f$ is clearly unitary. Since multiplication by Q^* maps QH_N^2 onto H_N^2 necessarily the image of $H(Q)$ is in $\bar{H}_{0,N}^2$. Conversely if $Q^*f \in H_{0,N}^2$ then $f = Qg \in H_N^2$ for some $g \in H_{0,N}^2$. Since f is in H_N^2 and clearly orthogonal to QH_N^2 it follows that $f \in H(Q)$.

Lemma 13-6 Let Q be inner. A function $f \in H(Q)$ can be continued analytically across any point λ, $|\lambda| = 1$, where Q has an analytic continuation.

PROOF Let $|\lambda| = 1$ and assume Q has an analytic continuation across λ. If $f \in H(Q)$ then $f = Qg$ for some $g \in \bar{H}_{0,N}^2$. $Q(z)$ is invertible for all points z in D that are sufficiently close to λ. Now the analytic continuation of Q to D_e is given by $Q(\bar{z}^{-1})^{*-1}$. Similarly g is the boundary function of an $\bar{H}_{0,N}^2$ function. Applying Theorem 12-26 we conclude that the function $Q(\bar{z}^{-1})^{*-1} g(z)$ is the analytic continuation of f across λ.

The following lemma is instrumental in the spectral analysis of restricted shifts.

Lemma 13-7 Let Q be an inner function and $|\lambda| < 1$. Given any unit vector $\xi \in N$ there exists a function $f \in H(Q)$ such that

$$\|(S(Q)^{*n} - \bar{\lambda}^n I) f\| \leq 4 \|Q(\lambda)^* \xi\| \cdot \|f\| \tag{13-18}$$

PROOF If $Q(\lambda)^* \xi = 0$ we choose $f(z) = (1 - |\lambda|^2)^{1/2} \xi/(1 - \bar{\lambda}z)$ which is an eigenfunction of $S(Q)^*$ corresponding to the eigenvalue $\bar{\lambda}$ and hence there is a trivial equality in (13-18). In the general case given

$$e(z) = (1 - |\lambda|^2)^{1/2} \xi/(1 - \bar{\lambda}z) \tag{13-19}$$

we write $e = f + g$ for the decomposition of e relative to the direct sum $H_N^2 = H(Q) \oplus QH_N^2$, with $f \in H(Q)$ and $g \in QH_N^2$. It is elementary to check that

$$f(z) = (1 - |\lambda|^2)^{1/2} (I - Q(z) Q(\lambda)^*) \, \xi /(1 - \bar{\lambda} z) \tag{13-20}$$

and

$$g(z) = (1 - |\lambda|^2)^{1/2} Q(z) Q(\lambda)^* \, \xi /(1 - \bar{\lambda} z) \tag{13-21}$$

Since Q is inner $\|g\|^2 = \|Q(\lambda)^* \, \xi\|^2$ and hence

$$\|f\|^2 = \|\xi\|^2 - \|Q(\lambda)^* \, \xi\|^2 \tag{13-22}$$

Now e is a null function of $S^* - \bar{\lambda} I$ and hence

$$0 = (S^{*n} - \bar{\lambda}^n I) e = (S(Q)^{*n} - \bar{\lambda} I) f + (S^{*n} - \bar{\lambda}^n I) g$$

Now $\|S^{*n} - \bar{\lambda}^n I\| \leq 2$ and consequently

$$\|(S(Q)^{*n} - \bar{\lambda}^n I) f\| \leq \min \{2\|g\|, 2\|f\|\}$$

For a unit vector $\xi \in N$ if $\|Q(\lambda)^* \, \xi\| \geq \frac{1}{2}$ then

$$\|(S(Q)^{*n} - \bar{\lambda}^n I) f\| \leq 2\|f\| \leq 4\|Q(\lambda)^* \, \xi\| \cdot \|f\|$$

On the other hand if $\|Q(\lambda)^* \, \xi\| < 1/2$ then from (13-22) we conclude that $\|f\| > 1/2$ and so

$$\|(S(Q)^{*n} - \bar{\lambda}^n I) f\| \leq 2\|g\| = 2\|Q(\lambda)^* \, \xi\| < 4\|Q(\lambda)^* \, \xi\| \cdot \|f\|$$

which proves the lemma.

We have now at hand all that is needed for the characterization of the spectra of the restricted shift operators.

Theorem 13-8
(c) The spectrum of $S(Q)^*$ consists of all points $\bar{\lambda}$, $|\lambda| < 1$, where $Q(\lambda)$ is not boundedly invertible and those points $\bar{\lambda}$, $|\lambda| = 1$, for which Q has no analytic continuation at λ.
(E) The spectrum of $S(Q)$ consists of all points λ, $|\lambda| < 1$, where $Q(\lambda)$ is not boundedly invertible and those points λ, $|\lambda| = 1$, at which Q has no analytic continuation.

PROOF In view of Theorem 13-2 it suffices to prove part (a). We assume therefore that λ is in D and $Q(\lambda)$, and with it $Q(\lambda)^*$, are boundedly invertible. By Theorem 13-3 $\bar{\lambda} I - S(Q)^*$ is injective and to show that it is surjective it is enough to show the solvability of the equation

$$(\bar{\lambda} I - S(Q^*)) f = g \qquad \text{for each} \qquad g \in H(Q) \tag{13-23}$$

Since $S(Q)^*$ acts as the left shift the above equation reduces to

$$\bar{\lambda} f(z) - \frac{f(z) - f(0)}{z} = g(\lambda) \tag{13-24}$$

or $f(z) = (f(0) - zg(z))/(1 - \bar{\lambda}z)$ and we have to show that $\xi = f(0)$ can be chosen so that f is in $H(Q)$. With Lemma 13-5 in mind f is in $H(Q)$ if and only if Q^*f is in $H^2_{0,N}$. Now $g \in H(Q)$ so $g = Qh$ for some $h \in H^2_{0,N}$ and so k defined by $k(e^{it}) = (Q^*f)(e^{it}) = (e^{-it}Q(e^{it})^* \xi - h(e^{it}))/(e^{-it} - \bar{\lambda})$ has a meromorphic extension to D_e, the exterior of the closed unit disc, the only possible pole located at $\bar{\lambda}^{-1}$. The extension is given by $k(z) = (Q(\bar{z}^{-1})^* \xi - zh(z))/(1 - \bar{\lambda}z)$ and is actually in $\bar{H}^2_{0,N}$ if the numerator vanishes at $\bar{\lambda}^{-1}$. This is equivalent to $Q(\lambda)^* \xi = \bar{\lambda}^{-1}h(\bar{\lambda}^{-1})$ and since we assumed $Q(\lambda)$ to be invertible $\xi = \bar{\lambda}^{-1}Q(\lambda)^{*-1} h(\bar{\lambda}^{-1})$ when substituted back in f provides a solution of (13-23). The same argument works for $|\lambda| = 1$. In this case, since $S(Q)^*$ is completely nonunitary, in fact $S(Q)^{*n}$ tends strongly to zero, $\bar{\lambda}I - S(Q)^*$ and its adjoint are injective. Again all we have to show is that (13-23) is solvable. By our assumption Q has, by applying Lemma 13-6, an analytic continuation at λ and so has the function h, this time into D. The function $k(z)$ defined as before has a numerator analytic at λ and it will be in $\bar{H}^2_{0,N}$ if $Q(\lambda)^* \xi = \lambda h(\lambda)$. By Theorem 12-27 $Q(\lambda)$ is invertible and so for $\xi = \lambda Q(\lambda)^{*-1} h(\lambda)$ is as before the key to the solution of (13-23).

To prove the converse assume first that $|\lambda| < 1$ and $Q(\lambda)$ is not boundedly invertible. We will show that in this case $\bar{\lambda} \in \sigma(S(Q)^*)$. Without loss of generality we may assume that $\operatorname{Ker} S(Q)$ and $\operatorname{Ker} S(Q)^*$ contain the zero vector only, the other alternatives have been settled already by Theorem 13-3. This leaves us with the assumption $0 \in \sigma_c(Q(\lambda)^*)$, a condition implying the existence of a sequence of unit vectors $\xi_n \in N$ for which $\lim_{n \to \infty} \|Q(\lambda)^* \xi_n\| = 0$. We will show the existence of a sequence of functions f_n in $H(Q)$ such that $\lim_{n \to \infty} \|f_n\| = 1$ and $\lim_{n \to \infty} \|(\bar{\lambda}^n I - S(Q)^{*n}) f_n\| = 0$ showing that $\bar{\lambda} \in \sigma_c(S(Q)^*)$.

Let μ be a complex number, $|\mu| > 1$ then

$$(\mu I - S(Q)^*)^{-1} = \mu^{-1}(I - \mu^{-1}S(Q)^*)^{-1} = \sum_{n=0}^{\infty} \frac{S(Q)^{*n}}{\mu^{n+1}}$$

Define an operator valued function Γ by

$$\Gamma(\mu) = (\mu I - S(Q)^*) - (\mu I - \bar{\lambda}I)^{-1} \tag{13-25}$$

$$= \sum_{n=0}^{\infty} \frac{S(Q)^{*n} - \bar{\lambda}^n I}{\mu^{n+1}}$$

On the other hand we have for Γ

$$\Gamma(\mu) = (\mu I - S(Q)^*)^{-1} - (\mu I - \bar{\lambda}I)^{-1}$$
$$= (\mu I - S(Q)^*)^{-1} (I - (\mu - \bar{\lambda})^{-1}(\mu I - S(Q)^*))$$
$$= -(\mu - \bar{\lambda})^{-1}(\mu I - S(Q)^*)^{-1}(\bar{\lambda}I - S(Q)^*)$$

and so $\Gamma(\mu)$ is invertible, for $|\mu| > 1$, if and only if $(\bar{\lambda}I - S(Q)^*)$ is. For $\xi_n \in N$ define f_n and g_n by (13-20) and (13-21), respectively, then equality (13-25) implies

$$\|\Gamma(\mu) f_n\| = \left\| \sum_{k=0}^{\infty} \frac{(S(Q)^{*k} - \bar{\lambda}^k I) f_n}{\mu^{k+1}} \right\| \le \sum_{k=0}^{\infty} \frac{\|(S(Q)^{*k} - \bar{\lambda}^k I) f_n\|}{|\mu|^{k+1}}$$

$$\le \frac{4 \|Q(\lambda)^* \xi_n\| \cdot \|f_n\|}{|\mu| - 1}$$

The last inequality is a consequence of Lemma 13-7. As $\lim\limits_{n \to \infty} \|Q(\lambda)^* \xi_n\| = 0$ and $\lim\limits_{n \to \infty} \|f_n\| = 1$ we have also $\lim\limits_{n \to \infty} \|\Gamma(\mu) f_n\| = 0$. Thus $\Gamma(\mu)$ is not invertible and so $\bar{\lambda} \in \sigma(S(Q)^*)$. Finally let $|\lambda| = 1$ and assume Q has no analytic continuation at λ. By Theorem 12-27 there exist points $\lambda_n \in D$ and unit vectors $\xi_n \in N$ such that $\lim\limits_{n \to \infty} \lambda_n = \lambda$ and $\lim\limits_{n \to \infty} \|Q(\lambda_n)^* \xi_n\| = 0$. Define f_n by (13-20) then by Lemma 13-7 $\lim\limits_{n \to \infty} \|f_n\| = 1$ and

$$\|(\bar{\lambda}I - S(Q)^*) f_n\| \le |\lambda - \lambda_n| \cdot \|f_n\| + \|(\bar{\lambda}_n I - S(Q)^*) f_n\|$$

$$\le |\lambda - \lambda_n| + 4 \|Q(\lambda_n)^* \xi_n\|$$

The right-hand side tends to zero which shows $\bar{\lambda} \in \sigma_c(S(Q)^*)$ and completes the proof.

14. FUNCTIONAL CALCULUS FOR CONTRACTIONS

The spectral theorem for self-adjoint and unitary operators in Hilbert space provided the key to the construction of a functional calculus for these operators. Thus for any unitary U and bounded measurable function on $\mathbb{T}$ the operator $f(U)$ was defined. Theorem 6-9 provided an extension of the calculus to $L^\infty(\nu)$ where ν was a scalar measure on $\mathbb{T}$ such that for all x $(E(\cdot) x, x) \ll \nu$. In particular if E is absolutely continuous with respect to σ, the normalized Lebesgue measure on $\mathbb{T}$, then $f(U)$ is well defined for all $f \in L^\infty = L^\infty(\sigma)$.

Let now T be an arbitrary contraction in a Hilbert space H. By Theorem 9-1 T decomposes into a direct sum $T = T_0 \oplus T_1$ relative to the direct sum decomposition $H = H_0 \oplus H_1$ of reducing subspace of T such that $T_0 = T|H_0$ is unitary whereas $T_1 = T|H_1$ is completely nonunitary. Questions of constructing a functional calculus for T_0 have already been settled and we are left with the task of producing a satisfactory functional calculus for T_1. Hence without loss of generality we may assume that T is a completely nonunitary contraction. We let U be the minimal unitary dilation of T acting in $K \supset H$.

Let now u be an analytic function in D having a Taylor expansion $u(z) = \sum_{n=0}^{\infty} a_n z^n$ satisfying $\sum_{n=0}^{\infty} |a_n| < \infty$. Since T is a contraction $\sum_{n=0}^{\infty} a_n T^n$ converges in the operator norm and we define $u(T)$ by

$$u(T) = \sum_{n=0}^{\infty} a_n T^n \tag{14-1}$$

Since U is the strong minimal unitary dilation of T we have by (9-24) $T^n = PU^n|H$ where P is the orthogonal projection of K on H. Substituting back in (14-1) we obtain

$$u(T) = \sum_{n=0}^{\infty} a_n PU^n|H = P\left(\sum_{n=0}^{\infty} a_n U^n\right)|H = Pu(U)|H$$

or

$$u(T) = Pu(U)|H \tag{14-2}$$

Using (14-2) as a guide it is easy to generalize this simple calculus. If T is completely nonunitary the spectral measure E of its minimal strong unitary dilation U is absolutely continuous with respect to the Lebesgue measure on $\mathbb{T}$. Thus $u(U)$ is well defined for all $u \in L^\infty$. However, we restrict ourselves to the sub-algebra H^∞ of L^∞.

Theorem 14-1 Let T be a completely nonunitary contraction in H and U its minimal strong unitary dilation acting in $K \supset H$ and P the orthogonal projection of K on H. The map $u \to u(T)$ defined by (14-2) is a continuous algebra homomorphism of H^∞ onto $B(H)$ that satisfies

$$\|u(T)\| \le \|u\|_\infty \tag{14-3}$$

$$u(T)^* = \tilde{u}(T^*) \tag{14-4}$$

where $\tilde{u}$ is defined by

$$\tilde{u}(z) = \overline{u(\bar{z})} \tag{14-5}$$

If $u_n \in H^\infty$ converge to u boundedly pointwise a.e. on $\mathbb{T}$ then $u_n(T)$ converges strongly to $u(T)$.

PROOF That the map $u \to u(T)$ is linear is obvious. To show that the map is a multiplicative homomorphism let $K_1 = \bigvee_{n=0}^{\infty} U^n H$ then clearly K_1 is an invariant subspace of U and, using Theorem 9-5, so is $\{\bigvee_{n=0}^{\infty} U^n H\} \ominus H$. Now for all $x \in H$

$$(uv)(T)(x) = P(uv)(U)x = Pu(U)v(U)x = Pu(U)\,Pv(U)x = u(T)v(T)x$$

Next the simple estimate

$$\|u(T)x\| = \|Pu(U)x\| \le \|u(U)x\| \le \|u\|_\infty \|x\|$$

proves (14-3).

As U^* is the minimal unitary dilation of T^* we have $u(T^*) = Pu(U^*)$. So if $x, y \in H$

$$\begin{aligned}
(u(T)^* x, y) &= (x, u(T) y) = (x, Pu(U) y) = (x, u(U) y) = (u(U)^* x, y) \\
&= (\tilde{u}(U^*) x, y) = (P\tilde{u}(U^*) x, y)
\end{aligned}$$

which implies (14-4).

Finally, if $\{u_n\}$ is a uniformly bounded sequence that converges to u pointwise a.e. on $\mathbb{T}$ then for $x \in H$

$$\begin{aligned}
\|(u_n(T) - u(T)) x\|^2 &= \|P(u_n(U) - u(U)) x\|^2 \leq \|(u_n(U) - u(U)) x\|^2 \\
&= \int |u_n(e^{it}) - u(e^{it})|^2 (E(dt) x, x)
\end{aligned}$$

and the right-hand side tends to zero by the Lebesgue dominated convergence theorem.

The restriction of the functional calculus to H^∞ rather than L^∞ provided us with the multiplicative property of the functional calculus, that is, with the equality $(uv)(T) = u(T) v(T)$ for all $u, v \in H^\infty$. This property is not generally satisfied for L^∞ functions but it should be noted that important classes of operators are introduced in this way. A particularly important one is the class of Toeplitz operators. If S is the right shift in H^2 then U, the bilateral right shift in L^2, is its minimal unitary dilation. For $u \in L^\infty$ the operator $u(U)$ is just the multiplication operator by u on L^2. Hence we define an operator T_u in H^2 by

$$T_u f = u(T) f = P_{H^2} u(U) f = P_{H^2}(uf) \tag{14-6}$$

The operator T_u is called a *Toeplitz operator*. For extensive studies of Toeplitz operators and their relation to Wiener–Hopf equations we refer to [24, 56].

Once a functional calculus has been constructed a natural question that arises is the characterization of $\sigma(u(T))$, that is, the question of finding the right spectral mapping theorem. We address ourselves now to this problem for the class of restricted shifts of finite multiplicity, which is the case of interest to us in view of the applications we have in mind. This turns out to be the case for which a satisfactory answer is available.

We first note that if Q is a matrix rigid function, that is N is assumed to be finite dimensional, then the availability of the determinant function simplifies the analysis considerably. Also the restriction to the class of restricted shifts yields a very concrete representation for $u(T)$.

Theorem 14-2 Let Q be a rigid function and $S(Q)$ the operator in $H(Q)$ defined by (13-4). For $u \in H^\infty$ $u(S(Q))$ is given by

$$u(S(Q)) f = P_{H(Q)}(uf) \qquad \text{for} \qquad f \in H(Q) \tag{14-7}$$

and

$$u(S(Q))^* f = P_{H(Q)}(\bar{u}f) \qquad (14\text{-}8)$$

PROOF The not necessarily minimal unitary dilation of $S(Q)$ is given by U the bilateral right shift in L_N^2. Since for $u \in L^\infty$ $u(U) f = uf$ for all $f \in L_N^2$ the result follows. We note that as H_N^2 is invariant under all $u(U)$ for $u \in H^\infty$ the $P_{H(Q)}$ may be taken in (14-7) as the orthogonal projection of H_N^2 onto $H(Q)$ whereas in (14-8) its interpretation is the projection of L_N^2 onto $H(Q)$.

If N is a finite dimensional Hilbert space and A is an operator in N then we define det A the determinant of A as the determinant of the matrix representation of A relative to any basis of N. It is a well-known fact that det A is well defined, that is, its definition is independent of the particular basis used.

Lemma 14-3 Let A be a linear operator in an n-dimensional Hilbert space N, then

$$|\det A| \geq \|A^{-1}\|^{-n} \qquad (14\text{-}9)$$

where we put $\|A^{-1}\|^{-1} = 0$ whenever A is not invertible.

PROOF If A is not invertible (14-9) reduces to a triviality. Otherwise det $A = \prod_{i=1}^{n} a_i$ where a_i are the eigenvalues of R. Since a_i^{-1} are the eigenvalues of A^{-1} and clearly $|a_i^{-1}| \leq \|A^{-1}\|$ then $|a_i| \geq \|A^{-1}\|^{-1}$ and by taking the product of these inequalities we obtain (14-9).

As was the case in the previous section we begin with the analysis of the point spectrum.

Theorem 14-4 Let N be finite dimensional, Q a rigid function, $S(Q)$ the restricted shift operator acting in $H(Q) = \{QH_N^2\}^\perp$ and $u \in H^\infty$.
(a) If Q is not inner, that is, det Q is identically equal to zero, then $u(S(Q)^*)$ is injective if and only if u is an outer function.
(b) If Q is inner $u(S(Q))$ is injective if and only if u and $q = \det Q$ are coprime, that is, have no nontrivial common inner factor.
(c) If Q is inner $u(S(Q)^*)$ is injective if and only if u and $\tilde{q}$ are coprime.

PROOF If Q is not inner then we will show that for each scalar inner function q there exists a function $f \in H(Q)$ such that $\bar{q}f$ is orthogonal, in L_N^2, to H_N^2. This will show that if u has an inner factor, that is, u is not outer, then it is not injective.

Now $Q(e^{it})$ is a.e. on $\mathbb{T}$ a partial isometry with a fixed initial space $M \subset N$ and the inclusion is proper since Q is assumed to be noninner. Let $\{e_1, \ldots, e_n\}$ be an orthonormal basis for N such that $\{e_1, \ldots, e_m\}$ is an orthonormal basis for M. Let $Qe_i = q_i$ for $e = 1, \ldots, m$. If given q there exists no f in $H(Q)$ for which $\bar{q}f \perp H_N^2$ then this is equivalent to $H(Q) \cap H(qI) = \{0\}$ or alternatively to $H(Q) \vee H(qI) = H_N^2$. Now $\{qe_1, \ldots, qe_m, q_1, \ldots, q_m\}$ is a set of generators for H_N^2. If we consider the invariant subspace spanned by $\{qe_1, \ldots, qe_m, q_1, \ldots, q_m\}$ then it is given by a rigid function Q_1 with the same initial space. If $q^{(1)}, \ldots, q^{(m)}$ are the nonzero columns of Q_1 then we would have that the $n \times n$ matrix function with columns $q^{(1)}, \ldots, q^{(m)}, qe_{m+1}, \ldots, qe_n$ corresponds to H_N^2 and hence is a constant unitary matrix. But that is impossible as $q^{(n-m)}$ is a factor of its determinant.

To prove the converse assume u is a nontrivial function in H^∞ and $u(S(Q)^*) f = 0$ for some nonzero f in $H(Q)$. The set $J = (v \mid v(S(Q)^*) f = 0)$ is a w^*-closed ideal in H^∞ hence of the form qH^∞ for some inner function q. Since $u \in J$ it is not outer. So we proved (a).

To prove (b) assume first that $S(Q)$ is not injective. Hence there exists a nontrivial f in $H(Q)$ for which $uf = Qg$ for some $g \in H_N^2$. If $q = \det Q$ then by Cramer's rule we can write $MQ = qI$ where M is the inner function whose entries are the cofactors of Q. Thus we have

$$uMf = qg$$

Now if q and u are coprime then $Mf = qh$ for some $h \in H_N^2$. In other words $f = Qh$ which means that $f \in QH_N^2$ contrary to an assumption. So q and u have a nontrivial common inner factor.

Conversely let us assume u and q have a nontrivial common inner factor. If that is the case then also u and m have a nontrivial factor, where m is the minimal function of Q. Let us put $m = \psi a$ and $u = \psi b$ where ψ is the greatest common inner division of m and u. Since we cannot have $H(Q) \perp aH_N^2$ there exists a $g \in H_N^2$ for which the decomposition $ag = f + Qh$ relative to the direct sum $H_N^2 = H(Q) \oplus QH_N^2$ yields a nontrivial f. Obviously f is in $\mathrm{Ker}(u(S(Q)))$ for

$$uf = \psi bf = \psi b(ag - Qh) = m(bg) - Q(uh)$$

So $uf \in QH_N^2$ and hence $u(S(Q)) f = 0$. Part (c) follows from (b) by an application of Theorem 13-2. $u(S(Q)^*)$ is unitarily equivalent to $u(S(\tilde{Q}))$ acting in $H(\tilde{Q})$. Hence $u(S(Q)^*)$ is injective if and only if u and $\tilde{q} = \det \tilde{Q}$ are coprime.

Theorem 14-5 Let N be finite dimensional, Q a rigid function, $S(Q)$ the restricted shift operator acting in $H(Q) = \{QH_N^2\}^\perp$ and $u \in H^\infty$. Then $u(S(Q))$ is boundedly invertible if and only if for some $\delta > 0$

$$|u(z)| + \|Q(z)^{-1}\|^{-1} \geq \delta \tag{14-10}$$

holds for all z in D.

PROOF If (14-10) holds for all z in D then by Lemma 14-3 there exists a $\delta' > 0$ such that

$$|u(z)| + |q(z)| \geq \delta' \tag{14-11}$$

for all z in D, that is, u and q are strongly coprime. By the Corona theorem of Carleson there exist two functions a and b in H^∞ such that $au + bq = 1$.

The functional calculus transforms this into

$$a(S(Q)\,u(S(Q))) + b(S(Q))\,g(S(Q)) = I$$

But $q(S(Q)) = 0$ so we are left with $u(S(Q))^{-1} = a(S(Q))$ and $u(S(Q))$ is boundedly invertible.

To prove the converse we argue as in Theorem 13-8. If condition (14-10) is not satisfied there exists a sequence of points λ_n in D and unit vectors ξ_n in N such that $\lim_{n \to \infty} \|u(\lambda_n)\| = 0$ and $\lim_{n \to \infty} \|Q(\lambda_n)^* \, \xi_n\| = 0$. Let e_n, f_n, and g_n be defined by (13-19), (13-20), and (13-21), respectively, with λ replaced by λ_n and ξ replaced by ξ_n. Since $\|g_n\|^2 + \|f_n\|^2 = \|e_n\|^2$ we have $\lim \|f_n\|^2 = 1 - \lim \|g_n\|^2 = 1$. For the left shift S^* in H_N^2 we have

$$u(S)^* \, e_n = \tilde{u}(S^*)\, e_n = \tilde{u}(\lambda_n)\, e_n = \overline{u(\lambda_n)}\, e_n$$

Hence

$$u(S(Q))^* \, f_n - \overline{u(\lambda_n)}\, f_n = u(S)^* \, f_n - \overline{u(\lambda_n)}\, f_n = -\big(u(S)^* \, g_n - \overline{u(\lambda_n)}\, g_n\big)$$

Consequently we obtain the following estimate

$$\|u(S(Q)^*) \, f_n\| \le 2|u(\lambda_n)| + \|u\|_\infty \cdot \|g_n\|$$

and the right-hand side tends to zero. If Q is not inner then for each $\lambda \in D$ there exists, by Theorem 13-3, at least one eigenfunction of $S(Q)^*$. No decomposition of the eigenfunctions as carried out above is necessary. This completes the proof.

Corollary 14-6 $\lambda \in \rho(u(S(Q)))$ if and only if for some $\delta > 0$

$$|\lambda - u(z)| + \|Q(z)^{-1}\|^{-1} \ge \delta \tag{14-12}$$

for all z in D.

Restricting ourselves to functions in the algebra A of functions analytic in D and continuous in $\bar{D}$ we obtain the classical spectral mapping theorem.

Theorem 14-7 Let $u \in A$ and Q be a rigid function. Then $\sigma(u(S(Q))) = u(\sigma(S(Q)))$.

PROOF Assume $\lambda \in \sigma(S(Q))$. By Theorem 14-6 there are points λ_n in D such that $\lim \lambda_n = \lambda$ and $\lim \|Q(\lambda_n)^{-1}\|^{-1} = 0$. It follows by continuity that

$$\lim |u(\lambda_n) - u(\lambda)| + \|Q(\lambda_n)^{-1}\|^{-1} = 0$$

or that $u(\lambda) \in \sigma(u(S(Q)))$.

Conversely assume, without loss of generality, that $0 \in \sigma(u(S(Q)))$. Then, by passing to a subsequence, we may assume $\lambda_n \to \lambda$ and $|u(\lambda_n)| + \|Q(\lambda_n)^{-1}\|^{-1} \to 0$. Clearly $\lambda \in \sigma(S(Q))$ and $u(\lambda) = 0$.

The analysis carried out above can be generalized and the same methods appl ed to operators intertwining two restricted shifts.

Let T_1 and T_2 be contractions and assume $T_i^{*n} \to 0$ strongly. This assumption simplifies significantly the construction of a functional model for the T_i. In fact by Theorem 12-1 each T_i can be represented by an operator of the form $S(Q_i)$ acting in a left invariant subspace $H(Q_i)$ of $H_{N_i}^2$ where Q_i is a rigid function. As of now we will assume the Q_i to be inner and the spaces N_i to be finite dimensional.

The lifting theorem, that is, Theorem 11-4, when translated into the language of vectorial function theory reads as follows.

Theorem 14-8 Let Q_i for $i = 1, 2$ be inner functions N_i finite dimensional Hilbert spaces and $S(Q_i)$ the restricted shifts acting in $H(Q_i) = \{Q_i H_{N_i}^2\}^{\perp}$. An operator $X: H(Q_1) \to H(Q)$ satisfies

$$XS(Q_1) = S(Q_2) X \tag{14-13}$$

if and only if there exist functions $\Xi, \Xi_1 \in H_{B(N_1, N_2)}^{\infty}$ satisfying

$$\Xi Q_1 = Q_2 \Xi_1 \tag{14-14}$$

$$\|\Xi\|_{\infty} = \|X\| \tag{14-15}$$

and for which

$$Xf = P_{H(Q_2)} \Xi f \tag{14-16}$$

PROOF If X is given by (14-16) then since by (14-14) multiplication by Ξ maps $Q_1 H_{N_1}^2$ into $Q_2 H_{N_2}^2$ we have for $f \in H(Q_1)$

$$XS(Q_1) f = P_{H(Q_1)} \Xi P_{H(Q_1)} \chi f = P_{H(Q_2)} \chi \Xi f = P_{H(Q_2)} \chi P_{H(Q_2)} \Xi f = S(Q_2) Xf$$

Moreover we have obviously $\|X\| \leq \|\Xi\|_{\infty}$.

Conversely assume X satisfies (14-13) that is X intertwines $S(Q_1)$ and $S(Q_2)$. The right shifts in $H_{N_1}^2$ and $H_{N_2}^2$, respectively, provide isometric dilations, which are not necessarily minimal. By Theorem 11-7 there exists an operator Y intertwining the right shifts for which $\|Y\| = \|X\|$. Since an operator intertwining the right shifts is necessarily a multiplication operator by a bounded operator valued analytic function Ξ we have (14-16). Now the operator Y satisfies $YQ_1 H_{N_1}^2 \subset Q_2 H_{N_2}^2$ which is equivalent to the existence of a function Ξ_1 such that (14-14) holds.

Theorem 14-9 Let $X: H(Q_1) \to H(Q_2)$ be the map defined in Theorem 14-1. Then its adjoint $X^*: H(Q_2) \to H(Q_1)$ is unitarily equivalent to the map $X : H(\tilde{Q}_2) \to H(\tilde{Q}_1)$ defined by

$$X_1 g = P_{H(\tilde{Q}_1)} \tilde{\Xi}_1 g \tag{14-17}$$

with Ξ_1 satisfying

$$\tilde{\Xi}_1 \tilde{Q}_2 = \tilde{Q}_1 \tilde{\Xi} \tag{14-18}$$

PROOF Let, for $i = 1, 2$, τ_{Q_i} be the unitary maps defined by (12-7). We will prove that $X^* = \tau_{Q_1}^* X_1 \tau_{Q_2}$. Note that (14-18) follows from (14-14) and is in turn equivalent to

$$\tilde{Q}_2 \tilde{\Xi}^* = \tilde{\Xi}_1^* \tilde{Q}_1 \tag{14-19}$$

The last equality implies that for each $g \in H(Q_1)$ we have $\tau_{Q_2} \Xi g = \tilde{\Xi}_1^* \tau_{Q_1} g$, also recall that we have $\tau_{Q_2} P_{H(Q_2)} = P_{H(\tilde{Q}_2)} \tau_{Q_2}$. Therefore for all $f \in F(Q_2)$ and $g \in H(Q_1)$

$$
\begin{aligned}
(X^* f, g) = (f, Xg) = (\tau_{Q_2} f, \tau_{Q_2} Xg) &= (\tau_{Q_2} f, \tau_{Q_2} P_{H(Q_2)} \Xi g) \\
&= (\tau_{Q_2} f, P_{H(\tilde{Q}_2)} \Xi g) = (\tau_{Q_2} f, \tau_{Q_2} \Xi g) \\
&= (\tau_{Q_2} f, \tilde{\Xi}_1^* \tau_{Q_1} g) = (\tilde{\Xi}_1 \tau_{Q_2} f, \tau_{Q_1} g) = (P_{H(\tilde{Q}_1)} \tilde{\Xi}_1 \tau_{Q_2} f, \tau_{Q_1} g) \\
&= (X_1 \tau_{Q_2} f, \tau_{Q_1} g)
\end{aligned}
$$

which proves the theorem.

In preparation for Theorem 14-11 we prove a matrix generalization of the Carleson corona theorem.

Theorem 14-10 Let N, N_i $i = 1, \ldots, p$ be finite dimensional Hilbert spaces.
(a) Given $A_i \in H_{B(N, N_i)}^\infty$ then a necessary and sufficient condition for the existence of $B_i \in H_{B(N_i, N)}^\infty$ satisfying

$$\sum_{i=1}^p B_i(z) A_i(z) = I_N \tag{14-20}$$

is that the strong coprimeness condition

$$[A_1, \ldots, A_p]_R = I_N \tag{14-21}$$

be satisfied.
(b) Given $A_i \in H_{B(N_i, N)}^\infty$ then a necessary and sufficient condition for the existence of $B_i \in H_{B(N, N_i)}^\infty$ satisfying

$$\sum_{i=1}^p A_i(z) B_i(z) = I_N \tag{14-22}$$

is that the strong coprimeness condition

$$[A_1, \ldots, A_p]_L = I_N \tag{14-23}$$

be satisfied.

PROOF
(a) The necessity part of the proof is simple. If (14-21) is not satisfied then there exist points $\lambda_n \in D$ and unit vectors $\xi_n \in N$ such that for $i = 1, \ldots, p$ $\lim \|A_i(\lambda_n) \xi_n\| = 0$. In that case (14-20) cannot hold for it implies $\xi_n =$

$\sum_{i=1}^{p} B_i(\lambda_n) A_i(\lambda_n) \xi_n$ and hence the following estimate

$$1 = \|\xi_n\| \le \sum_{i=1}^{p} \|B_i\| \|A_i(\lambda_n) \xi_n\|$$

and the right-hand side tends to zero.

To prove sufficiency we fix orthonormal bases in N, N_i, $i = 1, \dots, p$ and express A_i in matrix form, retaining the letters A for the corresponding matrices. Denote by $A_{jk}^{(i)}$ the elements of A_i. A_i is an $n_i \times n$ matrix where n and n_i are the dimensions of N, N_i, respectively.

Let W be the $\sum n_i \times n$ matrix composed of the rows of all A_i. Let $W_{i_1 \dots i_n}$ be the $n \times n$ matrix whose rows are the $i_1, \dots, i_n$ rows of W.

We claim that if (14-21) holds then the set of scalar functions $\det W_{i_1 \dots i_n}$, $1 \le i_1 < \cdots < i_n \le \sum_{i=1}^{p} n_i$ is strongly coprime, that is, there exists a $\delta > 0$ such that

$$\sum |\det W_{i_1 \dots i_n}(z)| \ge \delta \qquad (14\text{-}24)$$

for all $z \in D$.

The basic idea is that if for some $\lambda \in D$ we have $\sum |\det W_{i_1 \dots i_n}(\lambda)| = 0$ then the vectors represented by the rows of the matrices $A_i(\lambda)$ lie all in a proper subspace of N. If that is the case then there exists a nonzero vector orthogonal to all of them implying that for some $\xi \in N$ we have $\sum_{i=1}^{p} \|A_i(\lambda) \xi\| = 0$ in contradiction to (14-20). In the general case we have to argue differently. If (14-24) is not satisfied then there exists a sequence of points $\lambda_v \in D$ for which

$$\lim_{v \to \infty} \sum |\det W_{i_1 \dots i_n}(\lambda_v)| = 0 \qquad (14\text{-}25)$$

We will show that (14-25) contradicts (14-21) by proving that

$$\lim_{v \to \infty} \inf \left\{ \sum_{i=1}^{p} \|A_i(\lambda_v) x\| \, |x \in N|, \|x\| = 1 \right\} \qquad (14\text{-}26)$$

Let $\varepsilon_v = \sum |\det W_{i_1 \dots i_n}(\lambda_v)|$ and let $y_i^{(v)}$ be the ith row of $W(\lambda_v)$. There is one set of indices $i_1 \dots i_n$ such that for all $j_1 \dots j_n$ we have

$$|\det W_{j_1 \dots j_n}(\lambda_v)| \le |\det W_{i_1 \dots i_n}(\lambda_v)| \qquad (14\text{-}27)$$

By Lemma 14-3 there exists a unit vector $\xi \in N$ for which $|(\xi, y_{ij}^{(v)})| \le \varepsilon_v^{1/n}$. If $\varepsilon_v > 0$ then the vectors $y_{ij}^{(v)}$ are a basis for N. Each $y_k^{(v)}$ has a representation $y_k^{(v)} = \sum_{j=1}^{n} \beta_{kj} y_{ij}^{(v)}$ and (14-27) implies that $|\beta_{kj}| \le 1$ and with it that for each $y_k^{(v)}, |(x, y_k^{(v)})| \le n\varepsilon_v^{1/n}$. This estimate shows that (14-20) implies (14-24).

Invoke now the scalar corona theorem to deduce the existence of $a_{i_1 \dots i_n} \in H^\infty$, $1 \le i_1 < \cdots < i_n = \sum_{j=1}^{p} n_j$ such that

$$\sum a_{i_1 \dots i_n}(z) \det W_{i_1 \dots i_n}(z) = 1 \qquad (14\text{-}28)$$

To complete the proof we have to show the existence of $B_{ik}^{(l)}$ in H^∞

such that

$$\sum_{l=1}^{p} \sum_{k=1}^{n} B_{ik}^{(l)}(z) A_{kj}^{(l)}(z) = \delta_{ij} \tag{14-29}$$

holds.

From equality (14-28) we can, by collecting terms, define matrices $B^{(l)}$ with elements $B_{jk}^{(l)}$ in H^{∞} such that

$$\sum_{l=1}^{p} \sum_{k=1}^{n} B_{jk}^{(l)}(z) A_{kj}^{(l)}(z) = 1$$

Thus (14-29) is satisfied for $i = j$. If $i \neq j$ then

$$\sum_{i=1}^{p} \sum_{k=1}^{n_i} B_{ik}^{(l)}(z) A_{kj}^{(l)}(z) = \sum a_{i_1 \ldots i_n}(z) \det W_{i_1 \ldots i_n}'(z)$$

where $W_{i_1 \ldots i_n}'$ is equal to $W_{i_1 \ldots i_n}$ with the ith column replaced by the jth and thus $\det W_{i_1 \ldots i_n}'(z) = 0$. As a consequence (14-29) holds for all $i, j = 1, \ldots, n$.

(b) This follows from part (a) by noting that $\sum_{i=1}^{p} A_i(z) B_i(z) = I_N$ if and only if $\sum_{i=1}^{p} \tilde{B}_i(z) \tilde{A}_i(z) = I_N$ and that $[\tilde{A}_1, \ldots, \tilde{A}_n]_R = I_N$ is equivalent to (14-23).

Theorem 14-11 Let for $i = 1, 2$ N_i be finite dimensional Hilbert spaces, Q_i inner functions and $S(Q_i)$ the restricted shifts in $H(Q_i) = \{Q_i H_{N_i}^2\}^{\perp}$. Let $X: H(Q_1) = H(Q_2)$ be an operator intertwining $S(Q_1)$ and $S(Q_2)$ having the representation (14-16) with (14-14) satisfied. Then

(a) X is injective if and only if

$$(\Xi_1, Q_1)_R = I_{N_1} \tag{14-30}$$

(b) X^* is injective if and only if

$$(\Xi, Q_2)_L = I_{N_2} \tag{14-31}$$

(c) X has a bounded left inverse if and only if

$$[\Xi_1, Q_1]_R = I_{N_1} \tag{14-32}$$

(d) X has a bounded right inverse if and only if

$$[\Xi, Q_2]_L = I_{N_2} \tag{14-33}$$

PROOF Let us start with (b). X^* is injective if and only if the range of X is dense in $H(Q_2)$ and this occurs if and only if the invariant subspace spanned by $\Xi H_{N_1}^2$ and $Q_2 H_{N_2}^2$ is all of $H_{N_2}^2$. This is equivalent to (14-31). Next X is unitarily equivalent to X_1^* where X_1 is defined by (14-17). Applying part (b) X is injective if and only if $(\tilde{\Xi}_1, \tilde{Q}_1)_L = I_{N_1}$ and this condition is equivalent to (14-30) proving (a).

Assume now (14-33) holds. By Theorem 14-10 there exist $\Theta \in H^\infty_{B(N_1, N_2)}$ and $R \in H^\infty_{B(N_2)}$ such that $\Xi\Theta + Q_2 R = I_{N_2}$. Define a bounded operator $T: H(Q_2) \to H(Q_1)$ by $Tf = P_{H(Q_1)}\Theta f$ for all $f \in H(Q_2)$. For such f we have, recalling that $\Xi Q_1 H^2_{N_1} \subset Q_2 H^2_{N_2}$,

$$P_{H(Q_2)}\{\Xi\Theta f + Q_2 Rf\} = P_{H(Q_2)}\Xi P_{H(Q_1)}\Theta f = f$$

or $XT = I$ proving that T is a right inverse of X.

In the same manner, using Theorem 14-9, we prove that (14-32) is sufficient for the existence of a bounded left inverse for X. To prove the necessity part we argue in the way we did in the proof of Theorem 14-5. We will show that (14-33) is necessary for the existence of a bounded right inverse for X. Now X has a bounded right inverse if and only if X^* has a bounded left inverse. Since X^* and X_1 are unitarily equivalent it is enough to show that X_1 does not have a bounded left inverse. For this it suffices to exhibit a sequence of functions such that $\lim_{n\to\infty} \|F_n\| = 1$ and $\lim_{n\to\infty} \|X_1 F_n\| = 0$.

Since we assume that (14-33) does not hold there exists a sequence of points λ_n in D and unit vectors $\xi_n \in N$ such that $\lim \|\Xi(\lambda_n)^* \xi_n\| = \lim \|Q_2(\lambda_n)^* \xi_n\| = 0$. Given λ_n and ξ_n define the functions e_n, f_n, and g_n by (13-19), (13-20), and (13-21), respectively. The function f_n is the projection of e_n onto $H(Q_2)$. Apply the unitary transformation τ_{Q_2} to all three functions and let $E_n = \tau_{Q_2} e_n$, $F_n = \tau_{Q_2} f_n$, and $G_n = \tau_{Q_2} g_n$.

A simple calculation yields that

$$E_n(z) = (1 - |\lambda_n|^2)^{1/2}\, \tilde{Q}_2(z)\, \xi_n/(z - \bar{\lambda}_n) \tag{14-34}$$

$$F_n(z) = (1 - |\lambda_n|^2)^{1/2}\, (\tilde{Q}_2(z) - \tilde{Q}_2(\bar{\lambda}_n))\, \xi/(z - \bar{\lambda}_n) \tag{14-35}$$

$$G_n(z) = (1 - |\lambda_n|^2)^{1/2}\, \tilde{Q}_2(\bar{\lambda}_n)\, \xi_n/(z - \bar{\lambda}_n) \tag{14-36}$$

Moreover, since $\|F_n\|^2 + \|G_n\|^2 = 1$ and $\lim \|G_n\|^2 = 0$ we have $\lim \|F_n\|^2 = 1$. Note also that $E_n \perp \tilde{Q}_2 H^2_{N_2}$ and $F_n = P_{H(\tilde{Q}_2)}E_n$. To compute $X_1 F_n$ we note that

$$\begin{aligned}
\tilde{\Xi}_1(z)\, F_n(z) &= \tilde{\Xi}_1(z)(1 - |\lambda_n|^2)^{1/2}\, (\tilde{Q}_2(z) - Q_2(\bar{\lambda}_n))\, \xi_n/(z - \bar{\lambda}_n) \\
&= (1 - |\lambda_n|^2)^{1/2}\, \tilde{\Xi}_1(z)\, \tilde{Q}_2(z)\, \xi_n/(z - \bar{\lambda}_n) \\
&\quad - (1 - |\lambda_n|^2)^{1/2}\, \tilde{\Xi}_1(z)\, \tilde{Q}_2(\bar{\lambda}_n)\, \xi_n/(z - \bar{\lambda}_n) \\
&= (1 - |\lambda_n|^2)^{1/2}\, \tilde{Q}_1(z)\, \tilde{\Xi}(z)\, \xi_n/(z - \bar{\lambda}_n) \\
&\quad - (1 - |\lambda_n|^2)^{1/2}\, \tilde{\Xi}_1(z)\, Q_2(\lambda_n)^*\, \xi_n/(z - \bar{\lambda}_n) \\
&= \tilde{Q}_1(z)(1 - |\lambda_n|^2)^{1/2}\, (\tilde{\Xi}(z) - \tilde{\Xi}(\bar{\lambda}_n))\, \xi_n/(z - \bar{\lambda}_n) \\
&\quad + \tilde{Q}_1(z)(1 - |\lambda_n|^2)^{1/2}\, \tilde{\Xi}(\bar{\lambda}_n)\, \xi_n/(z - \bar{\lambda}_n) \\
&\quad - (1 - |\lambda_n|^2)^{1/2}\, \tilde{\Xi}_1(z)\, Q_2(\lambda_n)^*\, \xi_n/(z - \bar{\lambda}_n)
\end{aligned}$$

Since the first term in the last sum is obviously in $\tilde{Q}_1 H_{N_1}^2$ taking the projection on $H(\tilde{Q}_1)$ we obtain the following estimate.

$$\|X_1 F_n\| \leq \|\hat{\Xi}(\bar{\lambda}_n)\,\xi_n\| + \|\Xi_1\|_\infty \|Q_2(\lambda_n)^*\,\xi_n\|$$

$$= \|\Xi(\lambda_n)^*\,\xi_n\| + \|\Xi_1\|_\infty \|Q_2(\lambda_n)^*\,\xi_n\|$$

and so $\lim_{n\to\infty} \|X_1 F_n\| = 0$ and (d) is proved. The necessity of condition (14-32) for the existence of a bounded left inverse for X follows now by duality arguments.

15. JORDAN MODELS

In Chap. I we showed how the theory of equivalence of matrices over $F[\lambda]$ is connected with the theory of similarity of matrices over the field F. We proved that matrices A and A_1 are similar if and only if the polynomial matrices $\lambda I - A$ and $\lambda I - A_1$ are equivalent. As a consequence the reduction of $\lambda I - A$ to its Smith canonical form was the key to the reduction of A to its first canonical form, and from that the Jordan canonical form followed.

Our aim in this section is to develop the same theory for the case of restricted shifts of finite multiplicity. We will consider the set of all $n \times n$ matrices over H^∞, that is, the matrix ring $(H^\infty)^{n \times n}$. H^∞ is a ring, actually an algebra over the complex field. As a ring H^∞ is an integral domain, that is, has no zero divisors. However, H^∞ is not a principal idea domain, as the only principal ideals in H^∞ are the w^*-closed ones. Thus a straightforward application of the classical algebraic theory is impossible. We will have to relax our notion of equivalence to obtain a richer theory.

In H^∞, as in any ring, we have the natural division relation. We say that b divides a and write $b\,|\,a$ if for some c we have $a = bc$. By $\wedge\,f_\alpha$ we will denote the greatest common inner divisor of the f_α and by $\vee\,f_\alpha$ their least common inner multiple. Both $\wedge\,f_\alpha$ and $\vee\,f_\alpha$ are determined up to a constant of absolute value one.

We begin by proving some lemmas which are important for the development of the theory.

Lemma 15-1 Let f_1, f_2 be functions in H^∞ and let ω be an inner function. Suppose

$$\omega \wedge f_1 \wedge f_2 = 1 \tag{15-1}$$

then for every complex number α, with the exception of at most a countable number of values we have

$$\omega \wedge (f_1 + \alpha f_2) = 1 \tag{15-2}$$

PROOF For complex α let $r_\alpha = \omega \wedge (f_1 + \alpha f_2)$. If $\beta \neq \alpha$ then necessarily $r_\alpha \wedge r_\beta = 1$ for if $r = r_\alpha \wedge r_\beta$ then r divides both $f_1 + \alpha f_2$ and $f_1 + \beta f_2$ and hence their difference $(\alpha - \beta)\, f_2$. Since $\alpha \neq \beta$ r divides f_2 and as a consequence also f_1. Now r is an inner factor of ω and so must be equal to one otherwise (15-1) is contradicted.

Now we claim that at most a countable number of the r_α are nontrivial. Let us factor the inner function ω as BS where B is a Blaschke product corresponding to the zeros of ω and S a singular inner function which is associated with the singular measure μ. If r_α is an inner divisor of ω let $r_\alpha = B_\alpha S_\alpha$ be its corresponding factorization. The zeros of B_α form a subset of the zeros of B, the measure μ_α is also singular and $\mu - \mu_\alpha$ is positive. If $r_\alpha \wedge r = 1$ then $B_\alpha \wedge B_\beta = 1$ and $S_\alpha \wedge S_\beta = 1$. This means that the zero sets of B_α and B_β are nonintersecting and that the singular measures μ_α and μ_β are mutually singular. From this it is clear that at most a countable number of the r_α are nontrivial.

The preceding lemma has some immediate generalizations.

Lemma 15-2 Let $f_1, \ldots, f_m$ be functions in H^∞ and let ω be an inner function. Assume

$$\omega \wedge f_1 \wedge \cdots \wedge f_m = 1 \tag{15-3}$$

then there exist complex numbers $\alpha_2, \ldots, \alpha_m$ such that

$$\omega \wedge (f_1 + \alpha_2 f_2 + \cdots + \alpha_m f_m) = 1 \tag{15-4}$$

PROOF We prove the lemma by induction. Let us define $r_i = \omega \wedge f_1 \wedge \cdots \wedge f_i$. Obviously we have the division relations $1 = r_m | r_{m-1} | \ldots | r_1 | \omega$. For $i = 2$ we have $(\omega/r_2) \wedge (f_1/r_2) \wedge (f_2/r_2) = 1$ which, by the previous lemma guarantees the existence of a complex α_2 such that

$$\frac{\omega}{r_2} \wedge \left(\frac{f_1}{r_2} + \frac{\alpha_2 f_2}{r_2} \right) = 1$$

which is the same as $\omega \wedge (f_1 + \alpha_2 f_2) = r_2$.

Assume we proved the existence of $\alpha_2, \ldots, \alpha_k$ such that $\omega \wedge (f_1 + \alpha_2 f_2 + \cdots + \alpha_k f_k) = r_k$. Clearly we must have $\omega \wedge (f_1 + \alpha_2 f_2 + \cdots + \alpha_k f_k) \wedge f_{k+1} = r_{k+1}$. Applying the previous lemma once more there exists an α_{k+1} such that $\omega \wedge (f_1 + \alpha_2 f_2 + \cdots + \alpha_{k+1} f_{k+1}) = r_{k+1}$. Since $r_m = 1$ the lemma is proved.

Lemma 15-3 Let (f_{ij}) be an $n \times m$ matrix with H^∞ entries and let ω_i be inner functions. Suppose

$$\omega_i \wedge f_{i1} \wedge \cdots \wedge f_{im} = 1 \tag{15-5}$$

then there exist numbers $\lambda_1, \ldots, \lambda_m$ such that

$$\omega_i \wedge \sum_{j=1}^{m} \lambda_j f_{ij} = 1 \qquad i = 1, \ldots, n \tag{15-6}$$

PROOF The proof is analogous to the previous one. If $r_j^{(i)} = \omega_i \wedge f_{i1} \wedge \cdots \wedge f_{ij}$ at each step λ_j can be chosen arbitrarily except at most in the union of n countable sets. This means that the λ_j can be chosen to fit all i, $i = 1, \ldots, n$.

Corollary 15-4 Let (a_{ij}) be an $n \times m$ matrix with H^∞ entries and ω an inner function. Then there exist functions $h_i \in H^\infty$ such that

$$a_{i1} + \lambda_2 a_{i2} + \cdots + \lambda_m a_{im} = f_i(a_{i1} \wedge \cdots \wedge a_{im}) \tag{15-7}$$

and $f_i \wedge \omega = 1$.

A matrix X in $(H^\infty)^{n \times n}$ is a unit if X^{-1} is also in $(H^\infty)^{n \times n}$ which is equivalent to $\det X$ being an invertible element of H^∞. Two matrices A and B in $(H^\infty)^{n \times n}$ are *equivalent* if there exist units X and Y in $(H^\infty)^{n \times n}$ for which

$$XA = BY \tag{15-8}$$

A function X in $(H^\infty)^{n \times n}$ has a scalar multiple $\varphi \in H^\infty$ if there exists a matrix $X^a \in (H^\infty)^{n \times n}$ such that

$$X^a X = XX^a = \varphi I \tag{15-9}$$

Clearly $\det X$ is a scalar multiple of X. Given an inner function ω we denote by $\mathcal{N}_\omega(n)$ the set of all matrices in $(H^\infty)^{n \times n}$ that have a scalar multiple φ which is coprime with ω, that is for which $\varphi \wedge \omega = 1$. In particular all units in $(H^\infty)^{n \times n}$ belong to $\mathcal{N}_\omega(n)$.

Given matrices A and B in $(H^\infty)^{n \times n}$ and an inner function ω we say A and B are *ω-equivalent* if there exist X and Y in $\mathcal{N}_\omega(n)$ for which (15-8) holds. If A and B are ω-equivalent for every inner function ω then we say that A and B are *quasiequivalent*. All three notions of equivalence are bona fide equivalence relations, that is, they are reflexive, symmetric, and transitive. For equivalence the three properties are immediate, and for quasiequivalence it would follow from proving these properties for ω-equivalence. Reflexivity is trivial choosing $X = Y = I$. To show symmetry assume (15-8) holds and X, Y have scalar multiples φ and ψ, respectively, satisfying $\varphi \wedge \omega = \psi \wedge \omega = 1$. There exist X^a and Y^a such that (15-9) holds as well as the analogous condition for Y. From (15-8) we have $XAY^a = BYY^a = B\psi$ which implies $X^a XAY^a = \psi X^a B$ or $A(\varphi X^a) = (\psi X^a) B$. Since $\psi Y^a X = \varphi Y^a Y = \varphi \psi I$ and $\varphi \psi \wedge \omega = 1$ symmetry follows. Finally, if besides (15-8) we have also $ZB = CW$ with $Z^a Z = ZZ^a = \zeta I$ and $\zeta \wedge \omega = 1$ it follows that $(ZX)A = Z(BY) = C(WY)$. ZY has $\zeta \psi$ as a scalar multiple and clearly $\zeta \psi \wedge \omega = 1$. Similarly WY has a scalar multiple coprime with ω which shows transitivity.

It is clear that equivalence implies quasiequivalence, but generally the converse is not true. To give an example consider the diagonal matrices A and B given by

$$A = \begin{pmatrix} \varphi & 0 \\ 0 & \psi \end{pmatrix}, \qquad B = \begin{pmatrix} \varphi\psi & 0 \\ 0 & 1 \end{pmatrix}$$

where φ and ψ are the inner functions $\varphi(z) = e^{-(1+z)/(1-z)}$ and ψ the Blaschke product with zeros at the points $1 - (1/n^2)$. We will show that A and B are quasiequivalent but not equivalent.

For the equivalence of A and B we have to have unit matrices

$$X = \begin{pmatrix} \xi_{11} & \xi_{12} \\ \xi_{21} & \xi_{22} \end{pmatrix} \quad \text{and} \quad Y = \begin{pmatrix} \eta_{11} & \eta_{12} \\ \eta_{21} & \eta_{22} \end{pmatrix}$$

in $(H^\infty)^{n \times n}$ such that (15-8) holds. A simple calculation shows that necessarily $\psi | \xi_{11}$ and $\varphi | \xi_{12}$. Thus

$$A = \begin{pmatrix} \psi\xi'_{11} & \varphi\xi'_{12} \\ \xi_{21} & \xi_{22} \end{pmatrix}$$

and since $\det A = \psi\xi'_{11}\xi_{22} - \varphi\xi_{21}\xi_{12}$ is a unit in H^∞ we must have that φ and ψ satisfy the Carleson condition $\inf_{z \in D} \{|\varphi(z)| + |\psi(z)|\} \geq \delta$ for some $\delta > 0$. Our choice of φ and ψ rules that out so A and B are not equivalent. It is easy, however, to show that A and B are quasiequivalent. Let ω be an arbitrary inner function. By Lemma 15-1 we can choose a complex α so that $(\psi + \alpha\varphi) \wedge \omega = 1$. Define matrices X and Y by

$$X = \begin{pmatrix} \psi & -\alpha\varphi \\ 1 & 1 \end{pmatrix} \quad \text{and} \quad Y = \begin{pmatrix} 1 & -\alpha \\ \varphi & \psi \end{pmatrix}$$

then clearly $XA = BY$ and $\det X = \det Y = \psi + \alpha\varphi$. This implies ω-equivalence and as ω was arbitrary the quasiequivalence of A and B.

For a matrix A in $(H^\infty)^{n \times n}$ we introduce the determinant divisors $D_i(A)$ and invariant factors $E_i(A)$ in analogy with that of Chap. I. We let $D_0(A) = 1$ and define $D_i(A)$ to be the greatest common inner divisor of all $i \times i$ minors of A. We write $D_i(A) = 0$ if all $i \times i$ minors are identically equal to zero. By the expansion rules of determinants it follows that $D_i(A) | D_{i+1}(A)$ and in particular $D_i(A) = 0$ implies $D_{i+1}(A) = 0$. Next we define $E_i(A)$ by $E_i(A) = D_i(A)/D_{i-1}(A)$ if $D_{i-1}(A) \neq 0$ and $E_i(A) = 0$ if $D_{i-1}(A) = 0$. For $A \in (H^\infty)^{n \times n}$ we define the compound matrices of A, $A^{(p)}$ as in Chap. I. The results of Theorem I 2-17 holds equally well over H^∞. In particular if A and B are equivalent so are $A^{(p)}$ and $B^{(p)}$. The same is true for ω-equivalence, for if (15-8) holds and $\varphi = \det X$ and $\psi = \det Y$ are coprime with ω then

$$X^{(p)}A^{(p)} = B^{(p)}Y^{(p)} \tag{15-10}$$

holds and as $\det X^{(p)} = \varphi^{\binom{n-1}{p-1}}$ and $\det Y^{(p)} = \psi^{\binom{n-1}{p-1}}$ obviously $\det X^{(p)}$ and $\det Y^{(p)}$ are also coprime with ω. From this it also follows that quasiequivalence of A and B implies the quasiequivalence of $A^{(p)}$ and $B^{(p)}$.

We can prove now the analog of Lemma I 2-20.

Lemma 15-5 Let $A, B \in (H^\infty)^{n \times n}$ and let ω be an inner function. Then

$$D_i(A) = D_1(A^{(i)}) \tag{15-11}$$

If A and B are ω-equivalent then there exist inner functions α_i and β_i coprime with ω such that

$$D_i(A)\,|\,\alpha_i D_i(B), \qquad D_i(B)\,|\,\beta_i D_i(A), \qquad i = 1, \ldots, n \qquad (15\text{-}12)$$

If A and B are quasiequivalent then

$$D_i(A) = D_i(B), \qquad i = 1, \ldots, n \qquad (15\text{-}13)$$

PROOF That $D_i(A) = D_1(A^{(i)})$ is obvious from the definition of the determinant divisors. Assume now A and B are ω-equivalent, thus $XA = BY$ with $\varphi = \det X$ and $\psi = \det Y$ coprime with ω. From this we obtain

$$XAY^a = \psi B, \qquad X^a BY = \varphi A \qquad (15\text{-}14)$$

where X^a and Y^a are the classical adjoints of X and Y, respectively. From (15-14) it is clear that

$$D_1(A)\,|\,\psi D_1(B) \qquad \text{and} \qquad D_1(B)\,|\,\varphi D_1(A) \qquad (15\text{-}15)$$

From (15-14) we have immediately

$$X^{(p)}A^{(p)}(Y^a)^{(p)} = \psi^p B^{(p)}, \qquad (X^a)^{(p)} B^{(p)} Y^{(p)} = \varphi^p A^{(p)} \qquad (15\text{-}16)$$

and hence

$$D_1(A^{(p)})\,|\,\psi^p D_1(B^{(p)}), \qquad D_1(B^{(p)})\,|\,\varphi^p D_1(A^{(p)}) \qquad (15\text{-}17)$$

which is equivalent to (15-12) once we use the identity (15-11).

If A and B are quasiequivalent then for a fixed index i choose $\omega = D_i(A)\,D_i(B)$, then φ^p and ψ^p are coprime with $D_i(A)$ and $D_i(B)$ separately. In this case (15-17) implies (15-12).

Next we introduce some convenient notation. Let $u = (0, u_2, \ldots, u_n)$ and $(0, v_2, \ldots, v_n)$ be vectors with H^∞ entries. We define matrices $C(u)$ and $R(v)$ in $(H^\infty)^{n \times n}$ by

$$C(u) = \begin{pmatrix} 1 & & & \\ u_2 & 1 & & \\ \cdot & & \cdot & \\ \cdot & & & \cdot \\ u_n & & & \cdot & 1 \end{pmatrix}, \qquad R(v) = \begin{pmatrix} 1 & v_2 & \cdots & v_n \\ & 1 & & \\ & & \cdot & \\ & & & \cdot \\ & & & & 1 \end{pmatrix} \qquad (15\text{-}18)$$

Obviously $C(u)$ and $R(v)$ are units in $(H^\infty)^{n \times n}$. In fact we have

$$C(u)\,C(-u) = I \qquad R(v)\,R(-v) = I \qquad (15\text{-}19)$$

We are ready for the main theorem of diagonalization.

Theorem 15-6 Let $A \in (H^\infty)^{n \times n}$ then A is quasiequivalent to a diagonal matrix having the invariant factors $E_i(A)$ on the diagonal, and we have

$$E_i(A)\,|\,E_{i+1}(A) \qquad (15\text{-}20)$$

PROOF We will show that for any choice of inner function ω in H^∞ A is ω-equivalent to $\operatorname{diag}(E_1(A), \ldots, E_n(A))$. For $A = 0$ this is trivial so we assume A is nonzero and hence $D_1(A)$ is a nontrivial inner function. Let ω' denote the product of ω and all nontrivial determinant divisors of A. Thus a function f is coprime with ω' if and only if it is coprime with ω and all nonzero $D_i(A)$.

Let $a_i^\wedge = \wedge_{j=1}^n a_{ij}$ then $a_{i1}/a_i^\wedge, \ldots, a_{in}/a_i^\wedge$ are coprime and hence, by Lemma 15-3, there exist complex numbers $\alpha_2, \ldots, \alpha_n$ such that a_i defined by

$$a_i = a_{i1} + \alpha_2 a_{i2} + \cdots + \alpha_n a_{in} = a_i^\wedge h_i$$

and $h_i \wedge \omega' = 1$.

Since $\wedge_{i=1}^n a_i^\wedge = D_1(A)$ another application of Lemma 15-3 shows there exist complex $\beta_2, \ldots, \beta_n$ such that

$$a_1 + \beta_2 a_2 + \cdots + \beta_n a_n = h D_1(A) \qquad \text{and} \qquad h \wedge \omega' = 1$$

For the vectors $a = (0, \alpha_2, \ldots, \alpha_n)$ and $b = (0, \beta_2, \ldots, \beta_n)$ we define now the matrices $C(b)$ and $R(a)$ by (15-18). Let A' be given by

$$A' = R(b)\, A\, C(a) \tag{15-21}$$

Since $C(a)$ and $R(b)$ are units A' is equivalent to A and so has the same determinant divisors as A. Moreover by the special structure of $R(b)$ and $C(a)$ it follows that A' has the form

$$A' = \begin{pmatrix} h D_1(a) & a'_{12} & \cdots & a'_{1n} \\ a'_{21} & a_{22} & \cdots & a_{2n} \\ \vdots & \vdots & & \\ a'_{n1} & a_{n2} & \cdots & a_{nn} \end{pmatrix} \tag{15-22}$$

Define A'' by

$$A'' = \begin{pmatrix} D_1(a) & a'_{12} & \cdots & a'_{1n} \\ a'_{21} & h a_{22} & \cdots & h a_{2n} \\ \vdots & & & \\ a'_{n1} & h a_{n2} & \cdots & h a_{nn} \end{pmatrix} \tag{15-23}$$

then we have

$$A'\, \operatorname{diag}(1, h, \ldots, h) = \operatorname{diag}(h, 1, \ldots, 1)\, A''$$

Since both of the diagonal matrices have scalar factors, h^{n-1} and h, respectively, which are coprime with ω' it follows that A' and A'' are ω'-equivalent. Now as $D_1(A) = D_1(A')$ it follows that $D_1(A)$ divides all a'_{1j} and all a'_{i1}. Let $u_i = a'_{i1}/D_1(A)$ and $v_j = a'_{1j}/D_1(A)$. If $u = (0, u_2, \ldots, u_n)$ and $v = (0, v_2, \ldots, v_n)$ we form $C(u)$ and $R(v)$ by (15-18) and define A''' by

$$A''' = C(u)\, A''\, R(v) \tag{15-24}$$

then A''' is equivalent to A'' and has the form

$$A''' = \begin{pmatrix} D_1(A) & 0 \\ 0 & A_1 \end{pmatrix}$$ (15-25)

where A_1 is in $(H^\infty)^{(n-1) \times (n-1)}$. Now $D_1(A)$ divides all elements of A' by equivalence. A simple check shows that $D_1(A)$ divides all elements of A'' and by equivalence those of A'''. In particular all elements of A_1 are divisible by $D_1(A)$.

We proceed by induction to obtain the ω'-equivalence of A to a diagonal matrix $\Delta = \text{diag}(\delta_1, \ldots, \delta_n)$ with $\delta_i | \delta_{i+1}$ and $\delta_1 = D_1(A)$. Since A is ω'-equivalent to Δ it is also ω-equivalent and ω being arbitrary A and Δ are quasiequivalent. Therefore $D_i(A) = D_i(\Delta) = \delta_1 \ldots \delta_i$. Now the invariant factors of Δ are $\delta_1, \ldots, \delta_n$ and therefore also $E_i(A) = \delta_i$. This proves the theorem.

Corollary 15-7 Two matrices A and B in $(H^\infty)^{n \times n}$ are quasiequivalent if and only if they have the same invariant factors.

PROOF Assume A and B have the same invariant factors $\delta_1, \ldots, \delta_n$. Both A and B are quasiequivalent to $\text{diag}(\delta_1, \ldots, \delta_n)$ and since quasiequivalence is transitive A and B are quasiequivalent. The converse has already been proved in Lemma 15-5.

Now we proceed to the study of the relation between quasiequivalence and quasisimilarity and determine a quasisimilarity invariant for a class of restricted shifts.

To begin with we note that in one direction we can get the analog of the finite dimensional situation.

Theorem 15-8 Let Q_1 and Q_2 be $n \times n$ inner functions. If Q_1 and Q_2 are equivalent then the operators $S(Q_1)$ and $S(Q_2)$ defined by (13-4) in the left invariant subspaces $H(Q_1)$ and $H(Q_2)$ defined by (13-3) are similar.

PROOF By equivalence there exist units Ξ and Ξ_1 in $(H^\infty)^{n \times n}$ such that

$$\Xi Q_1 = Q_2 \Xi_1$$ (15-26)

Define an operator $X: H(Q_1) \to H(Q_2)$ by

$$Xf = P_{H(Q_2)} \Xi f \quad f \in H(Q_1)$$ (15-27)

then X satisfies $XS(Q_1) = S(Q_2) X$ and is boundedly invertible by Theorem 14-11.

Given inner functions $q_1, \ldots, q_n$ in H^∞ which satisfy $q_{i+1} | q_i$ we define a *Jordan operator* to be an operator of the form

$$S(q_1) \oplus \cdots \oplus S(q_n)$$ (15-28)

acting in the space

$$H(q_1) \oplus \cdots \oplus H(q_n) \tag{15-29}$$

Our aim will be to establish that a Jordan operator is a quasisimilarity invariant for the set of restricted shifts corresponding to matrix inner functions.

Theorem 15-9 Let Q_1 and Q_2 be inner functions in $\mathbb{C}^n$. If Q_1 and Q_2 are quasiequivalent then $S(Q_1)$ and $S(Q_2)$ are quasisimilar.

PROOF Since Q_1 and Q_2 are quasiequivalent then for each inner function ω there exist Ξ and Ξ_1 in $(H^\infty)^{n \times n}$ such that (15-26) holds and both $\det \Xi$ and $\det \Xi_1$ are coprime with ω. Choose ω to be equal to $\det Q_1 \cdot \det Q_2$ then necessarily we obtain the coprimeness relations

$$(\Xi, Q_2)_L = I \quad \text{and} \quad (\Xi_1, Q_1)_R = I$$

Define $X: H(Q_1) \to H(Q_2)$ by (15-27) then X satisfies $XS(Q_1) = S(Q_2) X$. Moreover by Theorem 14-11 both X and X^* are injective. Since $\{\operatorname{Range} X\}^\perp = \operatorname{Ker} X^*$ it follows that X has dense range and so X is quasi-invertible. By symmetry there exists also a quasi-invertible operator $Y: H(Q_2) \to H(Q_1)$ for which $YS(Q_2) = S(Q_2) Y$ and so quasisimilarity of $S(Q_1)$ and $S(Q_2)$ is proved.

Theorem 15-10 Let Q be an inner function in $\mathbb{C}^n$ and let $q_1, \ldots, q_n$ be its invariant factors ordered so that $q_{i+1}|q_i$. Then $S(Q)$ is quasisimilar to the Jordan operator $S(q_1) \oplus \cdots \oplus S(q_n)$.

PROOF As Q is quasiequivalent to $\operatorname{diag}(q_1, \ldots, q_n)$ the theorem follows from the previous one.

The question of whether the converse to Theorem 15-9 holds is connected to the question of whether two quasisimilar Jordan operators are necessarily equal. Both questions can be answered in the affirmative but we will have to establish some preliminary results before.

Theorem 15-11 Let q be an inner function in H^∞ and Q_1 an inner function in $\mathbb{C}^m$. Let $Q = qI_n$ and let $X: H(Q) \to H(Q_1)$ be a quasi-invertible operator satisfying

$$XS(Q) = S(Q_1) X \tag{15-30}$$

then necessarily $n \le m$.

PROOF Since $S(Q_1)$ is quasisimilar to its Jordan model we may assume without loss of generality that $Q_1 = \operatorname{diag}(q_1, \ldots, q_m)$ where q_i are the invariant factors of Q_1 and $q_{i+1}|q_i$.
  From (15-30) it follows that the equality

$$X\varphi\big(S(Q)\big) = \varphi\big(S(Q_1)\big) X \tag{15-31}$$

holds for all $\varphi \in H^\infty$. In particular the choice $\varphi = q$ yields $q(S(Q)) = 0$ and as the range of X is dense in $H(Q_1)$ this in turn implies $q(S(Q_1)) = q(S(q_1)) \oplus \cdots \oplus q(S(q_m)) = 0$ or $q_j | q$ for all j. Therefore there exist inner functions r_j for which

$$q = r_j q_j, \qquad j = 1, \dots, m \tag{15-32}$$

By Theorem 14-8 there exist matrices $\Xi = (\xi_{ij})$ and $\Theta = (\eta_{ij})$ in $(H^\infty)^{m \times n}$ such that

$$\Xi Q = Q_1 \Theta \tag{15-33}$$

$$X f = P_{H(Q_1)} \Xi f \quad \text{for} \quad f \in H(Q) \tag{15-34}$$

and the coprimeness relations $(\Xi, Q_1)_L = I$ and $(Q, \Theta)_R = I$ are satisfied. From (15-32) and (15-33) it follows that $\eta_{ij} = r_i \xi_{ij}$ or that

$$\Theta = R\Xi \tag{15-35}$$

where $R = \mathrm{diag}(r_1, \dots, r_m)$. Since $(Q, \Theta)_R = I$ it is impossible for all η_{ij} to be divisible by q. Let r be the order of a maximal minor of Θ whose determinant is not divisible by q. Without loss of generality we may assume it to be the minor (η_{ij}), $i, j = 1, \dots, r$. Obviously $1 \leq r \leq m$.

Assume now $n > m$ and we will show this leads to a contradiction. Define functions $u_1, \dots, u_{r+1}$ in H^∞ through the following determinant expansion

$$\begin{vmatrix} \eta_{11} & \cdots & \eta_{1r} & \eta_{1,r+1} \\ \vdots & & \vdots & \vdots \\ \eta_{r1} & \cdots & \eta_{rr} & \eta_{r,r+1} \\ x_1 & \cdots & x_r & x_{r+1} \end{vmatrix} = \sum_{i=1}^{r+1} x_i u_i \tag{15-36}$$

We note that

$$\sum_{j=1}^{r+1} \eta_{ij} u_j = \begin{cases} 0 & \text{for} \quad i = 1, \dots, r \\ \text{divisible by } q & \text{for} \quad i > r \end{cases} \tag{15-37}$$

as for $i \leq r$ this is the expansion of a determinant with two equal rows whereas for $i > r$ this is the determinant expansion of a minor of Θ of order $r + 1$ hence divisible by q. Let now $u \in H^2_{\mathbb{C}^n}$ have $u_1, \dots, u_{r+1}$ as its first $r + 1$ components, all others equal to zero. Since

$$u_{r+1} = \det \begin{pmatrix} \eta_{11} & \cdots & \eta_{1r} \\ \vdots & & \vdots \\ \eta_{r1} & \cdots & \eta_{rr} \end{pmatrix}$$

is not divisible by q the function u is not in $QH^2_{\mathbb{C}^n} = QH^2_{\mathbb{C}^n}$ and so $v = P_{H(Q)} u$ is nonzero. We will show now that $Xv = 0$. Indeed (15-37) shows that $\Theta u \in qH^2_{\mathbb{C}^m}$ and as $qI = RQ_1$ we obtain $\Theta u \in RQ_1 H^2_{\mathbb{C}^m}$. This coupled with (15-35) implies $\Xi u \in Q_1 H^2_{\mathbb{C}^m}$. Now $Xv = P_{H(Q_1)} \Xi v = P_{H(Q_1)} \Xi P_{H(Q)} u =$

$P_{H(Q_1)}\Xi u = 0$. This contradicts the assumption that X is quasi-invertible and in particular is injective. So $n \le m$ and the theorem is proved.

Lemma 15-12 Let $q = pr$ be a factorization of the inner function q into inner factors. Then

$$H(q) = H(r) \oplus rH(p) \tag{15-38}$$

PROOF We have $H^2 = H(p) \oplus pH^2$. Since multiplication by an inner function is an isometry in H^2 it preserves orthogonality. Hence

$$rH^2 = rH(p) \oplus rpH^2 = rH(p) \oplus qH^2$$

and as a consequence

$$H^2 = H(r) \oplus rH^2 = H(r) \oplus rH(p) \oplus qH^2$$

which is equivalent to (15-38).

Lemma 15-13 Let q and q' be inner functions in H^∞ and let $r = q \wedge q'$, $q = r\hat{q}$. Then

$$\overline{\text{Range}\, q'\,(S(q))} = rH(\hat{q}) \tag{15-39}$$

and there exists a quasi-invertible operator $K = H(\hat{q}) \to rH(\hat{q})$ such that

$$KS(q) = (S(q)|rH(\hat{q}))\, K \tag{15-40}$$

PROOF Define $X: H(\hat{q}) \to H(q)$ by

$$Xg = P_{H(q)}q'g, \quad \text{for} \quad g \in H(\hat{q}) \tag{15-41}$$

Clearly

$$XS(\hat{q}) = S(q)\, X \tag{15-42}$$

Let $f \in H(q)$ and $f = g + \hat{q}h$ be its decomposition with respect to the direct sum representation $H(q) = H(\hat{q}) \oplus \hat{q}H(r)$. Now

$$q'(S(q))\, f = P_{H(q)}q'f = P_{H(q)}(q'g + q'\hat{q}h) = P_{H(q)}q'g$$
$$= P_{H(q)}q'P_{H(\hat{q})}\, f = XP_{H(\hat{q})}\, f$$

Here we used the fact that q divides $q'\hat{q}$.

Define now $\hat{q}' = q'/r$ then $\hat{q}' \wedge q = 1$ and from (15-41) we have

$$Xg = P_{H(q)}q'g = P_{H(q)}r\hat{q}'g = rP_{H(\hat{q})}\hat{q}'g$$

By Theorem 14-5 X is injective and the range of $\hat{q}'(S(\hat{q}))$ is dense in $H(\hat{q})$ and so the range of X is dense in $rH(\hat{q})$. Define $K: H(\hat{q}) \to rH(\hat{q})$ by $Kg = Xg$ then K is quasi-invertible and (15-40) holds.

Corollary 15-14   Let $q = pr$ be a factorization of the inner function q into inner factors. Then the range of $p(S(q))$ is $rH(p)$ and the operators $S(p)$ and $S(q)|rH(p)$ are unitarily equivalent.

PROOF   The operator $X: H(p) \to H(q)$ defined by $Xg = rg$ is isometric with range equal to the invariant subspace $rH(p)$ of $H(q)$. If we restrict the range of X to $rH(p)$ it is unitary and provides the required unitary equivalence of $S(p)$ and $S(q)|(rH(p))$.

Lemma 15-15   If q_1, q_2, and q' are inner functions and $q_2|q_1$ then

$$\frac{q_2}{q_2 \wedge q'} \Bigg| \frac{q_1}{q_1 \wedge q'} \tag{15-43}$$

PROOF   Since $q_2|q_1$ there exists an inner function s such that $q_1 = q_2 s$. It follows that

$$q_1 \wedge q' = q_2 s \wedge q' = (q_2 \wedge q') \left(s \wedge \left(\frac{q'}{q_2 \wedge q'} \right) \right)$$

and as a consequence

$$\frac{q_1}{q_1 \wedge q'} = \left(\frac{q_2}{q_2 \wedge q'} \right) \frac{s}{s \wedge \left(\dfrac{q'}{q_2 \wedge q'} \right)} \tag{15-44}$$

But (15-44) clearly implies (15-43).

The next result shows that two quasisimilar Jordan operators are necessarily equal.

Theorem 15-16   Let $q_1, \ldots, q_n$ and $q'_1, \ldots, q'_m$ be inner functions in H^∞ such that $q_{i+1}|q_i$ and $q'_{i+1}|q'_i$. Let $Q = \text{diag}(q_1, \ldots, q_n)$ and $Q' = \text{diag}(q'_1, \ldots, q'_m)$. If $X: H(Q) \to H(Q')$ in a quasi-invertible operator for which

$$XS(Q) = S(Q') X \tag{15-45}$$

then $n = m$ and $q_i = q'_i$.

PROOF   Note that $H(Q) = H(q_1) \oplus \cdots \oplus H(q_n)$ and $H(Q') = H(q'_1) \oplus \cdots \oplus H(q'_m)$. Also $S(Q) = S(q_1) \oplus \cdots \oplus S(q_n)$ and $S(Q') = S(q'_1) \oplus \cdots \oplus S(q'_m)$. Since $q_n|q_j$ for $j = 1, \ldots, n$ we have $q_j = r_j q_n$ for some inner functions r_j, $j = 1, \ldots, n$, with $r_n = 1$. Consider the subspace $H_0 = r_1 H(q_n) \oplus \cdots \oplus H(q_n)$ of $H(Q)$ which is an invariant subspace of $S(Q)$. By Corollary 15-14 $S(Q)|H_0$ is unitarily equivalent to $S(q_n) \oplus \cdots \oplus S(q_n)$. X restricted to H_0 is a quasi-invertible operator from H_0 into $H'_0 = \overline{X H_0}$ which is a $S(Q')$-invariant subspace of $H(Q')$. Now $S(Q')|H'_0$ is unitarily equivalent to $S(Q'')$ for some $Q \times m$ inner function Q''. Applying Theorem 15-11 we have $n \leq m$.

Fix now $1 \leq k \leq m$. Since $X\varphi(S(Q)) = \varphi(S(Q'))X$ for all $\varphi \in H^\infty$ the choice $\varphi = q_k'$ yields

$$X[q_k'(S(q_1)) \oplus \cdots \oplus q_k'(S(q_n))]$$

$$= [q_k'(S(q_1')) \oplus \cdots \oplus q_k'(S(q_{k-1})) \oplus \{0\} \oplus \cdots \oplus \{0\}]X \qquad (15\text{-}46)$$

By Lemma 15-13 and Corollary 15-14

$$q_k'(S(q_j))H(q_j) = (q_j \wedge q_k')H\left(\frac{q_j}{q_j \wedge q_k'}\right)$$

and

$$S(q_j)\Bigg|(q_j \wedge q_k')H\left(\frac{q_j}{q_j \wedge q_k'}\right)$$

is unitarily equivalent to

$$S\left(\frac{q_j}{q_j \wedge q_k'}\right)$$

Similarly $q_k'(S(q_i'))H(q_i') = q_k'H(q_i'/q_k')$ and $S(q_i')|q_k'H(q_i'/q_k')$ is unitarily equivalent to $S(q_i'/q_k')$. Thus there exists a quasi-invertible operator X_1 for which

$$X_1\left[S\left(\frac{q_1}{q_1 \wedge q_k'}\right) \oplus \cdots \oplus S\left(\frac{q_n}{q_n \wedge q_k'}\right)\right]$$

$$= [S(q_1'/q_k') \oplus \cdots \oplus S(q_{k-1}'/q_k')]X_1$$

and Lemma 15-15 guarantees the division relations

$$\frac{q_{i+1}}{q_{i+1} \wedge q_k'}\Bigg|\frac{q_i}{q_i \wedge q_k'} \qquad \text{and} \qquad (q_{i+1}'/q_k')|(q_i'/q_k')$$

Let j be the maximal index for which $q_j/(q_j \wedge q_k')$ is nontrivial. Since

$$S\left(\frac{q_1}{q_1 \wedge q_k'}\right) \oplus \cdots \oplus S\left(\frac{q_j}{q_j \wedge q_k'}\right)$$

restricted to an invariant subspace is unitarily equivalent to the direct sum of j copies of

$$S\left(\frac{q_j}{q_j \wedge q_k'}\right)$$

then another application of Theorem 15-11 shows that $j \leq k - 1$. This implies $q_k = q_k \wedge q_k'$ or $q_k|q_k'$.

To complete the proof we take the adjoint of equality (15-45). Thus we have

$$X^*S(Q')^* = S(Q)^*X^* \qquad (15\text{-}47)$$

and X^* is also quasi-invertible. Now, by Theorem 13-2 $S(Q)^*$ is unitarily equivalent to $S(\tilde{Q})$ and $S(\tilde{Q}) = S(\tilde{q}_1) \oplus \cdots \oplus S(\tilde{q}_n)$. Similarly for $S(Q_1)^*$. Apply the first part of the theorem to obtain $m \leq n$ and $\tilde{q}_i | \tilde{q}_i$ which clearly implies $q_i' | q_i$ and this completes the proof.

By Theorem 15-10 given any matrix inner function Q then $S(Q)$ is quasi-similar to a Jordan operator. By Theorem 15-16 this Jordan operator is unique. We call it the *Jordan model* of the operator $S(Q)$.

Theorem 15-17 Let Q_1 and Q_2 be two matrix inner functions. Then $S(Q_1)$ and $S(Q_2)$ are quasisimilar if and only if they have the same Jordan model.

PROOF Assume $S(Q_1)$ and $S(Q_2)$ have the same Jordan model $S(\Delta)$. Since both $S(Q_i)$ are quasisimilar to $S(\Delta)$ the quasisimilarity of $S(Q_1)$ and $S(Q_2)$ follows by transitivity. Conversely let $S(\Delta_1)$ and $S(\Delta_2)$ be the Jordan models of $S(Q_1)$ and $S(Q_2)$, respectively. Again by transitivity $S(\Delta_1)$ and $S(\Delta_2)$ are quasisimilar and by Theorem 15-16 Δ_1 and Δ_2 coincide.

Theorem 15-18 Let Q_1 and Q_2 be two matrix inner functions. Then $S(Q_1)$ and $S(Q_2)$ are quasisimilar if and only if Q_1 and Q_2 are quasiequivalent.

PROOF The if part has been proved in Theorem 15-9. Let Δ_1 and Δ_2 be the diagonal inner functions with the invariant factors of Q_1 and Q_2 as their respective entries, then Q_i is quasiequivalent to Δ_i. As $S(\Delta_1)$ and $S(\Delta_2)$ are quasisimilar by the transitivity of the quasisimilarity relation we must have, by Theorem 15-16, that $\Delta_1 = \Delta_2$. By transitivity of quasiequivalence, Q_1 and Q_2 are quasiequivalent.

In a finite dimensional vector space a linear transformation is cyclic if and only if its characteristic and minimal polynomials coincide. An analogous situation holds for Jordan operators.

Let us denote by C_0 the class of all completely nonunitary contractions such that $\varphi(T) = 0$ for some nonzero function φ in H^∞.

Lemma 15-19 Let $T \in C_0$ then $J = \{\psi \, | \, \psi \in H^\infty, \psi(T) = 0\}$ is a w^*-closed ideal in H^∞.

PROOF That J is an ideal is clear from the properties and as $T \in C_0$ it is a nontrivial ideal of the functional calculus. To see that J is w^*-closed let φ_α be a net converging to φ in the w^*-topology of H^∞. Recall that the minimal unitary dilation U of T has a spectral measure which is absolutely continuous with respect to Lebesgue measure. Let x and y be arbitrary vectors in H, then we have

$$(\varphi(T)x, y) = (P\varphi(U)x, y) = (\varphi(U)x, y) = \int \varphi(e^{it})(E(dt)x, y)$$

Since by the Radon–Nykodim theorem $(E(dt)x, y) = k(e^{it}) \, dt$ for some $k \in L_1$

we have

$$\int \varphi(e^{it})\,(E(dt)\,x,\,y) = \lim_{\alpha} \int \varphi_\alpha(e^{it})\,(E(dt)\,x,\,y) = 0$$

This implies that $(\varphi(T)\,x,\,y) = 0$ for all x and y hence $\varphi(T) = 0$.

We use now the representation theorem for w^*-closed ideals in H^∞ to get $J = m_T H^\infty$ for some inner function m_T which is uniquely determined up to a constant factor of absolute value one. We call m_T the *minimal function* of T. If $T \in C_0$ so does T^* and we have $m_{T^*} = \tilde{m}_T$.

If q is an inner function in H^∞ then clearly for $S(q)$ the minimal function coincides with q. Now let $S(q_1) \oplus \cdots \oplus S(q_n)$ be a Jordan operator. Since $q_{i+1}\,|\,q_i$ for $i = 1, \ldots, n - 1$ we immediately infer that the minimal function of $S(q_1) \oplus \cdots \oplus S(q_n)$ is q_1.

The minimal function is a quasisimilarity invariant. Actually we prove a slightly stronger result.

Theorem 15-20 Let T, T_1 be two completely nonunitary contractions and X a quasi-invertible operator that intertwines them, i.e.

$$XT = T_1 X \tag{15-48}$$

Then if one of the operators is C_0 so is the other and their minimal functions coincide.

PROOF   From (15-48) it follows that

$$X\varphi(T) = \varphi(T_1)\,X \qquad \text{for all} \qquad \varphi \in H^\infty \tag{15-49}$$

Assume $T \in C_0$ then $m_T(T_1)\,X = Xm_T(T) = 0$ and as the range of X is dense we must have $m_T(T_1) = 0$, and so $T_1 \in C_0$ and $m_{T_1}\,|\,m_T$. Conversely if $T_1 \in C_0$ then $0 = m_{T_1}(T_1)\,X = Xm_{T_1}(T)$. Since X is injective $m_{T_1}(T) = 0$ which shows that $T \in C_0$ and $m_T\,|\,m_{T_1}$. Hence the two minimal functions coincide.

We have now two notions of minimal inner functions associated with a matrix inner function Q. One is the minimal inner function of Q while the other is the minimal function of $S(Q)$. Not surprisingly the two notions of minimality coincide.

Theorem 15-21 Let Q be an inner function in $\mathbb{C}^n$. Then m, the minimal inner function of Q, and m_1, the minimal function of $S(Q)$, coincide.

PROOF   Since $m_1(S(Q)) = 0$ it follows that $m_1 H(Q) \subset QH^2_{\mathbb{C}^n}$ and since clearly $m_1 Q H^2_{\mathbb{C}^n} = Q m_1 H^2_{\mathbb{C}^n} \subset Q H^2_{\mathbb{C}^n}$ we have $m_1 H^2_{\mathbb{C}^n} \subset Q H^2_{\mathbb{C}^n}$ or $m\,|\,m_1$. Conversely if $m H^2_{\mathbb{C}^n} \subset Q H^2_{\mathbb{C}^n}$ we have $m(S(Q)) = 0$ and so $m_1\,|\,m$. Thus m and m_1 coincide.

An operator T in H is *cyclic* and b a *cyclic vector* if the set of linear combinations of the vectors $T^j b, j = 0, 1, \ldots$ is dense in m. Cyclicity too is a quasisimilarity invariant.

Theorem 15-22 Let T and T_1 be contractions and X a quasi-invertible operator that intertwines T and T_1. Then if T is cyclic so is T_1. Consequently if T and T_1 are quasisimilar then T is cyclic if and only if T_1 is cyclic.

PROOF Let T be cyclic and b a cyclic vector. From (15-48) it follows that

$$XT^n b = T_1^n Xb = T_1^n b_1 \tag{15-50}$$

where $b_1 = Xb$. As the range of X is dense this shows that T_1 is cyclic and b_1 a cyclic vector.

We describe an important class of cyclic operators.

Lemma 15-23 Let $q \in H^\infty$ be inner. Then $S(q)$ is cyclic and a function $f \in H(q)$ is a cyclic vector for $S(q)$ if and only if $f \wedge q = 1$.

PROOF A function f in $H(q)$ is a cyclic vector for $S(q)$ if and only if the smallest right invariant subspace that contains f and qH^2 is H^2. This, by Beurling's theorem, is the case if and only if $f \wedge q = 1$. To see that $S(q)$ is cyclic it suffices to exhibit an outer function in $H(q)$. Let $k = P_{H(q)} 1 = 1 - \overline{q(0)} q$ then k is an outer function. In fact $|q(0)| < 1$ as a consequence of the maximum modulus theorem and hence k is actually an invertible element of H^∞.

As the result of the preceding development we can obtain a characterization of the restricted shifts of finite multiplicity which are cyclic.

Theorem 15-24 Let Q be an inner function in $\mathbb{C}^n$. Then the following statements are equivalent.
(a) $S(Q)$ is cyclic.
(b) $S(Q)^*$ is cyclic.
(c) $S(Q)$ is quasisimilar to $S(q)$ where $q = \det Q$.
(d) $m = q$ where m is the minimal function of $S(Q)$.

PROOF Since cyclicity is a quasisimilarity invariant we can replace $S(Q)$ by its Jordan model $S(q_1) \oplus \cdots \oplus S(q_n)$. We assume $q_{i+1} | q_i$ for $i = 1, \ldots, n - 1$. Since $S(q)$ is cyclic clearly $(c) \to (a)$. Also if $S(Q)$ is quasisimilar to $S(q)$ then $S(Q)^*$ is quasisimilar to $S(q)^*$. However, $S(q)^*$ is unitarily equivalent, by Theorem 13-2, to $S(\tilde{q})$ which is again cyclic, so $(c) \to (b)$. Minimal functions are a quasisimilarity invariant so since q is clearly the minimal function of $S(q)$ we have the implication $(c) \to (d)$. To see that $(d) \to (c)$ assume (d) holds, which is equivalent to $m = q = q_1 \cdots q_n$ or $q_2 = \cdots = q_n = 1$.

Finally assume (a) holds, that is, $S(Q)$ or $S(q_1) \oplus \cdots \oplus S(q_n)$ is cyclic and let $f = f_1 \oplus \cdots \oplus f_n$ be a cyclic vector for it. First we show that we may replace f by another cyclic vector $\xi_1 + \cdots + \xi_n$ with $\xi_i \in H^\infty$, $i = 1, \ldots, n$. To this end let S be the right shift in H_n^2 and let M be the invariant subspace spanned by $S^n f$. By Theorem 12-22, $M = \Omega H_{\mathbb{C}^2}^2$ for some rigid function Ω. Let $K \subset \mathbb{C}^n$ be the initial space of Ω. Since $f \in M$ $f = \Omega g$ for some $g \in H_{\mathbb{C}^n}^2$ and we may as well assume $g \in H_K^2$. Since Ω is isometric on H_K^2 clearly g spans H_K^2 which implies that dim $K = 1$. By a suitable choice of basis we may assume K coincides with the subspace spanned by e_1, the first element of the standard orthonormal basis of $\mathbb{C}^n$. Let $\xi_1, \ldots, \xi_n$ be the elements of the first column of Ω then $f_1 \oplus \cdots \oplus f_n = (\xi_1 \oplus \cdots \oplus \xi_n) g$ and $\sum_{j=1}^n |\xi_j(e^{it})|^2 = 1$. Moreover $\xi_1 \oplus \cdots \oplus \xi_n$ is also a cyclic vector for $S(q_1) \oplus \cdots \oplus S(q_n)$. If $k = P_{H(q_1)} 1 = 1 - \overline{q_1(0)} q_1$ then we define an operator $X : H(q_1) \to H(q_1) \oplus \cdots \oplus H(q_n)$ by

$$XS(q_1)^m k = S(q_1)^m \xi_1 \oplus \cdots \oplus S(q_n)^m \xi_n \tag{15-51}$$

Clearly X extends by linearity to a contraction. We have in fact

$$Xh = P_{H(q_1) \oplus \cdots \oplus H(q_n)} \begin{pmatrix} \xi_1 \\ \vdots \\ \xi_n \end{pmatrix} h \qquad \text{for} \qquad h \in H(q_1) \tag{15-52}$$

Since $\xi_1 \oplus \cdots \oplus \xi_n$ is a cyclic vector X clearly has dense range. If X is not injective then also the operator $Y : H(q_1) \to H(q_1)$ given by

$$Yh = P_{H(q_1)} \xi_1 h \tag{15-53}$$

is not injective. This is equivalent to ξ_1 and q_1 having a nontrivial inner factor. But this contradicts the fact that ξ_1 is cyclic for $S(q_1)$. Therefore the operator X defined by (15-52) is a quasi-invertible operator and it clearly satisfies

$$XS(q_1) = (S(q_1) \oplus \cdots \oplus S(q_n)) X \tag{15-54}$$

We apply Theorem 15-16 to infer that $q_2 = \cdots = q_n = 1$. So $S(q)$ is the Jordan model of $S(q)$ and so they are quasisimilar and the proof complete.

The previous theorem suggests the question as to when $S(Q)$ is actually similar to $S(q)$. Such questions are always associated with the Carleson corona theorem and this one is no exception.

Theorem 15-25 Let Q be an inner function in $\mathbb{C}^n$, and let $\Omega = (\omega_{ij})$ be the classical adjoint of Q, that is, $Q\Omega = \Omega Q = qI$ where $q = \det Q$. Then $S(Q)$ is similar to $S(q)$ if and only if there exist $h, u_i, v_i \in H^\infty$ such that

$$\sum_{i,j=1}^n Q_{ij} u_i v_j + qh = 1 \tag{15-55}$$

Proof Assume $S(Q)$ and $S(q)$ are similar. Then there exist boundedly invertible maps $X : H(Q) \to H(q)$ and $Y : H(q) \to H(Q)$ satisfying $XY = I_{H(q)}$,

$YX = I_{H(Q)}$

$$XS(Q) = S(q) X \qquad (15\text{-}56)$$

and

$$S(Q) Y = YS(q) \qquad (15\text{-}57)$$

By the representation theory for intertwining operators there exist $1 \times n$ and $n \times 1$ bounded matrix functions Φ and Ψ, respectively, such that

$$\Phi Q = q\Phi_1 \qquad \text{and} \qquad \Psi q = Q\Psi_1 \qquad (15\text{-}58)$$

$$Xf = P_{H(q)}\Phi f \qquad \text{for} \qquad f \in H(Q) \qquad (15\text{-}59)$$

$$Yg = P_{H(Q)}\Psi g \qquad \text{for} \qquad g \in H(q) \qquad (15\text{-}60)$$

and the coprimeness relations

$$[\Phi, qI]_L = I, \qquad [\Phi_1, Q]_R = I \qquad (15\text{-}61)$$

$$[\Psi, Q]_L = I, \qquad [qI, \Psi_1]_R = I \qquad (15\text{-}62)$$

Relation (15-58) implies that for $g \in H(q)$

$$g = P_{H(q)}\Phi P_{H(Q)}\psi g = P_{H(q)}\Phi\Psi g$$

and hence that

$$1 - \Phi\Psi = qh \qquad (15\text{-}63)$$

for some $h \in H^\infty$. Finally, from $\Phi Q = q\Phi_1$ and $qI = \Omega Q$, we get $\Phi = \Phi_1\Omega$. Thus

$$1 - \Phi_1\Omega\Psi = qh \qquad (15\text{-}64)$$

Let

$$\Phi_1 = (u_1, \ldots, u_n), \qquad \Psi = \begin{pmatrix} v_1 \\ \vdots \\ v_n \end{pmatrix} \qquad (15\text{-}65)$$

then (15-55) follows.

Conversely assume there exist $h, u_i, v_i \in H^\infty$ for which (15-55) holds. Define Φ and Ψ by (15-65) and let us define Φ and Ψ_1 by

$$\Phi = \Phi_1\Omega, \qquad \Psi_1 = \Omega\Psi \qquad (15\text{-}66)$$

From (15-66) we get, multiplying by Q on the right and left, respectively, that (15-58) holds. Therefore, the maps X and Y defined by (15-59) and (15-60), respectively, are bounded maps that satisfy (15-56) and (15-57), respectively. It remains to show that X and Y are invertible. It is clear that $XY = I_{H(q)}$ which follows from (15-55), and so X is surjective while Y is injective. To complete the proof we have to show that $YX = I_{H(Q)}$ or equivalently, since

$YXf = P_{H(Q)}\Psi\Phi f$ for $f \in H(Q)$ and $\Psi\Phi Q = \Psi q\Phi_1 = q\Psi\Phi_1 = Q\Omega\Psi\Phi_1$, that

$$(\Psi\Phi, Q)_L = I \qquad (15\text{-}67)$$

From (15-64) by multiplication by Ω we obtain

$$(\Phi_1\Omega\Psi)\,\Omega = \Omega + q\Omega h \qquad (15\text{-}68)$$

We will show now that q divides all elements of $\Gamma = (\Phi_1\Omega\Psi)\,\Omega - \Omega\Psi)(\Phi_1\Omega)$. Since Φ_1 and Ψ are given by (15-65) the k, l element of Γ is given by

$$\left(\sum_{j=1}^{n} \omega_{ij}u_iv_j \right)\omega_{kl} - \left(\sum_{j=1}^{n} \omega_{kj}v_j \right)\left(\sum_{j=1}^{n} u_i\omega_{il} \right) = \sum_{i,j=1}^{n} u_iv_j(\omega_{ij}\omega_{kl} - \omega_{kj}\omega_{il})$$

Now

$$(\omega_{ij}\omega_{kl} - \omega_{kj}\omega_{il}) = \begin{vmatrix} \omega_{ij} & \omega_{il} \\ \omega_{kj} & \omega_{kl} \end{vmatrix}$$

is the general element of the compound matrix $\Omega^{(2)}$, and we have to show its divisibility by q.

Observe that (15-55) implies that $S(Q)$ is cyclic for we have $D_{n-1}(Q) \wedge q = 1$, and as $D_{n-1}(Q) | D_n(Q) = \det\Omega = q$, necessarily $D_{n-1}(Q) = 1$. Thus the Jordan model for $S(Q)$ is $S(q)$ which is cyclic. Since $S(Q)$ and $S(q)$ are quasisimilar then also $S(Q)$ and $S(\hat{Q})$ are quasisimilar where $\hat{Q} = \text{diag}(q, 1, \ldots, 1)$. There exist therefore for each inner ω, Ξ and Θ in $\mathcal{N}_\omega(n)$ such that $\Xi\hat{Q} = Q\Theta$, $\Xi^a\Xi = \Xi\Xi^a = \xi I$, and $\Theta^a\Theta = \Theta\Theta^a = \eta I$, where Ξ^a and Θ^a are the classical adjoints of Ξ and Θ, respectively. We choose $\omega = q$ thus $\xi \wedge q = \eta \wedge q = 1$. From $\Omega Q = qI$ we have $q\eta I = \Theta^a\Omega Q\Theta = \Theta^a\Omega\Xi\hat{Q} = \hat{\Omega}\hat{Q}$ where $\hat{\Omega} = \Theta^a\Omega\Xi$. Now $\hat{\Omega}$ and Ω are clearly quasiequivalent as $\Theta\hat{\Omega} = \Omega(\eta\Xi)$ and $\hat{\Omega}\Xi^a = (\xi\Theta^a)\,\Omega$ and $\eta\Xi, \xi\Theta^a \in \mathcal{N}_\omega(n)$ if $\Xi, \Theta \in \mathcal{N}_\omega(n)$. Therefore $\hat{\Omega}$ and Ω have the same determinant divisors. Now from the equality $\hat{\Omega}\hat{Q} = q\eta I$ it follows that q divides all elements of the last $(n - 1)$ columns of $\hat{\Omega}$ This means that $q | D_2(\hat{\Omega}) = D_2(\Omega)$ and hence all elements of $\Omega^{(2)}$, and therefore of Γ, are divisible by q. Let $\Gamma = qD$ for some $n \times n$ matrix D over H^∞. Equality (15-68) can be rewritten as

$$\Omega = \Omega\Psi\Phi_1\Omega + q(D - \Omega h)$$

or, using the fact that $\Omega Q = qI$, as

$$I = \Psi\Phi_1\Omega + Q(D - \Omega h) = \Psi\Phi + Q(D - \Omega h)$$

By Theorem 14-10 this is equivalent to (15-67) and this completes the proof of the theorem.

The terminology concerning Jordan operators is not in total agreement with the finite dimensional situation. In fact a Jordan operator is the proper generalization of the first canonical form (I 4-21) which is derived from Theorem I 4-14. A natural question poses itself, that of finding conditions under which a further reduction is possible. The extreme simplification would be to find conditions under which $S(Q)$ is completely diagonalizable in the sense that there exists a basis consisting of eigenfunctions of $S(Q)$.

We restrict ourselves to the study of the scalar case only. Let us review the finite dimensional situation. Consider a polynomial $q \in F[\lambda]$ which has the zeros $\lambda_1, \ldots, \lambda_n$ all of them assumed to be simple. The quotient ring $F[\lambda]/(q)$ is isomorphic to the set of polynomials of degree $\leq n - 1$. Let us define n polynomials of degree $n - 1$ by

$$\pi_i(\lambda) = \prod_{j \neq i} \frac{\lambda - \lambda_j}{\lambda_i - \lambda_j} \tag{15-69}$$

Then we clearly have

$$\pi_i(\lambda_j) = \delta_{ij} \qquad \text{for} \quad 1 \leq i, j \leq n \tag{15-70}$$

The polynomials $\pi_1, \ldots, \pi_n$ are clearly linearly independent, as a consequence of (15-70), hence they form a basis for $F[\lambda]/(q)$. Now given any set of numbers $a_1, \ldots, a_n$ in F the polynomial $a(\lambda) = \sum_{i=1}^{n} a_i \pi_i(\lambda)$ is an interpolating polynomial in the sense that $a(\lambda_j) = a_j, j = 1, \ldots, n$. Finally we note that if $S(q)$ is defined by (I 4-8) then the polynomials π_i are eigenvectors of $S(q)$ corresponding to the eigenvalues λ_i. This follows from the fact that $(\lambda - \lambda_i) \pi_i(\lambda)$ is divisible by q. The moral of this simple example is that certain problems of interpolation are connected with the problem of existence of bases consisting of eigenfunctions.

We make a few definitions first. A sequence of points $\{\lambda_n\}$ in the open unit disc D is called *uniformly separated*, or equivalently a *Carleson sequence*, if there exists a $\delta > 0$ such that

$$\prod_{j \neq k} \left| \frac{\lambda_j - \lambda_k}{1 - \bar{\lambda}_j \lambda_k} \right| \geq \delta \qquad \text{for all} \quad k \geq 1 \tag{15-71}$$

It is quite easy to see that if $\{\lambda_i\}$ is a uniformly separated sequence then there exists a Blaschke product with the set $\{\lambda_i\}$ as its set of zeros. Without loss of generality we may assume $\lambda_0 = 0$ then

$$\prod_{j > 0} \left| \frac{\lambda_j - \lambda_0}{1 - \lambda_0 \bar{\lambda}_j} \right| = \prod_{j > 0} |\lambda_j| \geq \delta$$

The convergence of the last infinite product is equivalent to the convergence of the series $\sum_{j=1}^{\infty} (1 - |\lambda_j|)$.

We will say that a sequence of points $\{\lambda_n\}$ in D is a *p-interpolating sequence* if there exists a constant M such that for any sequence $a = \{a_n\}$ in l^p there exists a function $f \in H^p$ that satisfies

$$f(\lambda_i) = a_i/(1 - |\lambda_i|^2)^{1/p} \tag{15-72}$$

and

$$\|f\|_p \leq M \cdot \|a\|_p \tag{15-73}$$

that is, we can interpolate the values $a_i/(1 - |\lambda_i|^2)^{1/p}$ in a uniformly bounded way. An ∞-interpolating sequence is also referred to as an *interpolating sequence*. In this case (15-72) is replaced by

$$f(\lambda_i) = a_i \qquad (15\text{-}74)$$

We continue with the introduction of some notions concerning bases in a Hilbert space. Let $\{f_i\}$ be a sequence of vectors in a separable Hilbert space then $\{f_i\}$ is a *basis* if for each $f \in H$ there exist unique scalars $\alpha_i(f)$ such that $\lim_{n \to \infty} \|f - \sum_{i=1}^n \alpha_i(f) f_i\| = 0$. A basis $\{f_i\}$ is called a *bounded basis* if there exists a constant $m > 0$ for which $m^{-1} \le \|f_i\| \le m$, for all $i \ge 0$. A basis $\{f_i\}$ is called *Hilbertian* if for each sequence $\{\alpha_i\} \in l^2$ the series $\sum \alpha_i f_i$ converges. A basis $\{f_i\}$ is called *Besselian* if for each $g \in H$ we have $\sum |(g, \psi_i)|^2 < \infty$ where $\{\psi_i\}$ is a sequence which is biorthogonal to $\{f_i\}$, that is, which satisfies $\psi_i(f_j) = \delta_{ij}$.

It will turn out to be useful to have some additional information concerning bases in Hilbert space. Let us assume H is a separable Hilbert space and $\{\varphi_i\}$ a basis in H. Let $\{\varphi_i\}$ be the biorthogonal sequence that corresponds to $\{\varphi_i\}$, that is $(\varphi_i, \psi_j) = \delta_{ij}$. Finally let e_i be the standard orthonormal basis for l^2.

Theorem 15-26 The following statements are equivalent.
(a) $\{\varphi_i\}$ is a Besselian basis for H.
(b) There exists a bounded map $T: H \to l^2$ such that

$$T\varphi_i = e_i \qquad (15\text{-}75)$$

(c) $\{\psi_i\}$ is a Hilbertian basis for H.
(d) There exists a bounded map $R: l^2 \to H$ such that

$$Re_i = \psi_i \qquad (15\text{-}76)$$

PROOF Assume $\{\varphi_i\}$ is a Besselian basis for H. Thus $\sum_{i=1}^\infty |(x, \psi_i)|^2 < \infty$ for all x in H. Define a map $T: H \to l^2$

$$T \sum_{j=1}^\infty \alpha_j \varphi_j = \{\alpha_j\} \qquad (15\text{-}77)$$

Clearly T is linear, its domain of definition is all of H and it is easily checked to have a closed graph. So by the closed graph theorem T is bounded. In particular we have the inequality

$$\{\sum |(x, \psi_j)|^2\}^{1/2} \le \|T\| \cdot \|x\| \qquad (15\text{-}78)$$

Conversely if there exists a bounded linear map $T: H \to l^2$ that satisfies (15-75) then we have

$$\alpha_j = \left(\sum_{i=1}^n \alpha_i e_i, e_j \right) = \left(T \sum_{i=1}^n \alpha_i \varphi_i, e_j \right) = \sum_{i=1}^n \alpha_i(\varphi_i, T^*e_j)$$

We must have therefore that

$$\psi_j = T^* e_j \tag{15-79}$$

and so we obtain

$$\sum |(x, \psi_j)|^2 = \sum |(x, T^* e_j)|^2 = \sum |(Tx, e_j)|^2 = \|Tx\|^2 \leq \|T\|^2 \|x\|^2$$

which shows that $\{\varphi_i\}$ is a Besselian basis. We have proved so far the equivalence of (a) and (b). In a completely analogous way we can show the equivalence of (c) and (d). But equality (15-79) shows that (b) → (d). Similarly if R satisfies (15-76) then

$$R^* \varphi_i = e_i \tag{15-80}$$

and so (d) → (b) and this completes the proof.

For two subspaces M and N of a Hilbert space H we define the *angle between the two subspaces*, which we denote by $\alpha(M, N)$, by $\alpha(M, N) = $ arc cos sup $\{(x, y) \mid x \in M, y \in N, \|x\| = \|y\| = 1\}$ and we assume $0 \leq \alpha(M, N) \leq \pi/2$. It is obvious that inf $\{\|x - y\| \mid x \in M, y \in N, \|x\| = \|y\| = 1\} > 0$ if and only if $\alpha(M, N) > 0$.

Lemma 15-27 Let $R: H \to K$ be a boundedly invertible map between two Hilbert spaces. Then there exists a $\delta > 0$ such that for any pair of orthogonal subspaces M and N in H we have $\alpha(RM, RN) \geq \delta$.

PROOF It suffices to show that there exists a $d > 0$ such that for every pair of orthogonal subspaces M and N in H we have inf $\{\|m' - n'\| \mid m' \in RM, n' \in RN, \|m'\| = \|n'\| = 1\} \geq d$. We will show that we can take $d = 2^{1/2} \|R^{-1}\| \cdot \|R^{-1}\|^{-1}$.

Note the two elementary inequalities $\|R\|^{-1} \|y\| \leq \|R^{-1} y\|$ and $\|R^{-1}\|^{-1} \|x\| \leq \|Rx\|$. Let $m' \in RM$ and $n' \in RN$ be unit vectors and let $m = R^{-1} m'$, $n = R^{-1} n'$. Thus

$$\|m' - n'\| \geq \|R^{-1}\|^{-1} \|m - n\|$$

or

$$\begin{aligned}
\|m' - n'\|^2 &\geq \|R^{-1}\|^{-2} (\|m\|^2 + \|n\|^2) \\
&= \|R^{-1}\|^{-2} (\|R^{-1} m'\|^2 + \|R^{-1} n'\|^2) \\
&\geq \|R^{-1}\|^{-2} \|R\|^{-2} (\|m'\|^2 + \|n'\|^2) \\
&= 2 \|R^{-1}\|^{-2} \cdot \|R\|^{-2}
\end{aligned}$$

and this proves the lemma.

We can state now the main result about uniformly separated sequences.

Theorem 15-28 Let $\{\lambda_i\}$ be a sequence of points in the open unit disc D. Then the following statements are equivalent.

(a) $\{\lambda_i\}$ is a uniformly separated sequence satisfying

$$\prod_{j \neq k} \left| \frac{\lambda_k - \lambda_j}{1 - \bar{\lambda}_j \lambda_k} \right| \geq \delta \qquad \text{for all} \qquad k \geq 1 \tag{15-81}$$

(b) The map T defined by

$$Tf = \{ f(\lambda_i)(1 - |\lambda_i|^2)^{1/2} \} \tag{15-82}$$

is bounded map of H^2 onto l^2.

(c) Each of the two sequences of functions

$$f_j(z) = \frac{(1 - |\lambda_j|^2)^{1/2}}{1 - \bar{\lambda}_j z} \qquad j \geq 1 \tag{15-83}$$

and

$$g_j(z) = (1 - |\lambda_j|^2)^{1/2} \frac{B(z)}{z - \lambda_j} \tag{15-84}$$

is a Hilbertian and Besselian basis for $H(B)$ where B is the Blaschke product whose zeros are the points λ_i.

(d) The operator $S(B)$ defined in $H(B)$ by (13-4) is similar to a normal operator.

We note that if the operator T defined by (15-82) is bounded then $f \in H^2$ belongs to $\text{Ker } T$ if and only if $f(\lambda_i) = 0$ for all i which is equivalent to $f \in BH^2$ where B is the Blaschke product whose zeroes are the points λ_i. This means that T restricted to $H(B) = \{BH^2\}^\perp$ is a boundedly invertible map of $H(B)$ onto l^2. We can restate statement (b) as follows.

(b') There exists a constant M such that for all $f \in H^2$

$$\sum |f(\lambda_j)|^2 (1 - |\lambda_j|^2) \leq M \|f\|^2 \tag{15-85}$$

and there exists another constant K, which can be taken to be $2\delta^{-4}(1 - 2 \log \delta)$ such that for each sequence $\{\alpha_j\}$ in l^2 there exists a function f in H^1 for which

$$f(\lambda_j) = \frac{\alpha_j}{(1 - |\lambda_j|^2)^{1/2}} \tag{15-86}$$

and

$$\|f\|^2 \leq K \|\alpha\|^2 \tag{15-87}$$

Before we prove the theorem we state and prove some lemmas.

Lemma 15-29 Let (a_{ij}), $i, j = 1, 2, \ldots$ be a Hermitian matrix, that is, $a_{ij} = a_{ji}$ and suppose for some constant M we have

$$\sum_{j=1}^{\infty} |a_{ij}| \le M$$

Then for any sequence $\{\xi_j\}$ in l^2 we have

$$\left| \sum_{i,j=1}^{\infty} a_{ij} \xi_i \bar{\xi}_j \right| \le M \sum_{j=1}^{\infty} |\xi_j|^2 \tag{15-88}$$

PROOF We show first that

$$\left| \sum_{i,j=1}^{n} a_{ij} \xi_i \bar{\xi}_j \right| \le M \sum_{j=1}^{n} |\xi_j|^2 \tag{15-89}$$

for all n. For the finite Hermitian matrix $A_n = (a_{ij})$, $i, j = 1, \ldots, n$ we have

$$\| A_n \| = \sup \left\{ \left| \sum_{i,j=1}^{n} a_{ij} \xi_i \bar{\xi}_j \right| \, \middle| \, \sum_{j=1}^{n} |\xi_j|^2 \le 1 \right\} \tag{15-90}$$

and since $\| A_n \|$ is equal to the largest absolute value of any eigenvalue of A_n we have to get an estimate on these. If λ is an eigenvalue of A_n and $(\xi_1, \ldots, \xi_n)$ a corresponding eigenvector then

$$\lambda \xi_i = \sum_{j=1}^{n} a_{ij} \xi_j$$

and therefore

$$|\lambda| |\xi_i| \le \sum_{j=1}^{n} |a_{ij}| |\xi_j|$$

Summing up these inequalities for all i and dividing by $\sum_{j=1}^{n} |\xi_j|$ we obtain $|\lambda| \le M$. Since the norms $\| A_n \|$ are uniformly bounded we have $\lim(A_n x, y) = (Ax, y)$ for all x and y in l^2 if and only if this equality holds in a fundamental set. But for the standard orthonormal basis $\{e_i\}$ of l^2 we clearly have $\lim(A_n e_i, e_j) = a_{ji} = (A e_i, e_j)$ and so (15-88) holds.

Lemma 15-30 Let $\{\lambda_i\}$ be a uniformly separated sequence satisfying (15-81). Then

$$\sum_{j=1}^{\infty} \frac{(1 - |\lambda_j|^2)(1 - |\lambda_k|^2)}{|1 - \bar{\lambda}_j \lambda_k|^2} \le 1 - 2 \log \delta \qquad \text{for all} \qquad k \ge 1 \tag{15-91}$$

PROOF From (15-81) we obtain

$$-\sum_{j \ne k} \log \left| \frac{\lambda_k - \lambda_j}{1 - \bar{\lambda}_j \lambda_k} \right| \le -\log \delta^2 = -2 \log \delta \tag{15-92}$$

Applying the inequality $-\log y \geq 1 - y$ to the following identity

$$\frac{(1 - |\lambda_j|^2)(1 - |\lambda_k|^2)}{|1 - \bar{\lambda}_j \lambda_k|^2} = 1 - \left| \frac{\lambda_k - \lambda_j}{1 - \bar{\lambda}_j \lambda_k} \right|^2 \qquad (15\text{-}93)$$

we obtain, by summing up over the all $j \neq k$, that

$$\sum_{j \neq k} \frac{(1 - |\lambda_j|^2)(1 - |\lambda_k|^2)}{|1 - \bar{\lambda}_j \lambda_k|^2} \leq - \sum_{j \neq k} \log \left| \frac{\lambda_k - \lambda_j}{1 - \bar{\lambda}_j \lambda_k} \right|^2 \leq -2 \log \delta$$

which is equivalent, adding 1 to both sides of the inequality, to (15-91).

PROOF OF THEOREM 15-28 We will prove the implications $(a) \Rightarrow (b) \Rightarrow (c) \Rightarrow (d) \Rightarrow (a)$. Let us assume first that $\{\lambda_i\}$ is a uniformly separated sequence. We already know that there exists a Blaschke product B whose zero set is $\{\lambda_i\}$. Let us define also

$$B_n(z) = \prod_{j=1}^{n} \frac{z - \lambda_j}{1 - \bar{\lambda}_j z} \qquad (15\text{-}94)$$

$$B_{nj}(z) = B_n(z) \frac{1 - \bar{\lambda}_j z}{z - \lambda_j} \qquad \text{for} \qquad j \leq n \qquad (15\text{-}95)$$

and

$$b_{nj} = B_{nj}(\lambda_j) \qquad (15\text{-}96)$$

Let us solve the finite interpolation problem by finding a function $h_n \in H^2$ that satisfies $h_n(\lambda_j) = \alpha_j/(1 - |\lambda_j|^2)^{1/2}$ for $j = 1, \ldots, n$. Now the function $B_{nj}(z)/b_{nj}$ satisfies

$$B_{nj}(\lambda_k)/b_{nj} = \delta_{jk} \qquad 1 \leq j, k \leq n \qquad (15\text{-}97)$$

Therefore a solution to the finite interpolation problem is given by

$$h_n(z) = \sum_{j=1}^{n} \alpha_j B_{nj}(z)/b_{nj} (1 - |\lambda_j|^2)^{1/2}$$

Moreover any other solution of the finite interpolation problem differs from h_n by $B_n g$ for some g in H^2. To find the function of minimum norm we have to find $\inf_{g \in H^2} \| h_n - B_n g \|$. We rewrite $h_n - B_n g$ as follows

$$h_n(z) - B_n(z) g(z) = B_n(z) \left\{ \sum_{j=1}^{n} \frac{\alpha_j}{b_{nj}} \frac{1 - \bar{\lambda}_j z}{z - \lambda_j} \frac{1}{(1 - |\lambda_j|^2)^{1/2}} - g(z) \right\}$$

Since B_n is inner we have

$$\| h_n - B_n g \| = \left\| \sum_{j=1}^{n} \frac{\alpha_j}{b_{nj}} \frac{1 - \bar{\lambda}_j z}{z - \lambda_j} \frac{1}{(1 - |\lambda_j|^2)^{1/2}} - g \right\| \qquad (15\text{-}98)$$

If M is any subspace of a Hilbert space then we know

$$\inf_{m \in M} \left\| x - m \right\| = \sup_{\substack{\|y\| = 1 \\ y \in M}} \left| (x, y) \right|$$

Applying this identity to (15-98) we have

$$\inf_{g \in H^2} \| h_n - B_n g \|$$

$$= \sup \left\{ \frac{1}{2\pi} \int \sum_{j=1}^{n} \frac{\alpha_j}{b_{nj}} \frac{(1 - \bar{\lambda}_j e^{it})}{(e^{it} - \lambda_j)} \frac{\overline{k(e^{it})}}{(1 - |\bar{\lambda}_j|^2)^{1/2}} \, dt \, \middle| \, k \in \bar{H}_0^2, \|k\| = 1 \right\}$$

If $k \in \bar{H}_0^2$ then $\overline{k(e^{it})} = e^{it} f(e^{it})$ for some $f \in H^2$. Thus we obtain finally, evaluating the integral by the use of Cauchy's theorem, that

$$\inf_{g \in H^2} \| h_n - B_n g \| = \sup \left\{ \sum_{j=1}^{n} \frac{\alpha_j}{b_{nj}} f(\lambda_j)(1 - |\lambda_j|^2)^{1/2} \, \middle| \, f \in H^2, \|f\| = 1 \right\}$$

Given a sequence $a = \{\alpha_j\}$ in l^2 which satisfies $\|a\|^2 = \sum |\alpha_j|^2 \leq 1$ we define

$$m_n(a) = \sup \left\{ \sum_{j=1}^{n} \frac{\alpha_j}{b_{nj}} f(\lambda_j)(1 - |\lambda_j|^2)^{1/2} \, \middle| \, f \in H^2, \|f\| = 1 \right\}$$

then assuming (15-81) holds we easily obtain

$$\sup \left\{ \sum_{j=1}^{\infty} |f(\lambda_j)|^2 (1 - |\lambda_j|^2) \, \middle| \, f \in H^2, \|f\| = 1 \right\}$$

$$\leq [\sup_{n} \sup_{a} m_n(a)]^2 \leq \delta^{-2} \sup \left\{ \sum_{j=1}^{\infty} |f(\lambda_j)|^2 (1 - |\lambda_j|^2) \, \middle| \, f \in H^2, \|f\| = 1 \right\}$$

and we conclude that (15-85) holds for all $f \in H^2$ if and only if $\sup_{n} \sup_{a \in l^2} m_n(a) < \infty$. To complete the proof we will show that for $K = 2\delta^{-4}(1 - 2 \log \delta)$ (15-86) and (15-87) hold.

To this end define

$$g_{nj}(z) = \left(\frac{B_n(z)}{z - \lambda_j} \right)^2 (1 - |\lambda_j|^2)^{3/2}$$

We note that

$$g_{nj}(\lambda_k) = \frac{B_{nj}(\lambda_j)^2}{(1 - |\lambda_j|^2)^{1/2}} \delta_{jk}$$

therefore

$$\varphi_n(z) = \sum_{j=1}^{n} \frac{\alpha_j}{b_{nj}^2} g_{nj}(z)$$

satisfies

$$\varphi_n(\lambda_j) = \frac{\alpha_j}{(1 - |\lambda_j|^2)^{1/2}} \qquad (15\text{-}99)$$

Next we show that the set of functions φ_n is uniformly bounded in H^2 norm.

$$\|\varphi_n\|^2 = (\varphi_n, \varphi_n) = \sum_{i,j=1}^{n} \frac{\lambda_i \bar{\lambda}_j}{b_{ni}^2 \bar{b}_{nj}^2} (g_{ni}, g_{nj}) \qquad (15\text{-}100)$$

and since B_n is an inner function

$$(g_{ni}, g_{nj}) = (1 - |\lambda_i|^2)^{3/2} (1 - |\lambda_j|^2)^{3/2} \int \frac{dt}{(e^{it} - \lambda_i)^2 (e^{-it} - \lambda_j)^2}$$

$$= (1 - |\lambda_i|^2)^{3/2} (1 - |\lambda_j|^2)^{3/2} \cdot \frac{(1 + \lambda_i \bar{\lambda}_j)}{(1 - \lambda_i \bar{\lambda}_j)^3} \qquad (15\text{-}101)$$

From the trivial inequality $2 \operatorname{Re} \lambda_i \bar{\lambda}_j \le |\lambda_i|^2 + |\lambda_j|^2$ we obtain

$$(1 - |\lambda_i|^2)^{1/2} (1 - |\lambda_j|^2)^{1/2} \le |1 - \lambda_i \bar{\lambda}_j|$$

Substituting this back into (15-101) and using Lemma 15-30 we have

$$\sum_j (g_{ni}, g_{nj}) \le 2 \sum_j \frac{(1 - |\lambda_i|^2)(1 - |\lambda_j|^2)}{|1 - \lambda_i \bar{\lambda}_j|^2} \le 2(1 - 2\log\delta) \qquad (15\text{-}102)$$

Using Lemma 15-29 and (15-100) we have

$$\|\varphi_n\|^2 \le 2\delta^{-4}(1 - 2\log\delta) \sum_k |a_k|^2$$

Since the sequence φ_n is uniformly bounded there is a subsequence that converges weakly to a function f in H^2 whose norm is bounded by the same bound. Since weak convergence in H^2 implies pointwise convergence in the open unit disc the function f satisfies (15-86) and (15-87).

To prove that (b) implies (c) we assume the map T defined by (15-82) is a bounded map of H^2 onto l^2. As we saw this means that its restriction to $H(B) = \{\operatorname{Ker} T\}^{\perp}$ is boundedly invertible. Let $\{e_i\}$ be the standard orthonormal basis in l^2 then the functions $d_i \in H(B)$ defined by $d_i = T^{-1} e_i$ are uniformly bounded. Moreover d_i has to satisfy

$$d_i(\lambda_j) = \frac{\delta_{ij}}{(1 - |\lambda_j|^2)^{1/2}} \qquad (15\text{-}103)$$

We can easily derive an explicit expression for d_i. Indeed if B is the Blaschke product that corresponds to the set of zeroes $\{\lambda_i\}_{i=1}^{\infty}$ then we let

$B^{(j)}(z) = B(z)((1 - \bar{\lambda}_j z)/(z - \lambda_j))$, that is $B^{(j)}$ is the Blaschke product that corresponds to the same set of zeroes with λ_j excluded. It is easy to check that necessarily

$$d_j(z) = \frac{(1 - |\lambda_j|^2)^{1/2}}{B^{(j)}(\lambda_j)} \frac{B(z)}{z - \lambda_j} \qquad (15\text{-}104)$$

Using the fact that B is inner the norm of d_j turns out to be equal to $|B^{(j)}(\lambda_-)|^{-1}$. Since the d_j are uniformly bounded this means that $|B^{(j)}(\lambda_j)|$ is bounded away from zero. So in particular $\{\lambda_i\}$ is a uniformly separated sequence.

We note that the map $T: H(B) \to l^2$ defined by (15-82) is a bounded map for which $Td_i = e_i$ where d_i is defined by (15-104). By Theorem 15-26 we infer that $\{d_i\}$ is a Besselian basis for $H(B)$. Now it is easily seen that a sequence $\{\lambda_i\}$ is uniformly separated if and only if the sequence $\{\bar{\lambda}_i\}$ is. To the sequence $\{\bar{\lambda}_i\}$ corresponds to Blaschke product $\tilde{B}(z)$ and so the sequence of functions $\{\tilde{d}_j\}$ is a Besselian basis for $H(\tilde{B})$. We compute now the biorthogonal sequence that corresponds to $\{d_j\}$. Let $\{f_j\}$ be the sequence of functions defined by (15-83) then $(f, f_j) = (1 - |\lambda_j|^2)^{1/2} f(\lambda_j)$ for all $f \in H^2$ and moreover all f_j belong to $H(B)$. Therefore we have

$$(d_i, f_j) = (1 - |\lambda_j|^2)^{1/2} d_i(\lambda_j) = \delta_{ij}$$

and as a consequence of Theorem 15-26 it follows that $\{f_j\}$ is a Hilbertian basis for $H(B)$. Now let $\tau_B: H(B) \to H(\tilde{B})$ be the unitary map defined by (13-7) then by an elementary computation we have

$$\tau_B f_j = \tilde{g}_j \qquad \tau_{\tilde{B}} \tilde{g}_j = f_j \qquad (15\text{-}105)$$

This shows that $\{\tilde{g}_j\}$ is a Hilbertian and Besselian basis for $H(\tilde{B})$ and so $\{f_j\}$ is a Hilbertian and Besselian basis for $H(B)$. The result for $\{g_j\}$ follows by symmetry. Of course $\{g_j\}$ and $\{d_j\}$ differ only by factors of a bounded sequence which is also bounded away from zero. So one is Besselian, or Hilbertian, if and only if the other is.

Finally assume that B is an inner function in H^∞ for which $S(B)$ is similar to a normal operator. Since similarity preserves invariant subspaces and since for a normal operator every invariant subspace is reducing it follows that every invariant subspace M of $S(B)$ has a complementary invariant subspace N such that $H(B) = M + N$ and $M \cap N = \{0\}$. By Lemma 15-27 the angle between M and N is positive. If $M = H(B_1)$ and $N = H(B_2)$ it follows that $H(B) = H(B_1) + H(B_2)$ with $B = B_1 B_2$. The condition $H(B_1) \cap H(B_2) = \{0\}$ is equivalent to $B_1 \wedge B_2 = 1$. Since the angle between $H(B_1)$ and $H(B_2)$ is positive actually more is true as will be proved in the next chapter. However, since for an arbitrary factorization $B_1 B_2$ of B we have $B_1 \wedge B_2 = 1$ we conclude that B is a Blaschke product with simple zeroes. Let $\{\lambda_i\}$ be the set of zeroes of B. Clearly the functions $g_i(z) = (1 - |\lambda_i|^2)^{1/2} \{B(z)/(z - \lambda_i)\}$ are eigenfunctions of $S(B)$, corresponding to the eigenvalues λ_i, and they span $H(B)$. Let L be a normal operator in a Hilbert space H which is similar to $S(B)$. H has therefore an orthonormal basis

$\{e_i\}$ consisting of eigenvectors of L. Assume $RL = S(B) R$ and let M_i and N_i be the one-dimensional subspaces of H and $H(B)$ spanned by e_i and g_i, respectively. By Lemma 15-27 the angle between N_i and $\{B^{(i)}H^2\}^\perp$ is bounded away from zero, uniformly in i. We compute $\inf\{\|g_i - x\| \mid \|x\| = 1, \, x \in H(B^{(i)})\}$ which by the projection theorem is equal to $\|g_i - P_{H(B^{(i)})}g_i\|$. This projection is easily computed to be

$$(P_{H(B^{(i)})}g_i)(z) = (1 - |\lambda_i|^2)^{1/2} \frac{(B^{(i)}(z) - B^{(i)}(\lambda_i))}{z - \lambda_i}$$

and hence $\|g_i - P_{H(B^{(i)})}g_i\| = |B^{(i)}(\lambda_j)|$. Since

$$B^{(i)}(\lambda_i) = \prod_{j \neq i} \left| \frac{\lambda_i - \lambda_j}{1 - \bar{\lambda}_j \lambda} \right|$$

it follows that the sequence $\{\lambda_i\}$ is uniformly separated, which completes the proof.

As a corollary we obtain the solution to the H^∞-interpolation problem.

Theorem 15-31 A necessary and sufficient condition for $\{\lambda_i\}$ to be an ∞-interpolating sequence is that it be uniformly separated.

PROOF Assume $\{\lambda_i\}$ is an ∞-interpolating sequence. Consider the sequences $\{\delta_{kj}\}_{k=1}^\infty$ and let f_j be interpolating functions such that

$$f_j(\lambda_k) = \delta_{kj} \tag{15-106}$$

and

$$\|\delta_j\| \leq M \|\{\delta_{kj}\}\|_\infty = M \tag{15-107}$$

hold. If B_j is the Blaschke product corresponding to the zeroes $\{\lambda_k\}_{k \neq j}$ then B_j divides f_j. So $f_j + B_j h_j$ with $\|h_j\| = \|f_j\| \leq M$ and therefore

$$1 = f_j(\lambda_j) = B_j(\lambda_j) h_j(\lambda_j) \leq |B_j(\lambda_j)| \cdot \|h_j\|$$

or $|B_j(\lambda_j)| \geq \delta$ with $\delta = M^{-1}$. But

$$|B_j(\lambda_j)| = \prod_{i \neq j} \left| \frac{\lambda_j - \lambda_i}{1 - \bar{\lambda}_i \lambda_j} \right|$$

and so $\{\lambda_i\}$ is uniformly separated.

Conversely assume $\{\lambda_i\}$ is uniformly separated and let $\{\alpha_j\}$ be an l^∞ sequence. Define bounded operators A and L in l^2 by

$$Ae_i = \alpha_i e_i \quad \text{and} \quad Le_i = \lambda_i e_i$$

where $\{e_i\}$ is the standard orthonormal basis of l^2. Clearly $AL = LA$ and so if $R: l^2 \to H(B)$ is defined by

$$R \sum \alpha_i e_i = \sum \alpha_i f_i$$

then R is a boundedly invertible operator, $S(B) = RLR^{-1}$ and RAR^{-1} is in the commutant of $S(B)$. By the scalar version of Theorem 14-8 there exists a function a in H^∞ such that $a(S(B)) = RAR^{-1}$. Since $a(\lambda_i) = \alpha_i$, $\{\lambda_i\}$ is an interpolating sequence.

NOTES AND REFERENCES

The study of Hilbert spaces began with the work of Hilbert and others on integral equations. The axiomatic introduction of Hilbert spaces and much of the early theory is due to von Neumann. Among the other pioneers in the field one should mention Riesz, Fischer, Stone, and others. There exist several excellent introductions to Hilbert spaces and the theory of linear operators. Among others Stone [111], Akhiezer and Glazman [2], Riesz and Sz.-Nagy [99], Halmos [62, 64], Douglas [24] as well as the relevant parts of Dunford and Schwartz [29] should be mentioned.

Unbounded operators were studied by von Neumann who was the first to prove the spectral theorem for unbounded self-adjoint operators. To von Neumann is also due the Cayley transform technique.

The integral representations obtained in Sec. 4 have been the object of study of many mathematicians early in the century, among them Caratheodory, Herglotz, Toeplitz, Bochner, and Nevanlinna. For harmonic functions and the Poisson integral one can refer to Hoffman [73]. The exposition in the text of the relation between the representation theorems to the spectral theorem and moment problems follows Akhiezer and Glazman [2]. A short and beautifully written introduction to spectral theory is in Lorch [85]. The commutative B^*-algebra approach to the spectral theorem may be found in Dunford and Schwartz [29] and Douglas [24].

The theory of spectral representations amd multiplicity theory is developed in Halmos [62], Dunford and Schwartz [29], Beals [10], and Plessner [98] to cite a few. The exposition in Sec. 6 uses L^2 spaces of matrix measures which though less general than, say, the use of direct integrals of Hilbert spaces is more concrete. The writing of this section has been motivated in part by Brown [18] and Nelson [94].

The factorization lemma of Douglas appears in [23]. Embry's [34] contains a correct Banach space formulation. Theorem 7-5 is from [41]. Wold's decomposition is from [122] though it has been obtained earlier by von Neumann.

The study of shifts and models received its impetus from Beurling's fundamental paper [11] characterizing the invariant subspaces of the shift as well as Rota's elegant paper on models [101]. Other motivating sources were Livsic's introduction of characteristic functions [83], scattering theory as developed by Lax and Phillips [82], work in prediction theory by Wiener and Masani [121] and Helson and Lowdenslager [68]. An important influence was the work of de Branges and Rovnyak on the invariant subspace problem [14] and finally the large body of work of Sz.-Nagy and Foias on contractions [115, 116–118]. Three excellent references are Fillmore [37], Helson [67], and the survey of Douglas [25].

There is no attempt to credit all who contributed to the theory and only a few theorems are cited. Dilations and compressions were introduced by Halmos [61], minimal unitary dilations by Sz.-Nagy [112], they being a special case of Naimark's Theorem 9-8 on unitary representation of positive definite functions on groups [92]. The structure of the minimal dilation space has been elucidated by Schaffer [105] and Sz.-Nagy and Foias [116].

Semigroup theory is very completely covered in Hille and Phillips [72]. Other references are Yosida [125] and Dunford and Schwartz [29]. The proof of the Hille–Yosida theorem in the text follows Lax and Phillips [82]. Infinitesimal cogenerators were introduced by Sz.-Nagy and Foias [115].

For the Fourier–Plancherel transform basis references are Akhiezer and Glazman [2], Bochner and Chandrasekharan [13], and Katznelson [78]. In particular the proof of Theorem 10-13 can be found in [13]. Theorem 10-17 and its discrete analog are due to Sz.-Nagy and Foias. Theorems 10-18 and 10-23 are from Lax and Phillips [82]. The results on isometric semigroups are due to Cooper [21] and Sz.-Nagy [113].

The proof of the lifting theorem of Sec. 11 is due to Douglas, Muhly, and Pearcy [27]. The scalar version of the theorem has been proved first by Sarason [104] whereas the general case is due to Sz.-Nagy and Foias [115].

For H^p-theory the most readable account is Hoffman's excellent exposition [73]. Another comprehensive survey is Duren [30] which contains also a proof of Carleson's corona theorem. The proof of the Paley–Wiener theorem follows Dym and McKean [31]. The connection between the H^p spaces of the disc and the half plane appears in Hoffman [73]. Theorem 12-20 is due to Lax [81] in the vectorial case. The extension of Beurling's theorem given in Theorem 12-22 is due to Halmos [63]. For vectorial H^2-theory Helson [67] and Sz.-Nagy and Foias [115] are good sources. Our exposition follows in large part Helson. Theorem 12-27 is adapted from [82].

The content of Sec. 13 lies halfway between Rota's paper [101] and the Sz.-Nagy and Foias theory of characteristic functions and functional models [115]. The convenient unitary maps τ_Q introduced in Theorem 13-2 are from [39]. They have their counterpart in the Sz.-Nagy and Foias work. Lemma 13-7 is adapted from its continuous analog in Lax and Phillips [82]. Theorem 13-8 originated with Moeller [89], Lax and Phillips [82], Helson [67], and Sz.-Nagy and Foias [115].

The functional calculus constructed in Sec. 14 is due to Sz.-Nagy and Foias [115]. The spectral analysis of Theorems 14-4 and 14-5 is from [39]; so is the matrix version of the corona theorem. The extension to the spectral analysis of operators intertwining compressions of shifts is from [40]. Further extensions appear in Sz.-Nagy and Foias [118].

The study of diagonalization of matrices over H^∞ has been initiated by Nordgren [97] who also introduced quasiequivalence. The present exposition follows mainly Sz.-Nagy [114]. The fundamental Lemma 15-1 is due to Sherman [108]. Independently it has been proved by the author in [50] as an extension of a lemma of Wonham. Lemma 15-12 is from Ahern and Clark [1]. Theorem 15-18 is due to Moore and Nordgren [91]. Theorem 15-25 is from Sz.-Nagy and Foias [116] while generally the theory of Jordan models follows [114, 117]. The material on bases in left invariant subspaces is based on Shapiro and Shields [107] as well as Nikolskii and Pavlov [95].

THREE

LINEAR SYSTEMS IN HILBERT SPACE

1. FUNDAMENTAL CONCEPTS

As will become clear in this chapter there is a great similarity in the formalism of finite and infinite dimensional linear systems. However, the infinite dimensional cases abound with a variety of phenomena missing from the finite dimensional situations. In the following we will focus our attention on the theory of modeling infinite dimensional linear time invariant systems in state space form. We will consider solely systems with a finite number of inputs and outputs.

The need for such a theory is obvious inasmuch as realistic modeling of most physical systems must include the distributed effects. Whereas in some cases these effects can be safely ignored there are various situations when they have to be taken into account.

We note that while it is possible by experimental tests to conclude that a system is not finite dimensional it seems much harder, if not altogether impossible, to conclude that a system is finite dimensional. This indicates that it may be better to view finite dimensional systems as specializations of infinite dimensional ones as opposed to viewing infinite dimensional systems as extensions of finite dimensional ones. We note also that to specify an infinite dimensional system does not require an infinite number of parameters. Thus the nonrational transfer function $e^{-\alpha s}/(s + \beta)$ does not have a finite dimensional realization but can be specified by the two parameters α and β. Finally, even if one were to replace an infinite dimensional system by a finite dimensional approximation to do it the right way would necessarily mean the development of a complete state space theory for infinite dimensional systems.

We will restrict ourselves to systems which can be realized on state spaces which are Hilbert spaces. This is certainly not the most general framework discussed in the literature. One other alternative would be to use a distributional framework. Our assumption though somewhat more restrictive has the advantage of yielding a richer theory and a very explicit structure theory for the dynamical models. In fact it is this insistence on getting at a precise description of the internal structure which forces the various assumptions made in this chapter.

Our aim will be to develop a theory of infinite dimensional systems that encompasses both the discrete and continuous time case. As in the finite dimensional case, to specify a *system*, or rather a constant, or time invariant, linear system we need a quadruple (A, B, C, D) of operators and three spaces U, K, and Y. The input and output value spaces are U and Y and will be assumed finite dimensional. K denotes the state space and will be assumed to be a Hilbert space. The operators are assumed linear and $A: K \rightarrow K$, $B: U \rightarrow K$, $C: K \rightarrow Y$, and $D: U \rightarrow Y$.

In the discrete time case (A, B, C, D) stands for the set of dynamical equations

$$
\begin{aligned}
x_{n+1} &= Ax_n + Bu_n \\
y_n &= Cx_n + Du_n
\end{aligned}
\tag{1-1}
$$

whereas in the continuous time case the system equations will be

$$
\begin{aligned}
\dot{x}(t) &= Ax(t) + Bu(t) \\
y(t) &= Cx(t) + Du(t)
\end{aligned}
\tag{1-2}
$$

Of course these definitions are strictly formal and meaningless unless further assumptions are made on the spaces as well as the operators involved.

Even in a cursory acquaintance with modern operator theory, its extreme richness on the one hand and its essential incompleteness on the other, suggest immediately that there are only very few general results one could expect and it is only by restricting the setting that we can expect to develop an interesting theory. Here one must be guided both by physical intuition as well as by a sense of mathematical aesthetics. Our choice of Hilbert spaces could be justified in certain cases as ensuing naturally out of energy considerations, but mostly it is done for mathematical convenience. Within this general framework it is possible to develop a relatively complete and satisfying theory of systems.

To continue with our introduction we consider for the time being only discrete time systems. We assume all operators A, B, C, and D to be bounded and linear.

Let $u_{-j} \in U$ denote an input applied at time $t = -j$. Assuming the system to have been at rest in the remote past we obtain at time $t = 1$ the state

$$
x_1 = \sum_{j=0}^{\infty} A^j Bu_{-j}
\tag{1-3}
$$

Contrary to the development in Chap. I we may consider an infinite number of nonzero inputs but have to assume that the series in (1-3) converges. If as of time $t = 1$ no further inputs are applied we obtain a sequence of outputs

$$
y_k = \sum_{j=0}^{\infty} CA^{j+k} Bu_{-j}
\tag{1-4}
$$

Let us consider now the space of input string, or input functions, to be $l^2(-\infty, 0; U)$ and the space of output functions to be $l^2(1, \infty; Y)$. We define a map $f: l^2(-\infty, 0; U) \to l^2(1, \infty; Y)$ by

$$f(\{u_j\}_{j=-\infty}^0) = \{y_k\}_{k=1}^\infty \tag{1-5}$$

where the y_k are determined by (1-4).

We call f the restricted input/output map of the system. We note that the restricted input/output map is not dependent on the operator D. Of course initially f is not defined on all of $l^2(-\infty, 0; U)$ as the series (1-3) may fail to converge. Thus the initial domain of definition of f is the dense set of all finitely nonzero sequences in $l^2(-\infty, 0; U)$. The function f may or may not be extended by continuity to all of $l^2(-\infty, 0; U)$. Even in the later case its range may fail to be in $l^2(1, \infty; Y)$. Thus our first restriction will be to study only systems whose restricted input/output map f defined by (1-4) and (1-5) extends to a bounded linear map of $l^2(-\infty, 0; U)$ into $l^2(1, \infty; Y)$.

Let us denote by S_-^* and S_+^* the left shifts in $l^2(-\infty, 0; U)$ and $l^2(1, \infty; Y)$, respectively. Thus

$$S_-^* \{u_{-j}\}_{j=-\infty}^0 = \{v_{-j}\}_{j=-\infty}^0 \tag{1-6}$$

with

$$v_{-j} = \begin{cases} u_{-j+1} & j > 0 \\ 0 & j = 0 \end{cases} \tag{1-7}$$

whereas

$$S_+^* \{y_j\}_{j=1}^\infty = \{z_j\}_{j=1}^\infty \tag{1-8}$$

with

$$z_j = v_{j+1} \qquad j \geq 1 \tag{1-9}$$

Obviously S_-^* is unitarily equivalent to the right shift in $l^2(0, \infty; U)$.

It is clear from the definition of f that it necessarily satisfies the functional equation

$$f S_-^* = S_+^* f \tag{1-10}$$

We take this functional equation as our intrinsic definition of a restricted input/output map which for us is any bounded operator $f: l^2(-\infty, 0; U) \to l^2(1, \infty; Y)$ that satisfies (1-10).

It is of interest to derive a matrix representation for f. If we write $\{u_j\}_{j=-\infty}$ and $\{z_k\}_{k=1}^\infty$ as column vectors then (1-4) can be rewritten as

$$\begin{pmatrix} y_1 \\ y_2 \\ y_3 \\ \vdots \end{pmatrix} = \begin{pmatrix} CB & CAB & CA^2B & \cdots \\ CAB & CA^2B & & \cdots \\ CA^2B & & & \\ \vdots & \vdots & & \end{pmatrix} \begin{pmatrix} u_0 \\ u_{-1} \\ u_{-2} \\ \vdots \end{pmatrix} \tag{1-11}$$

Thus the matrix representation of f is a block Hankel matrix. This leads us to define a Hankel operator to be any bounded operator $H: l^2(-\infty, 0; U) \to l^2(1, \infty; Y)$ which satisfies (1-10).

We define an *extended causal input/output map* to be a bounded linear map $\tilde{f}: l^2(-\infty, \infty; U) \to l^2(-\infty, \infty; Y)$ which satisfies

$$\tilde{f}U = U\tilde{f} \tag{1-12}$$

and

$$\tilde{f}l^2(0, \infty; U) \subset l^2(0, \infty; Y) \tag{1-13}$$

where U denotes the bilateral right shift in $l^2(-\infty, \infty; U)$ and $l^2(-\infty, \infty; Y)$.

We say that $\tilde{f}$ is *strictly causal* if (1-13) is replaced by

$$\tilde{f}l^2(0, \infty; U) \subset l^2(1, \infty; Y) \tag{1-14}$$

It is natural to inquire whether a given restricted input/output map f can be extended to an extended causal input/output map $\tilde{f}$. A straightforward application of Theorem II 11-7 yields the following.

Theorem 1-1 Let $f: l^2(-\infty, 0; U) \to l^2(1, \infty; Y)$ be a restricted input/output map then there exists a map $\tilde{f}: l^2(-\infty, \infty; U) \to l^2(-\infty, \infty; Y)$ satisfying

$$\|\tilde{f}\| = \|f\| \tag{1-15}$$

as well as (1-12) and for which

$$f = P_{l^2(1, \infty; Y)}\tilde{f} \,|\, l^2(-\infty, 0; Y) \tag{1-16}$$

Of course Theorem 1-1 does not say anything about causality. In fact to study causal extensions it is convenient to work in a functional representation for f and $\tilde{f}$ and to this end we utilize the Fourier transform. Under the Fourier transform $\mathscr{F}$ we have $\mathscr{F}l^2(-\infty, \infty; U) = L_U^2$, $\mathscr{F}L^2(-\infty, \infty; Y) = L_Y^2$, $\mathscr{F}l^2(-\infty, 0; U) = \bar{H}_U^2$ and $\mathscr{F}l^2(1, \infty; Y) = H_{0,Y}^2$.

Since $\tilde{f}$ commutes with the bilateral shift its image under the Fourier transform $\bar{H} = \mathscr{F}\tilde{f}\mathscr{F}^{-1}$ commutes with all multiplication operators by scalar bounded measurable functions. It follows that $\bar{H}$ is a multiplication operator by a function in $L^\infty(B(U, Y))$. This implies the following result.

Theorem 1-2 Let $H: \bar{H}_U^2 \to H_{0,Y}^2$ be a bounded Hankel operator. Then there exists a function $T \in L^\infty(B(U, Y))$ such that

$$Hg = P_{H_{0,Y}^2} Tg \qquad \text{for all} \qquad g \in \bar{H}_U^2 \tag{1-17}$$

We say that H is the Hankel operator induced by T and write $H = H_T$.

Since T is in $L^\infty(B(U, Y))$ it has a Fourier expansion of the form $T(e^{it}) = \sum_{n=-\infty}^{\infty} T_n e^{int}$ with T_n being linear maps from U to Y. Since $\bar{H}$ is defined by $\bar{H}g = Tg$ for all $g \in L_U^2$ then the causality condition

$$\bar{H}(H_U^2) \subset H_Y^2 \tag{1-18}$$

is satisfied if and only if $T_n = 0$ for $n < 0$. Strict causality is equivalent to $T_n = 0$ for $n \leq 0$.

Corollary 1-3 Let $f: l^2(-\infty, 0; U) \to l^2(1, \infty; Y)$ be a restricted input/output map. Then f can be extended to a causal input/output map $\bar{f}$ if and only if the Hankel operator (T_{i+j}) is induced by a function

$$T(e^{it}) = \sum_{n=0}^{\infty} T_n e^{int} \quad \text{in} \quad H^\infty(B(U, Y))$$

Given a dynamical system (A, B, C, D) we define the *transfer function* T of the system by

$$T(z) = D + zC(I - zA)^{-1} B \tag{1-19}$$

T is analytic on the set $\{\lambda | \lambda^{-1} \in \rho(A)\}$ and for all values of z in a sufficiently small neighborhood of the origin, at least for all z such that $|z| < \|A\|^{-1}$, it is given by

$$T(z) = D + \sum_{j=0}^{\infty} CA^j B z^{j+1} \tag{1-20}$$

The system (A, B, C, D) has associated with it an extended causal input/output map if and only if its transfer function is in $H^\infty(B(U, Y))$. The transfer function can be considered as the Fourier transform of the sequence

$$(D, CB, CAB, \ldots) \tag{1-21}$$

which is called the *weighting pattern* or *impulse response* of the system.

Next we associate with any given system (A, B, C, D) a *reachability operator* R and *observability operator* O. R is defined on the set of finitely nonzero sequences $l^2(-\infty, 0; U)$ by

$$R(\{u_j\}_{j=-\infty}^0) = \sum_{j=0}^{\infty} A^j B u_{-j} \tag{1-22}$$

whereas O is defined by

$$O = \{CA^j x\}_{j=0}^{\infty} \tag{1-23}$$

for each x in the state space.

Again there is an ambiguity concerning the natural domain of definition of R and the range of O. Contrary to the finite dimensional situation there are various possibilities regarding the definition of reachability and observability all of which use the basic formulas (1-22) and (1-23).

The system (A, B, C), we omit D as it does not influence the dynamic behavior of the system, is called *reachable* if the reachability operator R has a range that is dense in the state space. This is clearly equivalent to

$$\bigcap_{i=0}^{\infty} \text{Ker } B^* A^{*i} = \{0\} \tag{1-24}$$

A system is *strongly reachable* if any state x is reachable from the origin through the application of a finite input string. That is for each x there exist $u_0, u_{-1}, \ldots, u_{-m}$ such that $x = \sum_{j=0}^{m} A^j B u_{-j}$. As a consequence of Theorem II 7-5 if a system is strongly reachable then there is a uniform bound on the length of the input string needed to reach any given state. This means that for some n the map $R_n: U^n \to K$ defined by

$$R_n(u_0, \ldots, u_{-n+1}) = \sum_{j=0}^{n-1} A^j B u_{-j} \tag{1-25}$$

is a surjective map. By Theorem II 7-1 this implies that $R_n R_n^*$ is a strictly positive operator or equivalently that for some $\delta > 0$

$$\sum_{i=0}^{n-1} A^i B B^* A^{*i} \geq \delta^2 I \tag{1-26}$$

Since we assume U to be finite dimensional BB^* is an operator of finite rank and so (1-26) can hold only if K is finite dimensional. Thus a finite input infinite dimensional system cannot be strongly reachable [41].

There are other definitions of reachability which are appropriate in the infinite dimensional situation. A system (A, B, C) is called *continuously reachable* if its reachability operator R extends to a bounded operator from $l^2(-\infty, 0; U)$ onto a dense subset of the state space K. We will say that (A, B, C) is *exactly reachable* if it is continuously reachable and R is a surjective map. In our definition continuous reachability refers to the space of input functions $l^2(-\infty, 0; U)$. Other input function spaces may be used and we will see one instance of this in section 7.

For observability we have analogous definitions. Thus (A, B, C) is, respectively, *observable, continuously observable,* and *exactly observable* if (A^*, C^*, B^*) is, respectively, reachable, continuously reachable, and exactly reachable. In particular the observability condition reduces to

$$\bigcap_{i=0}^{\infty} \operatorname{Ker} C A^i = \{0\} \tag{1-27}$$

If f is the restricted input/output map of the realization $\Sigma = (A, B, C)$ and if we assume Σ to be continuously reachable and continuously observable then we have $f = OR$.

Lemma 1-4 Two systems $\Sigma = (A, B, C, D)$ and $\Sigma_1 = (A_1, B_1, C_1, D_1)$ have the same transfer function if and only if

$$D = D_1 \quad \text{and} \quad CA^j B = C_1 A_1^j B_1 \quad \text{for} \quad j \geq 0 \tag{1-28}$$

If both systems are continuously reachable and continuously observable then they have the same transfer function if and only if

$$D = D_1 \quad \text{and} \quad OA^j R = O_1 A_1^j R_1 \quad \text{for} \quad j \geq 0 \tag{1-29}$$

PROOF The first part follows trivially from the definition of a transfer function. Equality (1-29) follows from (1-28) by observing that OA^jR has the matrix representation

$$OA^jR = \begin{pmatrix} CA^jB & CA^{j+1}B & \cdots \\ CA^{j+1}B & \cdot & \\ \cdot & \cdot & \\ \cdot & & \end{pmatrix} \tag{1-30}$$

which is an extension of (1-11).

One of the central problems we focus on in this chapter is the question of isomorphism theorems for systems. To this end we introduce some definitions and derive some elementary results.

Given two realizations $\Sigma = (A, B, C)$ and $\Sigma_1 = (A_1, B_1, C_1)$ with the same input and output spaces U and Y and state spaces K and K_1, respectively, we will say that a map $X: K \to K_1$ *intertwines* Σ and Σ_1 if the diagram

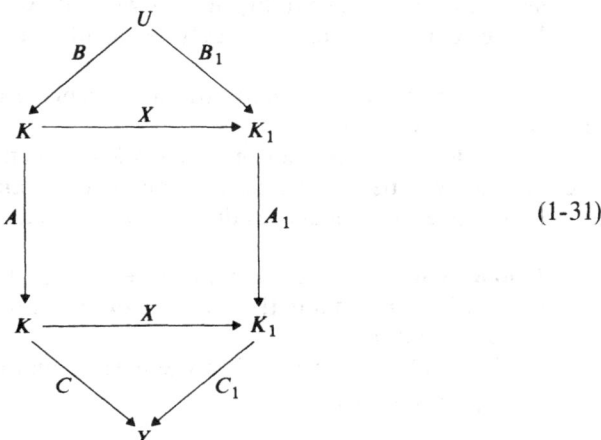

$$(1\text{-}31)$$

is commutative. If only the diagram

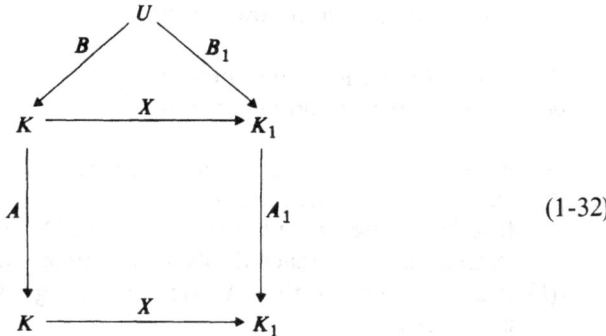

$$(1\text{-}32)$$

commutes then we say that X intertwines (A, B) and (A_1, B_1). Similarly we define an intertwining map of (A, C) and (A_1, C_1).

A system $\Sigma = (A, B, C)$ is a *quasi-invertible transform* of $\Sigma_1 = (A_1, B_1, C_1)$ if there exists a quasi-invertible operator $X: K \to K_1$ which intertwines Σ and Σ_1. Two systems Σ and Σ_1 are *quasisimilar* if each is a quasi-invertible transform of the other and *similar* if there exists a boundedly invertible operator X that intertwines the two systems.

For operators quasisimilarity does not imply similarity as we saw in Sec. II-15. However, for systems the intertwining relation is more rigid and hence the next result.

Lemma 1-5 Let $\Sigma = (A, B, C)$ and $\Sigma_1 = (A_1, B_1, C_1)$ be two reachable systems. Then Σ and Σ_1 are quasisimilar if and only if they are similar.

PROOF The if part is trivial. Thus let us assume $X: K \to K_1$ and $X_1: K_1 \to K$ are quasi-invertible maps that intertwine the two systems. Since intertwining is a transitive relation it follows that $X_1 X$ intertwines Σ with itself. Thus $X_1 X A^n B = A^n B$ for all $n \geq 0$ and analogously $X X_1 A_1^n B_1 = A_1^n B_1$. By the assumption of reachability it follows that $X_1 X = I_K$ and $X X_1 = I_{K_1}$ and hence X and X_1 are boundedly invertible and the two systems are similar.

A similar result holds naturally if we replace the condition of reachability by that of observability.

If no additional assumptions are made on the nature of the systems involved then nothing can be said concerning the existence and uniqueness of intertwining operators. One uniqueness result is the following.

Lemma 1-6 Given two systems $\Sigma = (A, B, C)$ and $\Sigma_1 = (A_1, B_1, C_1)$.
(a) Let Σ be reachable then if there exists an operator X intertwining Σ and Σ_1 it is unique.
(b) Let Σ_1 be observable then there exists an operator X intertwining Σ and Σ_1 it is unique.

PROOF Assume two intertwining operators X and X' are given. Then $Z = X - X'$ satisfies $Z A^i B = 0$ for all $i \geq 0$. Since Σ is reachable this implies $Z = 0$. Statement (b) follows by duality.

The next theorem relates properties of an intertwining operator to the properties of the corresponding systems.

Theorem 1-7 Let X be a bounded operator intertwining the systems $\Sigma = (A, B, C)$ and $\Sigma_1 = (A_1, B_1, C_1)$.
(a) If X has dense range the reachability of Σ implies that of Σ_1. If X is surjective the exact reachability of Σ implies that of Σ_1.
(b) If Σ_1 is reachable then X has dense range. If Σ_1 is exactly reachable X is surjective.

(c) If X is injective then the observability of Σ_1 implies that of Σ. If X^* is surjective and Σ_1 exactly observable then Σ is also exactly observable.

(d) If Σ is observable then X is injective. If Σ is exactly observable then X^* is surjective.

PROOF Let $z \in \bigcap_{n=0}^{\infty} \text{Ker } B_1^* A_1^{*n}$, then for all $n \geq 0$ we have $B_1^* A_1^{*n} z = 0$. Since X intertwines Σ and Σ_1 we have $B_1^* A_1^{*n} = B^* A^{*n} X^*$ and so $X^* z \in \bigcap_{n=0}^{\infty} \text{Ker } B^* A^{*n} = \{0\}$. If X has dense range then X^* is injective and so $z = 0$ which shows that $\bigcap_{n=0}^{\infty} \text{Ker } B_1^* A_1^{*n} = \{0\}$ or in other words the reachability of Σ_1. If X is surjective then $R_1 = XR$ shows that if R is onto so is R_1. Part (c) follows from (a) by duality. Statement (b) is a direct consequence of the definitions. Finally, to prove (b) we note that $C_1 A_1^n X z = C A^n z$ for each $z \in K$. Thus $z \in \bigcap_{n=0}^{\infty} \text{Ker } C A^n$ and hence $z = 0$. The second part of statement (d) follows from (b) by duality.

Lemma 1-8 Let X intertwine $\Sigma = (A, B, C)$ and $\Sigma_1 = (A_1, B_1, C_1)$. Then Σ and Σ_1 realize the same transfer function.

PROOF Since X intertwines Σ and Σ_1 we have $XA^j = A_1^j X$, $XB = B_1$ and $C_1 X = C$. Thus

$$C_1 A_1^j B_1 = C_1 A_1^j X B = C_1 X A^j B = C A^j B \qquad \text{for} \qquad j \geq 0$$

and hence the transfer functions, modulo the constant term, coincide.

The converse of this lemma is not generally true. For the finite dimensional case the additional assumption that both systems are canonical guarantees the similarity of the two systems. This is the content of Theorem I 8-4. In the infinite dimensional case we have to further strengthen the assumptions.

Theorem 1-9 Let $\Sigma = (A, B, C, D)$ and $\Sigma_1 = (A_1, B_1, C_1, D_1)$ be two realizations of the same transfer function T in the state spaces K and K_1, respectively.

(a) If Σ is continuously observable and exactly reachable and Σ_1 continuously observable and continuously reachable then there exists a quasi-invertible operator X which intertwines Σ and Σ_1.

(b) If Σ and Σ_1 are both continuously observable and exactly reachable then the two systems are similar.

PROOF Let R and R_1 be the respective reachability operators and O and O_1 the respective observability operators. By Lemma 1-4 we have

$$O A^j R = O_1 A_1^j R \tag{1-33}$$

For $j = 0$ this implies, O and O_1 being injective, that $\text{Ker } R = \text{Ker } R_1$.

If Σ is exactly reachable then $\bar{R} = R | \{\text{Ker } R\}^{\perp} \to K$ is a boundedly invertible operator. Let $X = R_1 \bar{R}^{-1}$ then X is a bounded quasi-invertible operator from K to K_1. Now from (1-33) restricted to $\{\text{Ker } R\}^{\perp} = \{\text{Ker } R_1\}^{\perp}$

we have $OA^j = O_1 A_1^j X$. For $j = 0$ this reduces to $O = O_1 X$ which implies $C = C_1 X$. Substituting $O = O_1 X$ back into (1-33) yields $O_1 X A^j R = O_1 A_1^j R_1$ and since O_1 is injective that $X A^j R = A_1^j R_1$ for all $j \geq 0$. For $j = 0$ this implies $XR = R_1$ and so also $XB = B_1$. Hence from $X A^j R = A_1^j R_1$ we obtain $X A^j R = A_1^j X R$ for all $j \geq 0$. As R has dense range this implies $X A^j = A_1^j X$ for all $j \geq 0$. Thus X intertwines the two systems which proves (a). If Σ_1 is also exactly reachable then X as defined before is boundedly invertible and hence the two systems are similar.

The previous theorem is an instance of an infinite dimensional state space isomorphism theorem. The proof depended crucially on the assumption of exact reachability. Of course exact observability might have been used instead. Another version of an isomorphism will be encountered in Sec. 7 where instead of strengthening the reachability or observability properties use is made of some symmetry assumptions.

That some extra assumptions need to be made to guarantee similarity is clear from the following example.

Let $A: l^2(0, \infty) \to l^2(0, \infty)$ be given by $Ae_i = \lambda_i e_i$ where $\{e_i\}_{i=0}^\infty$ is the standard orthonormal basis in $l^2(0, \infty)$, the λ_i are distinct numbers with $|\lambda_i| \leq 1$. Let b and c be vectors in $l^2(0, \infty)$ with real coordinates $\{\beta_i\}_{i=0}^\infty$ and $\{\gamma_i\}_{i=0}^\infty$, respectively.

Consider next the system $\Sigma = (A, b, c)$ and (A, c, b). Since $(A^j b, c) = \sum \lambda_i^j \beta_i \gamma_i = (A^j c, b)$ the two systems realize the same transfer functions. Let X be a bounded operator that intertwines Σ and Σ_1. As $XA = AX$ it follows from the cyclicity of A that $X = \varphi(A)$ for some bounded measurable φ. Let $\varphi_i = \varphi(\lambda_i)$ then $Xe_i = \varphi_i e_i$. Since $Xb = c$ we must have $\varphi_i \beta_i = \gamma_i$ which implies that

$$\text{Sup}_{i \geq 0} \left| \frac{\gamma_i}{\beta_i} \right| = \text{Sup}_{i \geq 0} |\varphi_i| = \|X\|$$

But for arbitrary b and c in $l^2(0, \infty)$

$$\text{Sup}_{i \geq 0} \left| \frac{\gamma_i}{\beta_i} \right|$$

need not be bounded. As a case in point we might take $\gamma_n = n^{-1}$ and $\beta_n = n^{-2}$. So the two systems cannot be similar in this case.

This example of nonisomorphic systems had still the same generators. We will see later that there exist canonical realizations of the same transfer functions whose generators have widely differing spectra.

2. HANKEL OPERATORS AND REALIZATION THEORY

It became clear in the previous section that the input/output behavior of a system is, except for the map D, completely determined by its Hankel operator. Thus given an impulse response $(T_0, T_1, T_2, \ldots)$ or its associated transfer function

$T(z) = \sum_{j=0}^{\infty} T_j z^j$ a system (A, B, C, D) is said to realize T if $T_0 = D$ and $T_j = CA^{j-1}B$ for $j \geq 1$.

Assume the function T is $H^{\infty}(B(U, Y))$ and let $H_T : \overline{H}_U^2 \to H_{0,Y}^2$ be its induced Hankel operator. Taking the finite dimensional theory as a guide we expect that the construction of a state space model should use $\overline{H}_U^2/\mathrm{Ker}\,H_T$ or $\overline{\mathrm{Range}\,H_T}$ as possible state spaces. This expectation turns out to be justified.

Theorem 2-1 Let $T \in H^{\infty}(B(U, Y))$. Then there exists a reachable and exactly observable Hilbert space realization of T.

PROOF It is easy to construct a realization of T. Let S_+ be the unilateral right shift in $H_{0,Y}^2$. Define operators $A_1, B_1,$ and C_1 by $A_1 = S_+^*$, $B_1\xi = H_T\xi$, $C_1 f = (S_+^* f)(0)$ and $D_1 = T_0$ then the system (A_1, B_1, C_1, D_1) is a realization of T. This realization may or may not be reachable, depending on whether $\overline{\mathrm{Range}\,H_T} = H_{0,Y}^2$ but it is exactly observable and $O^* : H_{0,Y}^2 \to H_{0,Y}^2$ is just the identity map. To obtain a reachable realization of T all we have to do is to replace the state space $H_{0,Y}^2$ by $\overline{\mathrm{Range}\,H_T}$. Since H_T satisfies the functional equation

$$H_T S_-^* = S_+^* H_T \tag{2-1}$$

it follows that $\mathrm{Ker}\,H_T$ is S_-^*-invariant whereas $\overline{\mathrm{Range}\,H_T}$ is an S_+^*-invariant subspace. Define now $A = S_+^* | \overline{\mathrm{Range}\,H_T}$, $B = B_1$, $C = C_1$, and $D = T_0$ then (A, B, C, D) is a reachable and exactly observable realization of T. We call this realization the *shift realization* of T. The reachability operator R of the shift realization of T coincides with H_T. This follows from the fact that for all $n \geq 0$ and $\xi \in U$

$$R(\xi \overline{\chi}^n) = A^n B\xi = S_+^{*n} H_T \xi = H_T S_-^{*n} \xi = H_T(\xi \overline{\chi}^n)$$

On the other hand the observability of operator of the system O: $\overline{\mathrm{Range}\,H_T} \to H_{0,Y}^2$ is given by $Of = f$ and so its adjoint $O^* : H_{0,Y}^2 \to \overline{\mathrm{Range}\,H_T}$ has the representation $O^* = P_{\overline{\mathrm{Range}\,H_T}}$.

It follows immediately from the previous theorem that a function T in $H^{\infty}(B(U, Y))$ has an exactly reachable and exactly observable Hilbert space realization if $\mathrm{Range}\,H_T$ is closed. We will defer to the next section the characterization of those functions admitting such a realization.

The realizability criterion given by Theorem 2-1 provides a sufficient condition only. In the discrete time case it is easy to characterize all weighting patterns having Hilbert space realizations. We say that a sequence $\{T_n\}_{n=0}^{\infty}$ of operators in $B(U, Y)$ is of exponential type if there exist constants M and w such that

$$\|T_n\| \leq Mw^n \qquad n \geq 0 \tag{2-2}$$

Theorem 2-2 A weighting pattern $\{T_n\}_{n=0}^{\infty}$ has a canonical Hilbert space realization if and only if it is of exponential type.

PROOF Assume (A, B, C, D) is a realization of $\{T_n\}_{n=0}^{\infty}$. Since $T_n = CA^{n-1}B$ it follows that $\|T_n\| \leq \|C\| \cdot \|B\| \cdot \|A\|^{n-1}$ and so $\{T_n\}_{n=0}^{\infty}$ is of exponential type.

Conversely if $\{T_n\}_{n=0}^{\infty}$ is of exponential type then for some nonzero λ, sufficiently small, the sequence $\{\lambda^n T_n\}_{n=0}^{\infty}$ is the sequence of Taylor coefficients of an $H^{\infty}(B(U, Y))$ function. By Theorem 2-1 it has a canonical Hilbert space realization (A, B, C, D). It is clear then that $\{\lambda^{-1}A, B, C, D\}$ is a realization of $\{T_n\}_{n=0}^{\infty}$ and the rescaling does affect neither reachability nor observability.

It will be seen later that while rescaling of a weighting pattern does not affect realizability the fine structure of the realization constructed in the proof of Theorem 2-1 is sensitive to rescaling.

The proof of the realizability result of Theorem 2-1 was asymmetrical inasmuch as it used only $\overline{\text{Range}\,H_T}$ as state space, not considering the possibility of using $\bar{H}_U^2 \ominus \text{Ker}\,H_T$. To get a more symmetric theory we have to use adjoints.

Given a system $\Sigma = (A, B, C, D)$ we define the *adjoint system* Σ^* by $\Sigma^* = (A^*, C^*, B^*, D^*)$. Thus dynamic equations for the adjoint systems are

$$x_{n+1} = A^*x_n + C^*y_n$$
$$u_n = B^*x_n + D^*y_n \tag{2-3}$$

If T_* denotes the transfer function of the adjoint system then it is clear that we obtain the following relation between it and the transfer function T of the original system, namely

$$T_*(z) = \tilde{T}(z) = T(\bar{z})^*$$

Now given a function $T \in H^{\infty}(B(U, Y))$, let Σ_T denote its shift realization. Since $\tilde{T}$ is in $H^{\infty}(B(Y, U))$ whenever T is in $H^{\infty}(B(U, Y))$ we can apply the construction of Theorem 2-1 to obtain the shift realization of $\tilde{T}$ $\Sigma_{\tilde{T}} = (\tilde{A}, \tilde{B}, \tilde{C}, \tilde{D})$. Thus $\Sigma_{\tilde{T}}$ has $\overline{\text{Range}\,H_{\tilde{T}}}$ as state space. Clearly by the definition of the adjoint system and the remarks following it the system $(\Sigma_T)^* = (\tilde{A}^*, \tilde{C}^*, \tilde{B}^*, \tilde{D}^*)$ is a realization of the transfer function T. We call this realization the *-shift realization* of T.

It is of interest to inquire about the relation between the shift and the *-shift realizations of a given transfer function. Contrary to the finite dimensional situation the two realizations, though both canonical, need not be isomorphic. Some obviously necessary conditions for isomorphism arise out of spectral considerations. If we consider the state operators $A = S_+^* | \overline{\text{Range}\,H_T}$ and $\tilde{A}^* = (S_+^* | \overline{\text{Range}\,H_{\tilde{T}}})^*$ then their spectra are determined by Theorem II 13-8 as well as the representation theorem for right invariant subspaces, namely Theorem II 12-22. If we assume that $\{\text{Range}\,H_T\}^{\perp} = QH_{0,Y}^2$ and $\{\text{Range}\,H_{\tilde{T}}\}^{\perp} = Q_1 H_{0,U}^2$ for some rigid functions Q and Q_1 then it is a consequence of Lemma II 13-4 that, U and Y being finite dimensional, that if Q and Q_1 fail to be inner then the spectra of A and $\tilde{A}^*$ coincide with the closed unit disc. However, while the point spectrum of A coincides with the open unit disc $\tilde{A}^*$ has no point spectrum at all. Thus the possibility of the similarity of A and $\tilde{A}^*$ is excluded.

A spectrum of the kind exhibited by the operators above in the case that Q and Q_1 fail to be inner is of a kind that operators arising in applications rarely have. This is an indication that the theory of the shift realization should be restricted to transfer functions T for which $\{\text{Range } H_T\}^\perp$ is an invariant subspace of full range. This theme is the subject of the next sections.

Before proceeding in that direction we elucidate the connection between the Hankel operator induced by $\tilde{T}$, namely $H_{\tilde{T}}$, and H_T^*.

Theorem 2-3 Let $T \in H^\infty(B(U, Y))$ and let H_T and $H_{\tilde{T}}$ be the Hankel operators induced by T and $\tilde{T}$, respectively. Then the operators $H_{\tilde{T}}$ and H_T^* are unitarily equivalent.

PROOF It is simple to check that if $H_T: H_U^2 \to H_{0,Y}^2$ is defined by (1-17) then its adjoint $H_T^*: H_{0,Y}^2 \to \bar{H}_U^2$ is given by

$$H_T^* g = P_{\bar{H}_U^2} T^* g \qquad \text{for} \qquad g \in \bar{H}_U^2 \tag{2-4}$$

Define now a map $\sigma: L_U^2 \to L_U^2$ by

$$(\sigma f)(e^{it}) = e^{it} f(e^{-it}) \tag{2-5}$$

σ is unitary and satisfies $\sigma^* = \sigma^{-1} = \sigma$ as well as $\sigma H_{0,U}^2 = \bar{H}_U^2$.

For every $g \in \bar{H}_Y^2$ we have

$$\sigma(H_{\tilde{T}} g) = P_{H_{0,U}^2} \tilde{T} g = P_{\bar{H}_U^2} \sigma(\tilde{T} g) = P_{\bar{H}_U^2} T^*(\sigma g) = H_T^*(\sigma g)$$

and hence the equality

$$\sigma H_{\tilde{T}} = H_T^* \sigma \tag{2-6}$$

follows.

3. RESTRICTED SHIFT SYSTEMS

It was indicated in the previous section that the shift realization might be a useful tool for the study of state space models of transfer functions T for which the induced Hankel operator H_T has a range whose orthogonal complement is an invariant subspace of full range.

In general we define a system (A, B, C) to be a *restricted shift system* whenever its generator, or state transition operator A, is unitarily equivalent to the restriction of a left shift to a left invariant subspace whose orthogonal complement is of full range.

Thus A is unitarily equivalent to $S_0(Q)^*$ where

$$S_0(Q)^* = S_+^* \,|\, H_0(Q) \tag{3-1}$$

and

$$H_0(Q) = \{QH_{0,N}^2\}^\perp \tag{3-2}$$

It is a consequence of Theorem II 13-2 that (A, B, C) is a restricted shift system if and only if (A^*, C^*, B^*) is. We further assume that the inner function Q has a scalar multiple. This is certainly the case whenever $\dim N < \infty$ when $\det Q$ provides a scalar multiple of Q. As before the spaces U and Y of input and output values are assumed to be finite dimensional.

With this definition we have greatly restricted the class of systems under consideration and our first object is the characterization of the class of input/output maps that have realizations by means of restricted shift systems.

Applying B to a fixed vector ξ in U we have $B\xi \in H_0(Q)$ and hence $(B\xi)(z) = b_\xi(z)$. Since the vector function b_ξ depends linearly on ξ there exists a $B(U, N)$ valued function D, analytic in the open unit disc, for which

$$(B\xi)(z) = b_\xi(z) = D(z)\,\xi \tag{3-3}$$

Similarly there exists a $B(Y, N)$-valued analytic function E for which

$$(C^*\eta)(z) = E(z)\eta \qquad \text{for all} \qquad \eta \in Y \tag{3-4}$$

The functions D and E need not necessarily be norm bounded in the open unit disc.

If we compute $(CA^n B\xi, \eta)$ for arbitrary $\xi \in U$, $\eta \in Y$ and $n \geq 0$ we obtain

$$(CA^n B\xi, \eta) = (CS_0(Q)^* B\xi, \eta) = (S_0(Q)^* B\xi, C^*\eta)$$

$$= (P_{H_0(Q)}\bar{\chi}^n D\xi, E\eta) = (\bar{\chi}^n D\xi, E\eta)$$

$$= \frac{1}{2\pi} \int e^{-int} (E(e^{it})^* D(e^{it})\,\xi, \eta)\, dt$$

Letting

$$(E^* D)(e^{it}) = \sum_{n=-\infty}^{\infty} T_n e^{int}$$

we get $T_n = CA^{n-1}B$ for $n > 0$. Thus the transfer function of (A, B, C) coincides with $\sum_{n=1}^{\infty} T_n z^n$.

The fact that the multiplication operator B defined by (3-3) has its range in $H_0(Q)$ implies a certain factorization representation given next.

Theorem 3-1 Let Q be an inner function in N and let $B: U \to H_0(Q)$ be a bounded linear operator defined by (3-3) for some $B(U, N)$-valued function D which is analytic in the open disc. Then D is factorable on the unit circle in the form

$$D(e^{it}) = Q(e^{it})\, F(e^{it})^* \tag{3-5}$$

where F is another $B(N, U)$-valued analytic function in the unit disc.

PROOF   We use Lemma II 13-5. Since $D\xi$ is in $H_0(Q)$ for each $\xi \in U$ it follows that $Q^* D\xi$ is in $\bar{H}_N^2$ for each ξ. Since the dependence on ξ is linear there exists a function F analytic in the unit disc such that $Q^* D\xi = F^*\xi$ which proves the theorem.

We can give now a first characterization of transfer functions realizable by restricted shift systems.

Theorem 3-2 An $H^\infty(B(U, Y))$ function T is realizable by a restricted shift system if and only if T is factorable on the unit circle as

$$T(e^{it}) = E(e^{it})^* \, Q(e^{it}) \, F(e^{it})^* \tag{3-6}$$

for some inner function Q acting in a Hilbert space N and where E and F are, respectively, $B(Y, N)$- and $B(N, U)$-valued analytic functions in the open unit disc that induce bounded multiplication operators through (3-3) and (3-4).

PROOF We saw that if T is realizable by a restricted shift system then it can be factored as $T = E^*D$. Theorem 3-1 and the factorization (3-5) of D imply the result.

Conversely if T admits a factorization of the form (3-6) then the system (A, B, C) defined by (3-1), (3-3), and (3-4) is clearly a restricted shift system. Moreover (A, B, C, T_0) realizes T.

We would like to relate the possibility of factoring a function as in Theorem 3-2 to some intrinsic property of the function. To this end we introduce some definitions.

We say that a function $T \in L^\infty(B(U, Y))$ is *cyclic* (cyclicity here is relative to the left shift in $H^2_{0,Y}$) if Range H_T is dense in $H^2_{0,Y}$ and *noncyclic* otherwise. A function T is called *strictly noncyclic* if $\{\text{Range } H_T\}^\perp$ is an invariant subspace of full range. In case dim $Y = 1$ noncyclicity and strict noncyclicity coincide but for shifts of multiplicity greater than one the notions differ. Of course strict noncyclicity implies noncyclicity. We note also that Range H_T cannot equal $H^2_{0,Y}$. This is excluded by the functional equation (2-1) of the Hankel operator and the fact that the left shift S_+^* in $H^2_{0,Y}$ is similar neither to the right shift nor to its compression to a left invariant subspace.

Let Ω be a domain in the complex plane. A $B(U, Y)$-valued function F is *meromorphic of bounded type* in Ω if $F = G/g$ where G and g are, respectively, bounded $B(U, Y)$-valued and scalar-valued analytic functions in Ω.

If a function T in $H^\infty(B(U, Y))$ is the strong radial limit of a function $\hat{T}$ meromorphic function and of bounded type in $D_e = \{\lambda \mid 1 < |\lambda| \leq \infty\}$ then we say that $\hat{T}$ is a *meromorphic pseudocontinuation of bounded type* of T. Clearly a meromorphic continuation of bounded type of T is at the same time a pseudocontinuation but the converse is generally false. There are functions in H^∞ for which the unit disc is the natural domain of analyticity but which still admit a meromorphic pseudocontinuation.

Lemma 3-3 Let $T \in L^\infty(B(U, Y))$. Suppose there exists a nonzero function $\varphi \in H^\infty$ such that $\varphi T \in H(B(U, Y))$ then there exists an inner function q for which $qT \in H^\infty(B(U, Y))$.

PROOF The set $J = \{\psi | \psi \in H^\infty, \psi T \in H^\infty(B(U, Y))\}$ is a nontrivial w^*-closed ideal in H^∞ hence of the form $J = qH^\infty$.

Consequently if F is a meromorphic function of bounded type in D_e then in a representation $F = G/g$ the denominator g can be taken to be an inner function in D_e.

Inner functions acting in finite dimensional Hilbert spaces have meromorphic pseudocontinuations of bounded type to D_e. Actually a somewhat stronger result holds.

Lemma 3-4 Let Q be an inner function in N. Then Q has a meromorphic pseudocontinuation of bounded type if and only if it has a scalar multiple.

PROOF Assume Q has a scalar multiple q. Thus

$$QQ^a = Q^a Q = qI \tag{3-7}$$

holds for some function Q^a. Since clearly qQ^* is analytic it follows from Lemma 3-3 that without loss of generality we may assume q, and therefore also Q^a, are inner. For $z \in D_e$ we define $\hat{Q}(z) = \tilde{Q}^a(z^{-1})/\tilde{q}(z^{-1})$. Clearly $\tilde{Q}^a(z^{-1})$ and $q(z^{-1})$ are bounded analytic functions in D_e. Thus $\hat{Q}$ is meromorphic and of bounded type in D_e. From the definition of $\hat{Q}$ it follows that a.e. on the unit circle

$$\lim_{R \to 1^+} \hat{Q}(Re^{it}) = \lim_{R \to 1^+} \tilde{Q}^a\left(\frac{1}{R}e^{-it}\right) \Big/ \tilde{q}\left(\frac{1}{R}\bar{e}^{it}\right) = \lim_{r \to 1^-} Q^a(re^{it})^*/\overline{q(re^{it})}$$

$$= Q^a(e^{it})^*/\overline{q(e^{it})} = q(e^{it})Q^a(e^{it})^*$$

But from (3-7) we obtain $q(Q^a)^* = Q$ and so $\lim_{R \to 1^+} \hat{Q}(Re^{it}) = \lim_{r \to 1^-} Q(re^{it})$ a.e. and $\hat{Q}$ is a pseudocontinuation of Q.

Conversely if Q has a meromorphic pseudocontinuation of bounded type $\hat{Q}$ in D_e then $\hat{Q} = G/g$ and g may be taken to be inner in D_e. This implies that $\bar{g}$ is a scalar multiple of Q.

We note that, as every inner function Q acting in a finite dimensional space has a scalar multiple, the construction in the lemma yields a pseudomeromorphic continuation of bounded type for Q. Moreover since (3-3) implies that, whenever $q(z) \neq 0$, $Q(z)^{-1} = Q^a(z)/q(z)$ then the function $\hat{Q}$ can be written also as $\hat{Q}(z) = \tilde{Q}(z^{-1})^{-1}$.

Theorem 3-5 Let $T \in L^\infty(B(U, Y))$ where U and Y are finite dimensional Hilbert spaces. Then the following statements are equivalent
(a) T is strictly noncyclic.
(b) T is a strong radial limit a.e. of a meromorphic function of bounded type in D_e.

c) On the unit circle T is factorable as

$$T = PC^* = C_1^* P_1 \tag{3-8}$$

where P and P_1 are inner functions in Y and U, respectively, C and C_1 are in $H^\infty(B(Y, U))$, and the coprimeness relations

$$(P, C)_R = I_Y \quad \text{and} \quad (P_1, C_1)_L = I_U \tag{3-9}$$

hold.

PROOF Assume T is strictly noncyclic, thus $\{\text{Range}\, H_T\}^\perp \supset PH_{0,Y}^2$ for some inner function P acting in Y. Equivalently $\overline{\text{Range}\, H_T} \subset H_0(P)$. Let $\xi \in U$ then $H_T\xi = P_{H_{0,Y}^2} T\xi$ and, by applying Lemma II 13-5, $P^* P_{H_{0,Y}^2} T\xi \in H_Y^2$. Since $P^* \bar{H}_Y^2 \subset \bar{H}_Y^2$ we have $P^* T\xi \in \bar{H}_Y^2$ for all $\xi \in U$. This implies that $P^* T$, which is in $L^\infty(B(U, Y))$, has a Fourier expansion in which all positively indexed Fourier coefficients vanish. Thus $P^* T = C^*$ for some C in $H^\infty(B(Y, U))$ which implies the factorization $T = PC^*$.

Assume now $T \in L^\infty(B(U, Y))$ admits a factorization of the form $T = PC^*$ with P and C as before. Define a function $\hat{T}$ in D_e by $\hat{T}(z) = \hat{P}(z)\tilde{C}(z^{-1}) = \tilde{P}(z^{-1})^{-1}\tilde{C}(z^{-1})$. Since $\tilde{C}(z)$ is analytic in the unit disc $\tilde{C}(z^{-1})$ is analytic in D_e. Also $\hat{P}(z) = \tilde{P}(z^{-1})^{-1}$ is, by Lemma 3-4, meromorphic of bounded type in D_e and hence so is $\hat{T}$. Moreover

$$\lim_{R \to 1^+} \hat{T}(Re^{it}) = \lim_{R \to 1^+} \tilde{P}\left(\frac{1}{R}e^{-it}\right)^{-1}\tilde{C}\left(\frac{1}{R}e^{-it}\right) = P(e^{it})C(e^{it})^* \quad \text{a.e.}$$

Thus (*a*) implies (*b*) and the first factorization in (3-8).

Next assume $T(e^{it})$ is a.e. the strong radial limit of a meromorphic function of bounded type $\hat{T}$ in D_e. If $\hat{T} = G/g$ with g an inner function then necessarily $G = H^*$ and $g = \bar{q}$ where $H \in H^\infty(B(Y, U))$ and $q = \bar{g}$ is inner. This last representation implies that $\bar{q}T\xi \in \bar{H}_Y^2$ for all $\xi \in U$. But $\bar{q}T\xi = \bar{q}\{P_{\bar{H}_Y^2} T\xi + P_{H_{0,Y}^2} T\xi\} = \bar{q}\{P_{\bar{H}_Y^2} T\xi + H_T\xi\}$. As $\bar{q}\bar{H}_Y^2 \subset \bar{H}_Y^2$ it follows that $\bar{q}H_T\xi \in \bar{H}_Y^2$ for all ξ in U and this implies that $\overline{\text{Range}\, H_T} \subset H_0(qI) = qH_{0,Y}^2\}^\perp$. Since $qH_{0,Y}^2$ is a subspace of full range so is $\{\text{Range}\, H_T\}^\perp$ which includes $qH_{0,Y}^2$. Thus T is strictly noncyclic and for this the coprimeness relations (3-9) are irrelevant.

If T is a radial limit of $\hat{T}$ which is meromorphic and of bounded type in D_e then $\tilde{T}$ is the radial limit of $\tilde{T}$. Thus T is strictly noncyclic if and only if $\tilde{T}$ is. Hence if T is strictly noncyclic $\tilde{T} = R_1 D_1^*$ with R_1 inner in U and $D_1 \in H^\infty(B(U, Y))$. It follows that $T = C_1^* P_1$ with $P_1 = \tilde{R}_1$ and $C_1 = \tilde{D}_1$.

Finally, assume $\overline{\text{Range}\, H_T} = H_0(P)$ which implies that $T = PC^*$. Apply the map $\tau_Q' : L_Y^2 \to L_Y^2$ defined by

$$\tau_Q' f = \chi \tilde{Q}(Jf) \tag{3-10}$$

where $(Jf)(e^{it}) = f(e^{it})$. τ_Q' is closely related to the map τ_Q introduced in Sec. II-13 and all duality results obtained there are easily adapted to the present setting. τ_P' maps $H_0(P)$ unitarily onto $H_0(\tilde{P})$ and $\tau_P' P_{H_0(P)} = P_{H_0(\tilde{P})}\tau_P'$.

It follows that for $f \in \bar{H}_U^2$

$$\tau_P' H_T f = \tau_P' P_{H_{0,Y}^2} T f = \tau_P' P_{H_0(P)} T f = P_{H_0(\tilde{P})} \tau_P' T f$$

$$= P_{H_0(\tilde{P})} \chi \tilde{P} \tilde{P}^* \tilde{C}(Jf) = P_{H_0(\tilde{P})} \chi \tilde{C}(Jf)$$

Hence $\overline{\text{Range } H_T} = H_0(P)$ if and only if the operator $M_{\tilde{C}} : H_U^2 \to H_0(\tilde{P})$ given by

$$M_{\tilde{C}} g = P_{H_0(\tilde{P})} \chi \tilde{C} g \qquad \text{for} \qquad g \in H_U^2$$

has dense range. By Theorem II 14-11 this is equivalent to the coprimeness condition $(\tilde{C}, \tilde{P})_L = I_U$ or, alternatively, to $(C, P)_R = I_U$. In a completely analogous fashion $\overline{\text{Range } H_T} = H_0(\tilde{P}_1)$ if and only if $(P_1, C_1) = I_Y$. This completes the proof.

As a straightforward corollary we obtain

Corollary 3-6
(a) Let T_1 and T_2 be strictly noncyclic functions in $H^\infty(B(U, Y))$ then $T_1 + T_2$ is strictly noncyclic.
(b) Let $T_1 \in H^\infty(B(U, W))$, and $T_2 \in H^\infty(B(W, Y))$ be strictly noncyclic then $T_2 T_1$ is strictly noncyclic.
(c) Let $T \in H^\infty(B(U, Y))$ be strictly noncyclic then $\tilde{T}$ is strictly noncyclic in $H^\infty(B(Y, U))$.

Corollary 3-7 A function $T \in H^\infty(B(U, Y))$ is realizable by a restricted shift system if and only if it is strictly noncyclic.

PROOF By Theorem 3-2 realizability by a restricted shift system is equivalent to a factorization of the form (3-6) on the unit circle. Since the factorizations (3-8) are special cases of (3-6) it is clear that strictly noncyclic functions are realizable by restricted shift systems. Conversely if $T = E^* Q F^*$ then, applying the previous theorem, $QF^* = F_1^* Q_1$ and $T = E^* F_1^* Q_1 = (F_1 E)^* Q_1$ is strictly noncyclic.

If we consider the restricted shift systems as generalizations of finite dimensional systems then strictly noncyclic functions take the place of rational functions. In fact the coprime factorizations (3-6) can be viewed as replacing the description of rational functions as quotients of polynomials. In this connection we note the following result.

Theorem 3-8 Let $T \in H^\infty(B(U, Y))$. Then Range H_T is finite dimensional if and only if T is rational.

PROOF Assume Range $H_T = H_0(Q)$ is finite dimensional. Thus Q is a finite Blaschke function, that is, $q = \det Q$ is a finite Blaschke product. Since the pseudomeromorphic continuation of T is an actual analytic continuation

T has only a finite number of poles on the Riemann sphere and hence is rational. Conversely if T is rational then $T(z) = G(z)/g(z)$ where $g(z)$ is a polynomial of degree k with zeroes in D_e. Let $g_1(z) = g(z)/z^k \tilde{g}(z^{-1})$ then g_1 is inner and we can write $T(z) = z^k \tilde{g}(z^{-1}) G(z)/g_1(z)$. This implies that $\text{Range}\, H_T \subset \{q_1 H_{0,Y}^2\}^\perp$ where $q_1 = \overline{g_1}$. But $\{q_1 H_{0,Y}^2\}^\perp$ is of finite dimension equal to $k \cdot \dim Y$.

Given a restricted shift system (A, B, C) its reachability and observability properties are completely determined by the analytic functions induced by the operators B and C and by the corresponding inner function. The next theorem describes the various possibilities in terms of coprimeness.

Theorem 3-9 Let $T \in H^\infty(B(U, Y))$ be strictly noncyclic admitting the factorization $T = E^*QF^*$ on the unit circle where E and F are, respectively, $B(Y, N)$- and $B(N, U)$-valued analytic functions in the unit disc that induce bounded multiplication operators. The realization (A, B, C, D) of T where

$$A = S_+^* \,|\, H_0(Q) \qquad (3\text{-}11)$$

$$B\xi = QF^*\xi \qquad (3\text{-}12)$$

$$Cf = (S_+^* E^* f)(0) \qquad (3\text{-}13)$$

and

$$D = T_0 \qquad (3\text{-}14)$$

is

(*a*) reachable if and only if

$$(Q, F)_R = I_N \qquad (3\text{-}15)$$

(*b*) exactly reachable if and only if

$$[Q, F]_R = I_N \qquad (3\text{-}16)$$

(*c*) observable if and only if

$$(E, Q)_L = I_N \qquad (3\text{-}17)$$

and

(*d*) exactly observable if and only if

$$[E, Q]_L = I_N \qquad (3\text{-}18)$$

PROOF The reachability and observability operators of the system are given by $R: \overline{H}_U^2 \to H_0(Q)$ and $O: H_{0,Y}^2 \to H_0(Q)$ where

$$Rf = P_{H_0(Q)} QF^* f = H_{QF^*} f, \qquad f \in \overline{H}_U^2 \qquad (3\text{-}19)$$

and

$$O^* g = P_{H_0(Q)} \chi E g, \qquad g \in H_{0,Y}^2 \qquad (3\text{-}20)$$

For O^* the result follows directly from Theorem II 14-11. To get the result for the reachability operator we apply the map τ'_Q defined by (3-10). Since

$$\Sigma A^i B \xi_i = \Sigma S_0(Q)^{*i} QF^* \xi_i$$

we have

$$\tau'_Q \Sigma S_0(Q)^{*i} QF^* \xi_i = \Sigma S_0(\tilde{Q})^i \tau'_Q (QF^* \xi_i) = P_{H_0(\tilde{Q})} \chi \, F \Sigma \chi \xi_i$$

Hence $\tau'_Q R f = P_{H_0(\tilde{Q})} \chi \bar{F}(Jf)$ for $f \in \bar{H}_U^2$. Since $J\bar{H}_U^2 = H_U^2$ we apply Theorem II 14-11 once again and the proof is complete.

Next we characterize the class of functions in $H^\infty(B(U, Y))$ whose induced Hankel operators have closed range.

Theorem 3-10 Let $T \in H^\infty(B(U, Y))$. Then the following statements are equivalent.
(a) H_T has closed range.
(b) $H_{\bar{T}}$ has closed range.
(c) On the unit circle T factors as

$$T = QH^* \qquad (3\text{-}21)$$

where Q is inner in Y and $H \in H^\infty(B(Y, U))$ and

$$[Q, H]_R = I_Y \qquad (3\text{-}22)$$

holds.
(d) On the unit circle T factors as

$$T = H_1^* Q_1 \qquad (3\text{-}23)$$

and

$$[Q_1, H_1]_L = I_U \qquad (3\text{-}24)$$

holds.

PROOF That (c) implies (a) follows from the previous theorem. Suppose conversely that H_T has closed range. As H_T satisfies the functional equation

$$S_+^* H_T = H_T S_-^* \qquad (3\text{-}25)$$

we deduce that $\text{Range}\, H_T$ is S_+^*-invariant and $\text{Ker}\, H_T$ is S_-^*-invariant. The restriction of H_T to $\{\text{Ker}\, H_T\}^\perp$ is a boundedly invertible operator of $\{\text{Ker}\, H_T\}^\perp$ onto $\text{Range}\, H_T$ which moreover satisfies

$$(S_+^* \,|\, \text{Range}\, H_T)\, H_T = H_T (P_{\{\text{Ker}\, H_T\}^\perp} S_-^* \,|\, \{\text{Ker}\, H_T\}^\perp)$$

Thus the operators $S_+^* \,|\, \text{Range}\, H_T$, and $P_{\{\text{Ker}\, H_T\}^\perp} S_-^* \,|\, \{\text{Ker}\, H_T\}^\perp$ are similar. If T is not strictly noncyclic then $\text{Range}\, H_T = \{QH_{0,Y}^2\}^\perp$ for some rigid function Q which is not inner. In that case it follows from Lemma II 13-4 that every point of the open unit disc is an eigenvalue of $S_+^* | \text{Range} H_T$. On the other hand the operator $P_{\{Ker H_T\}^\perp} S_-^* | \{Ker H_T\}^\perp$, which is unitarily equivalent to a compression of the right shift in H_U^2 to a left invariant subspace, can have at most a countable number of eigenvalues. Thus necessarily T is strictly noncyclic and $\text{Range} H_T = H_0(Q)$ for some inner function Q. So the factorization (3-21) holds and the strong coprimeness condition (3-22) follows from the previous theorem. By similar reasoning statements (b) and (d) are equivalent. Finally, we use the fact that a bounded operator A has closed range if and only if its adjoint A^* has closed range. So $\text{Range} H_T$ is closed if and only if $\text{Range} H_T^*$ is closed. But H_T^* and $H_{\tilde{T}}$ are unitarily equivalent, the equivalence given by Eqs (2-5) and (2-6). Thus (a) and (b) are equivalent.

Corollary 3-11 Let $T \in H^\infty(B(U, Y))$, U, Y finite dimensional. Then T is realizable by an exactly reachable and exactly observable system if and only if $\text{Range} H_T$ is closed.

PROOF If $\text{Range} H_T$ is closed then T is strictly noncyclic and factors on the unit circle as $T = QH^*$ where Q and H are strongly right coprime. Theorem 3-9 provides a realization which by Theorem 3-10 is exactly reachable and exactly observable.

Conversely assume T is realizable by an exactly reachable and exactly observable system. Since the shift realization of T is reachable and exactly observable it follows from Theorem 1-9 that the two systems are isomorphic. In particular the reachability operators are similar. But the reachability operator of the shift realization is H_T and so necessarily H_T has closed range.

4. SPECTRAL MINIMALITY OF RESTRICTED SHIFT SYSTEMS

In the absence of a general state space isomorphism theorem in the infinite dimensional context we are faced with a situation, and the last example in this section shows that this is a reality, that there may exist canonical realizations of the same transfer function which besides being nonisomorphic have generators with widely differing spectra.

From an intuitive point of view it seems clear that a state space model should, through the spectrum of the state operator, reflect the singularities of the transfer function in a faithful way. In some sense we should look for realization where the state operator has the smallest possible spectrum required to model the singularities of the transfer function.

To make this more precise let (A, B, C, D) be a realization of a transfer function T that is, in the neighborhood of the origin we have

$$T(z) = D + zC(I - zA)^{-1} B \tag{4-1}$$

Clearly this representation is analytic at all points z where $|z|^{-1}$ is larger than the spectral radius of A. But the formula (4-1) can be used as a basis for the analytic continuation of T at least to all points z where $z^{-1} \in \rho_0(A)$, $\rho_0(A)$ being the principal component of $\rho(A)$ the resolvent set of A. If $\sigma_0(z)$ denotes the complement of $\rho_0(A)$ we clearly have

$$\{\lambda \,|\, \lambda^{-1} \in \sigma(T)\} \subset \sigma_0(A) \tag{4-2}$$

$\sigma(T)$ denotes the set of nonanaliticity of T. We call (4-2) the *spectral inclusion property* and say that (A, B, C, D) is a *spectrally minimal* realization if there exists an analytic continuation of T for which equality holds in (4-2).

This section is devoted to a finer spectral analysis of the shift realization for strictly noncyclic functions, the ultimate aim being the proof of the spectral minimality of such realizations. In the process we prove some results concerning inner functions that, interesting in themselves, turn out to be useful also for the degree theory developed in the next section.

We recall, Lemma II 12-25, that given any inner function Q acting in an n-dimensional Hilbert space N there exists a unique, up to a constant factor of absolute value one, scalar inner function σ which satisfies $\sigma H_N^2 \subset QH_N^2$ and $\sigma \,|\, a$ for every $a \in H^\infty$ for which $aH_N^2 \subset QH_N^2$. The function σ is the minimal inner function of Q. We have $\sigma \,|\, \det Q$ and $\det Q \,|\, \sigma^n$.

Let now P and R be two inner functions acting in a finite dimensional Hilbert space N. Since $PH_N^2 \cap RH_N^2$ is an invariant subspace of full range, by the remarks following Theorem II 12-24, we have $PH_N^2 \cap RH_N^2 = QH_N^2$ for some inner function Q.

Theorem 4-1 Let P, R, and Q be inner functions such that $QH_N^2 = PH_N^2 \cap RH_N^2$ and let π, ρ, and σ be their respective minimal inner functions. Then

(a) $\pi \,|\, \sigma$, $\rho \,|\, \sigma$, and $\sigma \,|\, \pi\rho$.

(b) $\sigma = \pi\rho$, equality up to a constant factor of absolute value one, if and only if π and ρ are coprime.

PROOF Since $QH_N^2 = PH_N^2 \cap RH_N^2$ we have $QH_N^2 \subset PH_N^2$. Thus $\sigma H_N^2 \subset QH_N^2 \subset PH_N^2$ and hence $\pi \,|\, \sigma$ and similarly $\rho \,|\, \sigma$. Now as $\pi H_N^2 \subset PH_N^2$ and $\rho H_N^2 \subset RH_N^2$ it follows that

$$(\pi \vee \rho) H_N^2 = \pi H_N^2 \cap \rho H_N^2 \subset PH_N^2 \cap RH_N^2 = QH_N^2$$

and hence $\sigma \,|\, \pi \vee \rho$. This together with the division relations $\pi \,|\, \sigma$ and $\rho \,|\, \sigma$ implies that $\sigma = \pi \vee \rho$ and clearly $\pi \vee \rho \,|\, \pi\rho$ which proves (a). Statement (b) follows from the observation that $\pi \vee \rho = \pi\rho$ if and only if $\pi \wedge \rho = 1$, that is, if and only if π and ρ are coprime.

We proceed to discuss some closely related notions of minimality. Given an inner function Q acting in N, assumed to be finite dimensional, we have associated with it its minimal inner functions.

A contraction operator X is said to belong to the class C_0 if it is a completely nonunitary contraction for which there exists a nontrivial function a in H^∞ for which $a(X) = 0$. Clearly given the matrix inner function Q then the operators $S(Q)$ and $S_0(Q)$ defined by (II 13-4) and (3-1) are C_0 contractions. An annihilating function can be taken to be $q = \det Q$. Given a C_0 contraction X then $J_X = \{\varphi \in H^\infty \,|\, \varphi(X) = 0\}$ is a w^*-closed ideal in H^∞ and hence has the representation $J_X = m_X H^\infty$ for some inner function m_X. The function m_X is called the *minimal inner function* of X.

Finally, with each strictly noncyclic function $T \in L^\infty(B(U, Y))$ we associate the minimal inner function μ_T which makes $\bar{\mu}_T T$ have an analytic extension to the exterior of the closed unit disc.

Theorem 4-2 Let $T \in H^\infty(B(U, Y))$ be strictly noncyclic having the coprime factorizations

$$T = QH^* = H_1^* Q_1 \tag{4-3}$$

on the unit circle.

Let σ, m, and μ be the minimal inner functions of Q, $S_0(Q)$, and T, respectively. Then, up to a constant factor of absolute value one, σ, m, and μ coincide.

PROOF For each $f \in H_{0,N}^2$ we have $P_{H_0(Q)}\sigma f = 0$ as $\sigma H_{0,N}^2 \subset QH_{0,N}^2$. This implies that $m \,|\, \sigma$. Conversely if $m(S(Q)) = 0$, it follows by the invariance of $QH_{0,N}^2$ under multiplication by m that $P_{H_0(Q)}mf = 0$ for all $f \in H_{0,N}^2$. Thus $mH_{0,N}^2 \subset QH_N^2$ and so $\sigma \,|\, m$ and the coincidence of σ and m follows.

Since $\bar{\sigma}Q$ has an analytic extension to D_e so has $\bar{\sigma}T = \bar{\sigma}QH^*$ which shows that $\mu \,|\, \sigma$. Conversely since $\bar{\mu}T$ extends analytically to D_e we have $\bar{\mu}T = G^*$ for some G in $H^\infty(B(Y, N))$. Thus $T = \mu G^*$ and Range $H_T \subset (\mu H_{0,N}^2)^\perp$ which is equivalent to $\mu H_{0,N}^2 \subset QH_{0,N}^2$. Thus $\sigma \,|\, \mu$ which completes the proof.

Coprime factorizations of the minimal inner function σ of an inner function Q induce factorizations of Q itself.

Theorem 4-3 Let N be an n-dimensional Hilbert space and let Q be an inner function acting in N. Let σ be the minimal inner function of Q and $\sigma = \pi\rho$ any coprime factorization of σ. Then there exist inner functions P and R, having π and ρ as their respective minimal inner functions, such that

$$QH_N^2 = PH_N^2 \cap RH_N^2 \tag{4-4}$$

and

$$\det Q = \det P \cdot \det R \tag{4-5}$$

Furthermore there exist inner functions P_1 and R_1 for which

$$Q = PR_1 = RP_1 \tag{4-6}$$

PROOF Assume $\sigma = \pi\rho$ is a coprime factorization of σ. Let $M_\pi = \{f \in H_N^2 \,|\, \rho f \in QH_N^2\}$ and $M_\rho = \{f \in H_N^2 \,|\, \pi f \in QH_N^2\}$. Clearly M_π and M_ρ are invariant subspaces of H_N^2 and moreover $QH_N^2 \subset M_\pi \cap M_\rho$. Since QH_N^2 is of full range so are M_π and M_ρ and they have therefore representations of the form $M_\pi = PH_N^2$ and $M_\rho = RH_N^2$ for some inner functions P and R. To prove the converse inclusion $M_\pi \cap M_\rho \subset QH_N^2$ it suffices to show that $H(Q) = \{QH_N^2\}^\perp \subset (PH_N^2 \cap RH_N^2)^\perp$ or that $H(Q)$ is orthogonal to $PH_N^2 \cap RH_N^2$. To this end let f and g be arbitrary elements in $H(Q)$ and $PH_N^2 \cap RH_N^2$, respectively. Define $J = [\varphi \in H^\infty | \int \chi^n \varphi(g, f)\, dt = 0, n \geq 0]$. Obviously J is an invariant subspace of H^∞ which is nontrivial as both π and ρ are in J. Since π and ρ are coprime this implies that $J = H^\infty$. Letting $n = 0$ we obtain $\int (g(e^{it}), f(e^{it}))\, dt = 0$. Since f was arbitrary in $H(Q)$ necessarily $g \in QH_N^2$, or $PH_N^2 \cap RH_N^2 \subset QH_N^2$ and equality follows. Since $PH_N^2 \cap RH_N^2 \subset PH_N^2$ we have the existence of an inner function R_1 such that $Q = PR_1$ and analogously there exist an inner function P_1 such that $Q = RP_1$.

To see that π is the minimal inner function of P we note that for each $f \in H_N^2$, $\rho(\pi f) = (\rho \pi) f = \sigma f \in QH_N^2$, hence from the definition of M_π it follows that $\pi H_N^2 \subset M_\pi = PH_N^2$. If α is any inner function for which $\alpha H_N^2 \subset M_\pi$ then $(\alpha \rho) H_N^2 \subset M_\pi \subset QH_N^2$ which implies that $\sigma | \alpha \rho$ or that $\pi | \alpha$. Thus π is the minimal inner function of P.

Finally we prove equality (4-5). From the factorizations (4-6) we obtain $\det P | \det Q$ as well as $\det R | \det Q$. Since $\det P$ and $\det R$ are coprime it follows that $\det P \cdot \det R | \det Q$. To prove the converse we note that $\sigma | \det Q$ and $\det Q | \sigma^n$. Since $\sigma^n = \pi^n \rho^n$ it follows that $\det Q$ can be factored as $\det Q = pr$ where $p | \pi^n$ and $r | \rho^n$. Now $\rho M_\pi = \rho PH_N^2 \subset QH_N^2$ implies that $\det Q | \rho^n \det P$ or that $p | \det P$. Similarly $r | \det R$ and the two division relations taken together with the coprimeness of p and r yield $\det Q | \det P \cdot \det R$, which proves (4-5).

The importance of the minimal inner function μ of a strictly noncyclic function T stems from the fact that it gives a parametrization of the singularities of $\hat{T}$ the analytic extension of T to D_e. Since T is an operator valued function this description is insufficient and this is the motivation for deriving the next results concerning ideals in $H^\infty(B(U, Y))$.

Let V_1 and V_2 be two Hilbert spaces. We denote by $TC(V_1, V_2)$ and $HS(V_1, V_2)$ the trace class, and Hilbert–Schmidt class of operators from V_1 to V_2, respectively. The trace class norm of an operator $T \in B(V_1, V_2)$ is denoted by $\|T\|_1$. We let $L^1(TC(V_1, V_2))$ be the space of all weakly measurable $TC(V_1, V_2)$-valued functions F on the unit circle that satisfy

$$\|F\|_{L^1(TC(V_1, V_2))} = \int_0^{2\pi} \|F(e^{it})\|_1\, dt < \infty$$

$TC(V_1, V_2)$ considered as a Banach space has $B(V_1, V_2)$ as its dual under the pairing

$$\langle T, X \rangle = tr(X^* T)$$

As a consequence the dual of $L^1(TC(V_1, V_2))$ is given by $L^\infty(B(V_1, V_2))$ where the pairing is

$$\langle F, G \rangle = \frac{1}{2\pi} \int_0^{2\pi} tr\big(G(e^{it})^* \, F(e^{it})\big) \, dt$$

The Hilbert–Schmidt class $HS(V_1, V_2)$ becomes a Hilbert space under the inner product $(F_1, F_2) = tr(F_2^* F_1)$. Thus it is natural to consider the corresponding Hardy space $H^2_{HS(V_1,V_2)}$. A representation theorem extending the Beurling–Lax Theorem II 12-22, is the following.

Theorem 4-4

(a) A subspace of $H^2_{HS(V_1,V_2)}$ is invariant under right multiplication by all $H^\infty(B(V_1))$ functions if and only if it is of the form $QH^2_{HS(V_1,V_2)}$ for some rigid function Q.

(b) A subspace of $H^2_{HS(V_1,V_2)}$ is invariant under left multiplication by all $H^\infty(B(V_2))$ functions if and only if it is of the form $H^2_{HS(V_1,V_2)}Q_1$ for some rigid function Q_1.

PROOF Since $HS(V_1, V_2)$ is invariant under right multiplication by $B(V_1)$ operators it follows that given a rigid function Q the space $QH^2_{HS(V_1,V_2)}$ is invariant under right multiplication by all $H^\infty(B(V_1))$ functions. It is closed since it is the range of a partial isometry in $H^2_{HS(V_1,V_2)}$ induced by Q. The initial space of this partial isometry is the set of all $H^2_{HS(V_1,V_2)}$ functions whose values lie almost everywhere in the initial subspace of Q.

Conversely let $M \subset H^2_{HS(V_1,V_2)}$ be invariant under right multiplication by all $H^\infty(B(V_1))$ functions. Let M_0 be the subspace of $H^2_{V_2}$ spanned by all functions of the form $T\xi$ where $T \in M$ and $\xi \in V_1$. Clearly M_0 is an invariant subspace of $H^2_{V_2}$ and hence, by Theorem II 12-22, there exists a rigid function Q for which $M_0 = QH^2_{V_2}$. Given $T \in M$ and $\xi \in V_1$ we have $T\xi = Q\varphi_\xi$ for some $\varphi_\xi \in H^2_{V_2}$. The function φ_ξ is not uniquely determined but we can make it so by the additional requirement that φ_ξ is almost everywhere in the initial space of Q. Thus we may define an analytic operator valued function Φ by $\Phi(z)\xi = \varphi_\xi(z)$. For each $\xi \in V_1$ we have $\|T(z)\xi\|^2 = \|Q(z)\varphi_\xi(z)\|^2 = \|Q(z)\Phi(z)\xi\|^2$ which implies that almost everywhere on the unit circle $\|T(e^{it})\xi\| = \|\Phi(e^{it})\xi\|$. If $\{\xi_\alpha\}$ is an orthonormal basis in V_1 it follows that $\sum_\alpha \|T(e^{it})\xi_\alpha\|^2 = \sum_\alpha \|\Phi(e^{it})\xi_\alpha\|^2$ which shows that $\Phi(e^{it}) \in HS(V_1, V_2)$, $\Phi \in H^2_{HS(V_1,V_2)}$ and $M \subset QH^2_{HS(V_1,V_2)}$.

For a fixed ξ, $\{T\xi \mid T \in M\}$ is an invariant subspace included in $QH^2_{V_2}$. Thus for an arbitrary $e \in V_2$ there exists an $f \in V_1$ such that $Tf = Qe$ and this in turn implies that $Qe \otimes \bar{f}$ belongs to M for all $f \in V_1$ and $e \in V_2$, $e \otimes \bar{f}$ being defined by $(e \otimes \bar{f})x = (z, f)e$. From this we infer directly that $M \supset QH^2_{HS(V_1,V_2)}$ which proves (a). Part (b) follows by duality.

In the case of finite dimensional spaces V_1 and V_2 the trace class $TC(V_1, V_2)$ and the Hilbert–Schmidt class $HS(V_1, V_2)$ both coincide with $B(V_1, V_2)$ and this identification will be used in the sequel.

Let now T be a strictly noncyclic function in $H^\infty(B(U, Y))$. Given a complex

$\lambda, |\lambda| > 1$, we define $I_L(T; \lambda)$ and $I_R(T; \lambda)$ by

$$I_L(T; \lambda) = \{P \in H^\infty(B(Y)) | P^*T \text{ extends analytically to } \lambda\} \quad (4\text{-}7)$$

and similarly

$$I_R(T; \lambda) = \{P \in H^\infty(B(U)) | TP^* \text{ extends analytically to } \lambda\} \quad (4\text{-}8)$$

Clearly $I_L(T; \lambda)$ and $I_R(T; \lambda)$ are right and left ideals in $H^\infty(B(Y))$ and $H^\infty(B(U))$, respectively. Moreover $I_L(T; \lambda) = H^\infty(B(Y))$ if and only if T is analytic at λ, and likewise for $I_R(T; \lambda)$. If σ is the minimal function of T then σI belongs to $I_L(T; \lambda)$ as well as to $I_R(T; \lambda)$ which shows that both are subspaces of full range. Since they are also clearly w^*-closed, Theorem 4-5 can be applied in order to get the representations $I_L(T; \lambda) = S_\lambda H^\infty(B(Y))$ and $I_R(T; \lambda) = H^\infty(B(U)) S_\lambda'$ where S_λ and S_λ' are inner functions.

The ideals $I_L(T; \lambda)$ and $I_R(T; \lambda)$ serve as a local measure of the singularities of a strictly noncyclic function T. To get a global measure we introduce $I_L(T)$ and $I_R(T)$ through

$$I_L(T) = \{P \in H^\infty(B(Y)) | P^*T \text{ extends analytically to } D_e\} \quad (4\text{-}9)$$

and

$$I_R(T) = \{P \in H^\infty(B(U)) | TP^* \text{ extends analytically to } D_e\} \quad (4\text{-}10)$$

The spaces $I_L(T)$ and $I_R(T)$ are w^*-closed right and left ideals in $H^\infty(B(Y))$ and $H^\infty(B(U))$, respectively, which are of full range and hence have representations $I_L(T) = SH^\infty(B(Y))$ and $I_R = H^\infty(B(U)) S_1$, respectively, where S and S_1 are inner functions.

With the above definitions we can introduce some equivalence relations in the class of strictly noncyclic functions. We say that two strictly noncyclic functions T_1 and T_2 have equivalent left singularities at a point λ if $I_L(T_1; \lambda) = I_L(T_2; \lambda)$ and similarly for right singularities. T_1 and T_2 have globally equivalent left singularities if $I_L(T_1) = I_L(T_2)$ and similarly for globally equivalent right singularities. Thus the inner functions S_λ, S_λ', S, and S_1 parametrize the local and global singularities. Next we show that they are essentially related to the coprime factorizations of a strictly noncyclic function.

Theorem 4-5 Let $T \in H^\infty(B(U, Y))$ be strictly noncyclic having the coprime factorizations $T = QH^* = H_1^*Q_1$ on the unit circle and let $I_L(T) = SH^\infty(B(Y))$ and $I_R(T) = H^\infty(B(U)) S_1$. Then Q and S are equal up to a constant right unitary factor and Q_1 and S_1 are equal up to a constant left unitary factor.

PROOF Since $T = QH^*$ it follows that $Q^*T = H^*$ extends analytically to D_e and so $Q \in I_L(T)$ or $Q = SR$ for some, necessarily inner, function R. This means that $QH_Y^2 \subset SH_Y^2$.

Conversely since $S^*T\xi$ is orthogonal to H_Y^2 for every $\xi \in U$ and since $L_Y^2 \ominus H_Y^2$ is invariant under multiplication by S^* it follows that

$$P_{H_Y^2} S^* P_{H_Y^2} \bar{\chi}^n T\xi = P_{H_Y^2} \chi^n S^* T\xi = 0$$

for all $\xi \in U$. Since the vectors of the form $P_{H_Y^2} \chi^n T \xi$ span $H(Q) = \overline{\text{Range}\, H_T}$ it follows that $S^* H(Q) \perp H_Y^2$. By Lemma II 13-5 we have $H(Q) \subset H(S)$ or $SH_Y^2 \subset QH_Y^2$. Together with the previously obtained inverse inclusion we have $QH_Y^2 = SH_Y^2$ and hence Q and S differ by at most a constant right unitary factor.

Corollary 4-6

_a)_ Given two strictly noncyclic functions $T \in H^\infty (B(U, Y))$ and $T_1 \in H^\infty (B(U_1, Y))$ then T and T_1 have equivalent singularities if and only if $\overline{\text{Range}\, H_T} = \overline{\text{Range}\, H_{T_1}}$.

_b)_ Given two strictly noncyclic functions $T \in H^\infty (B(U, Y))$ and $T_1 \in H^\infty (B(U, Y_1))$ then T and T_1 have equivalent right singularities if and only if $\text{Ker}\, H_T = \text{Ker}\, H_{T_1}$.

Corollary 4-7 If $T \in H^\infty (B(U, Y))$ is strictly noncyclic and has the coprime factorization $T = QH^*$ then T is left equivalent to Q.

Next we pass on to the analysis of the local singularities. For every λ such that $|\lambda| > 1$ it is obvious that $I_L(T) \subset I_L(T; \lambda)$ and hence $S = S_\lambda S_\Lambda$ for some inner functions S_Λ. Assuming T admits the coprime factorization $T = QH^*$ we let σ be the minimal inner function of Q. Let $\sigma = \sigma_\lambda \sigma_\Lambda$ where σ_λ is the Blaschke factor that corresponds to the zeros of σ at λ^{-1}, thus $\sigma_\Lambda(\lambda^{-1}) \neq 0$. This means that σ_λ and σ_Λ are coprime. By Theorem 4-3 this factorization induces factorizations

$$Q = Q_\lambda Q_\Lambda = Q'_\Lambda Q'_\lambda \qquad (4\text{-}11)$$

of Q on the unit circle. σ_λ is the minimal inner function of Q_λ and Q'_λ whereas σ_λ is the minimal inner function of Q_Λ and Q'_Λ.

Theorem 4-8 Let $T \in H^\infty (B(U, Y))$ be strictly noncyclic admitting the coprime factorization $T = QH^*$ on the unit circle. Let Q be factored as in (4-11) and let $I_L(T; \lambda) = S_\lambda H^\infty (B(Y))$. Then Q_λ and S_λ coincide up to a constant right unitary factor.

PROOF Since $T = QH^* = Q_\lambda Q_\Lambda H^*$ it follows that $Q_\lambda^* T = Q_\Lambda H^*$ which has an extension to the exterior of the unit disc given by $\tilde{Q}_\Lambda(z^{-1}) H(z^{-1})$ that is analytic at λ as $Q_\Lambda(\lambda^{-1})$ is invertible. Thus $Q_\lambda \in I_L(T; \lambda)$ and $Q_\lambda = S_\lambda R$ for some inner function R.

Conversely assume $Q_\lambda = S_\lambda R$, R being a nontrivial inner factor. Obviously the minimal inner function of R is a factor of σ_λ. Thus the only singularity of the analytic extension of R to D_e is a pole at λ. Now $S_\lambda^* T$ extends meromorphically to D_e and the extension is analytic at λ. From the coprime factorizations of T and (4-11) we have

$$T' = S_\lambda^* T = S_\lambda^* Q_\lambda Q_\Lambda H^* = RQ_\Lambda H^*$$

Since the minimal inner functions of R and Q_Λ are coprime there exist, by Theorem 4-3, inner functions R'' and Q''_Λ satisfying $RQ_\Lambda = Q''_\Lambda R''$, $\det R =$

$\det R''$ and $\det Q_\Lambda = \det Q''_\Lambda$. Now $T' = RQ_\Lambda H^* = Q''_\Lambda R'' H^*$ and so $(Q''_\Lambda)^* T = R'' H^*$ extends analytically to D_e with the possible exception of λ. But it must be analytic at λ too since T extends analytically to λ. Thus the range of the Hankel operator induced by $R'' H^*$ is trivial which shows that R'' is constant and hence so is R.

Lastly we pass on to the study of the boundary behavior of a strictly non-cyclic function T and its associated inner function.

Theorem 4-9 Let $C \in H^\infty(B(U, Y))$, P an inner function in $H^\infty(B(Y))$ and assume that $(P, C)_L = I$. If $P(z)^{-1} C(z)$ has an analytic continuation at a point λ of the unit circle then P has an analytic continuation at λ.

PROOF Let U be a disc centered at λ where $P(z)^{-1} C(z)$ is analytic and assume $\|P(z)^{-1} C(z)\| \leq M$ for $z \in U \cap D$. Let V be a disc centered at λ and properly contained in U. Let π be the minimal inner function of P which we factor as $\pi = \pi_\lambda \pi_\Lambda$ where $\pi_\lambda \wedge \pi_\Lambda = 1$ and assume both $\pi_\lambda(z) \neq 0$ for $z \in D - V$ as well as that the singular measure in the integral representation of the singular factor of π_λ is supported on the intersection of the unit circle with $\overline{V}$. Let $\pi_\Lambda(z) \neq 0$ for $z \in V \cap D$ and assume its singular measure is supported on $T - V$. It follows that $\inf\{|\pi_\lambda(z)| \,|z \in D - U\} > 0$ and in particular π_λ extends analytically at all points of $T - U$, by an application of Theorem II 12-27.

Corresponding to the factorization of π we have a factorization $P = P_\lambda P_\Lambda$ with π_λ and π_Λ being the minimal inner functions of P_λ and P_Λ, respectively.

Consider now the function A defined by $A(z) = P_\Lambda(z) P(z)^{-1} C(z) = P_\lambda(z)^{-1} C(z)$. As π_λ is the minimal inner function of P_λ there exists an inner function Π_λ for which $P_\lambda \Pi_\lambda = \pi_\lambda I$. Therefore $P_\lambda(z)^{-1} = \pi_\lambda(z)^{-1} \Pi_\lambda(z)$ from which the boundedness of $P_\lambda(z)^{-1}$, and hence also that of $A(z)$, in $D - U$ follows. For $z \in D \cap U$, $P(z)^{-1} C(z)$ is bounded by assumption and hence also $A(z)$ is bounded. Thus $A \in H^\infty(B(U, Y))$ and so $C = P_\lambda A$ which together with the factorization $P = P_\lambda P_\Lambda$ contradicts the coprimeness condition $(P, C)_L = I_Y$. Thus necessarily P_λ is trivial, that is P_λ is a constant unitary operator and the minimal inner function of P is π_Λ which has an analytic continuation at λ and therefore P itself is analytically continuable at λ.

Theorem 4-10 Let T be strictly noncyclic in $H^\infty(B(U, Y))$ admitting the coprime factorization $T = QH^*$ on the unit circle. Then A has an analytic continuation at a point λ, $|\lambda| = 1$, if and only if Q has.

PROOF If Q is analytically continuable at λ then, by Lemma II 13-6, so is every function in $H_0(Q)$. Since for all $\xi \in U$ the function $H_T \xi = P_{H^2_{0,Y}} T \xi$ is in $H_0(Q)$ then $(T(z) \xi - T(0) \xi)/z$ has an analytic continuation at λ and so has $T(z) \xi$.

Conversely assume T has the coprime factorization $T = QH^*$ on the unit circle. Thus $\tilde{Q}(z^{-1})^{-1} \tilde{H}(z^{-1})$ is the meromorphic extension of T to D_e which by our assumption extends analytically to D at λ. This is equivalent to $\tilde{Q}(\zeta)^{-1} H(\zeta)$ having an analytic continuation at $\bar{\lambda}$. Now $(Q, H)_R = I_Y$ implies $(\tilde{Q}, \tilde{H})_L = I_Y$ and the result follows from the previous theorem.

In conclusion we are ready to put everything together and state the central result of this section.

Theorem 4-11 Let $T \in H^\infty(B(U, Y))$ be strictly noncyclic then the shift realization of T is spectrally minimal.

PROOF Let $T = QH^*$ be the coprime factorization of T on the unit circle. The state space of the shift realization is $H_0(Q)$ and the generator or state operators of the shift realization is given by $S_0(Q)^*$. By Theorem II 13-8 the spectrum of $S_0(Q)^*$ is completely determined by Q and is equal to the set of all points $\bar{\lambda}, |\lambda| < 1$, where $Q(\lambda)$ is not invertible as well as the points $\bar{\lambda}, |\lambda| = 1$, where Q is not analytically continuable to D_e.

Now T extends meromorphically to D_e with the exception of at most a countable number of poles located at the points λ where $Q(\bar{\lambda}^{-1})$ is not invertible. Similarly T and Q are analytically continuable at the same points of the unit circle, thus T has no analytic extension at $\lambda, |\lambda| = 1$, if and only if $\bar{\lambda} = \lambda^{-1}$ belongs to the continuous spectrum of $S(Q)^*$. This completes the proof.

For functions which are not strictly noncyclic the shift realization does not provide a useful tool for analysis. This is the setting for a striking counterexample to the state space isomorphism theorem. For simplicity we restrict ourselves to the scalar case.

Let $T \in H^\infty$ be a nonrational noncyclic function relative to the left shift operator. Let $T_\rho(z) = T(\rho z)$ where $0 < \rho < 1$. Obviously if $T \in H^\infty$ so do all T_ρ. In fact T_ρ is analytic in the region $|z| < \rho^{-1}$ and since T_ρ is not rational it is necessarily a cyclic function for the left shift. Let $\Sigma = (A, B, C)$ and $\Sigma_\rho = (A_\rho, b_\rho, c_\rho)$ be the shift realizations of T and T_ρ, respectively. Clearly $\Sigma'_\rho = (\rho^{-1}A_\rho, b_\rho, c_\rho)$ is a realization of T. Now $A = S^* | \operatorname{Range} H_T$ whereas $\rho^{-1}A_\rho = \rho^{-1}S^*$ as $\overline{\operatorname{Range} H_{T_\rho}} = H^2$. Since $\operatorname{Range} H_T$ is a proper left invariant subspace of H^2 the spectrum of A has at most a countable number of points inside the open unit disc. The spectrum of $\rho^{-1}A_\rho$ on the other hand coincides with the closed disc of radius ρ^{-1}. This excludes the possibility of the isomorphism of the two realizations as similarity preserves the spectrum.

5. DEGREE THEORY FOR STRICTLY NONCYCLIC FUNCTIONS

Although a general degree theory, extending the McMillan degree of rational functions, is not available it is possible to develop a complete degree theory for the case of strictly noncyclic transfer functions. The main difference is that while in the finite dimensional case degree is defined by use of the dimension function and hence is essentially an additive function, in the infinite dimensional case we will use a multiplicative analogue.

Given a finite dimensional vector space V then the dimension is a function from the set of all subspaces of V into the monoid Z_+, which has the following properties

$$(a) \; M \subset M_1 \Rightarrow \dim M \leq \dim M$$

$$(b) \; \dim(M + N) \leq \dim M + \dim N \tag{5-1}$$

with equality if and only if $M \cap N = \{0\}$.

Let now N denote a finite dimensional Hilbert space. Let L_N denote the set of all left invariant subspaces of H_N^2 whose orthogonal complement has full range. Thus L_N coincides with the set of all subspaces of H_N^2 of the form $H(Q) = \{QH_N^2\}^\perp$ where Q is an inner function. Let l_N denote the multiplicative monoid of all inner functions in N. Define a map $d : L_N \to l_N$ by

$$d(H(Q)) = \det Q \tag{5-2}$$

then the next theorem shows that the function d is a suitable generalization of the dimension function.

Theorem 5-1 Let P and R be inner functions in N. Then the following statements are true.
(a) $H(P) \subset H(R)$ implies $d(H(P)) | d(H(R))$
(b) $d(H(P) \vee H(R)) | d(H(P)) \cdot d(H(R))$
(c) $d(H(P) \vee H(R)) = d(H(P)) \cdot d(H(R))$

if and only if $(P, R)_L = I_N$.

PROOF (a) $H(P) \subset H(R)$ if and only if $RH_N^2 \subset PH_N^2$ which is equivalent to the factorization $R = PS$ for some inner function S. By taking determinants we obtain $\det P | \det R$, which is the same as $d(H(P)) | d(H(R))$.

Next we prove (c). Assume $H(P) \cap H(R) = \{0\}$ which is equivalent to $PH_N^2 \vee RH_N^2 = H_N^2$ or to $(P, R)_L = I_N$. Let Q be an inner function for which $QH_N^2 = PH_N^2 \cap RH_N^2$, Q exists since for N of finite dimensions the intersection of invariant subspaces of full range has full range. We note that Q is determined only up to a constant unitary factor on the right.

Since $QH_N^2 \subset PH_N^2$ we have $Q = PR_1$ and similarly $Q = RP_1$ for some inner functions P_1 and R_1. Define $A = P^*R$ then A is obviously a strictly noncyclic function in $L^\infty(R(N))$. By Theorem 3-5 A has also a factorization

$A = R_2 P_2^*$ on the unit circle and the coprimeness condition $(R_2, P_2)_R = I_N$ is satisfied. Since R_2 is inner the same must be true of P_2 as A is a.e. unitary on the unit circle.

Now the equality $P^*R = R_2 P_2^*$ implies

$$RP_2 = PR_2 \tag{5-3}$$

We can apply now Theorem II 14-11 to infer that $S(P_2)$ and $S(P)$ are quasisimilar, and by the same token also $S(R)$ and $S(R_2)$ are quasisimilar. By Theorem II 15-17, P and P_2 have the same Jordan model and the same is true of R and R_2. Since P and P_2 have the same invariant factors it follows in particular that $\det P = \det P_2$ and similarly that $\det R = \det R_2$. We will show that, modulo a constant unitary factor on the right, P_1 is equal to P_2 and the same holds for R_1 and R_2.

From (5-3) it follows that $RP_2 H_N^2 \subset RH_N^2$ and also $RP_2 H_N^2 = PR_2 H_N^2 \subset PH_N^2$ so $RP_2 H_N^2 \subset RH_N^2 \cap PH_N^2 = QH_N^2$. Hence there exists an inner function Z for which $RP_2 = QZ = RP_1 Z$, or $P_2 = P_1 Z$. By similar reasoning $R_2 = R_1 Z$ and so Z is a common right inner factor of P_2 and R_2 and by the assumption of the right coprimeness of R_2 and P_2, Z is constant. This implies that, up to a constant of absolute value one

$$\det Q = \det P \cdot \det R \tag{5-4}$$

If P and Q are not left coprime let S be a greatest common left inner divisor. Thus $P = SP'$ and $R = SR'$ with $(P', R')_L = I_N$. Since $SP'H_N^2 \cap SR'H_N^2 = S(P'H_N^2 \cap R'H_N^2) = SQ'H_N^2 = QH_N^2$ we can apply the first part of the proof to obtain $\det Q' = \det P' \cdot \det R'$. Hence $\det Q = \det SQ' = \det S \cdot \det Q' = \det S \cdot \det P' \cdot \det R' | (\det S)^2 \cdot \det P' \cdot \det Q' = \det SP' \cdot \det SR' = \det P \cdot \det R$. This proves (b).

Corollary 5-2 Let P and R be inner functions in N and let Q be an inner function for which $QH_N^2 = PH_N^2 \cap RH_N^2$. Then there exist inner functions P_1 and R_1 such that $Q = PR_1 = RP_1$ and moreover $(P, R)_L = I_N$ if and only if $(P_1, R_1)_R = I_N$.

PROOF That the factorizations $Q = PR_1 = RP_1$ hold follows from the inclusions $PH_N^2 \cap RH_N^2 \subset PH_N^2$ and $PH_N^2 \cap RH_N^2 \subset RH_N^2$. Assume $(P, R)_L = I_N$ and hence $H(P) \cap H(R) = \{0\}$ and with it the equalities $\det R = \det R_1$ and $\det P = \det P_1$. Now $\tilde{Q} = \tilde{R}_1 \tilde{P} = \tilde{P}_1 \tilde{R}$ and since $\det \tilde{Q} = \det Q = \det P \cdot \det R = \det \tilde{P}_1 \cdot \det \tilde{R}_1$ it follows that $(\tilde{P}_1, \tilde{R}_1)_L = I_N$ which is the same as $(P_1, R_1)_R = I_N$.

From Theorem 5-1 it is clear that the determinant of an inner function Q provides a suitable generalization of the concept of dimension for subspaces of the form $H(Q)$. This will also be the key for the generalization of McMillan degree theory of rational functions to the case of strictly noncyclic functions. Equality (5-4) is equivalent to $H(Q) = H(P) + H(R)$ and $H(P) \cap H(R) =$

$\{0\}$. This is certainly satisfied whenever

$$H(Q) = H(P) \dotplus H(R) \tag{5-5}$$

where $\dotplus$ denotes the, not necessarily orthogonal, direct sum of $H(P)$ and $H(R)$. For (5-5) to hold the condition $(P, R)_L = I_N$ is generally insufficient. One expects that $(P, R)_L = I_N$ should be replaced by the stronger coprimeness condition $[P, R]_L = I_N$ and this turns out to be true. In preparation we prove some necessary lemmas.

If R is an inner function acting in N then H_R, the Hankel operator induced by R and defined by (1-17), is a partial isometry from $\bar{H}_N^2$ into $H_{0,N}^2$. Its range is given by $H_0(R) = H_{0,N}^2 \ominus R H_{0,N}^2$. Applying Theorem II 2-3 the orthogonal projection of $H_{0,N}^2$ onto $H_0(R)$ is given by

$$P_{H_0(R)} = H_R H_R^* \tag{5-6}$$

Lemma 5-3 Let P and R be inner functions in N. If $(P, R)_L = I_N$ then $P_{H_0(R)}\{PH_{0,N}^2\}$ is dense in $H_0(R)$.

Proof Since for $f \in H_{0,N}^2$

$$P_{H_0(R)}Pf = H_R H_R^* Pf = P_{H_{0,N}^2} R P_{\bar{H}_{\tilde{N}}^2} R^* Pf$$

a simple adaptation of Theorem 3-5 shows that the coprimeness condition $(R, P)_L = I_N$ implies that the map $f \to P_{H_{\tilde{N}}^2} R^* Pf$ which is just $H_{R^* \bullet P}^*$ has range dense in $\bar{H}_N^2 \ominus R^* \bar{H}_N^2$. Since $\ker H_R = R^* \bar{H}_N^2$ the result follows.

Using our available information concerning range closure of Hankel operators we can strengthen the previous lemma to obtain.

Lemma 5-4 Let P and R be inner functions in N. If $[P, R]_L = I_N$ then $P_{H_0(R)}\{PH_{0,N}^2\} = H_0(R)$.

Proof By Theorem 3-10 the range of $H_{P^* \bullet R}^*$ is $\bar{H}_N^2 \ominus R^* \bar{H}_N^2$ given that $[P, R]_L = I_N$ is satisfied. But $\bar{H}_N^2 \ominus R^* \bar{H}_N^2$ is just the initial space of H_R and hence is mapped isometrically onto a closed subspace of $\bar{H}_{0,N}^2$ which by the previous lemma has to coincide with $H_0(R)$.

We can relate now strong coprimeness of inner functions to the geometry of left invariant subspaces in $H_{0,N}^2$.

For any two subspaces M_1 and M_2 of a Banach space X which satisfy $M_1 \cap M_2 = \{0\}$ the sum $M_1 + M_2$ is closed if and only if for some $d > 0$

$$\inf\{\|x_1 - x_2\| \,|\, x_i \in M_i, \|x_i\| = 1, i = 1, 2\} \geq d \tag{5-7}$$

In a Hilbert space condition (5-7) is equivalent to

$$\sup\{|(x_1, x_2)| \,|\, x_i \in M_i, \|x_i\| = 1, i = 1, 2\} < 1 \tag{5-8}$$

This last condition has the interpretation that the angle between M_1 and M_2 is positive.

Theorem 5-5 Let P and R be inner functions in N. The angle between the left invariant subspaces $H_0(P)$ and $H_0(R)$ is positive if and only if $[P, R]_L = I_N$.

Equivalently there exists an inner function Q such that

$$H_0(Q) = H_0(P) + H_0(R) \qquad (5\text{-}9)$$

and

$$\det Q = \det P \cdot \det R \qquad (5\text{-}10)$$

holds if and only if $[P, R]_L = I_N$.

PROOF Assume $[P, R]_L = I_N$. Let $PH^2_{0,N} \cap RH^2_{0,N} = QH^2_{0,N}$ then by Theorem 3-10 and Corollary 5-2, $Q = PR_1 = RP_1$ and also $[P_1, R_1]_R = I_N$. We apply now Theorem II 14-10 to infer the existence of Φ and Ψ in $H^\infty(B(N, N))$ for which $\Phi P_1 + \Psi R_1 = I_N$. In turn this implies $Q^* = \Phi P_1 Q^* + \Psi R_1 Q^* = \Phi R = \Psi P$ and by taking adjoints we obtain

$$Q = R\Phi^* + P\Psi^* \qquad (5\text{-}11)$$

We saw that already the weaker condition $(P, R)_L = I_N$ implied $H_0(Q) = H_0(P) \vee H_0(R)$. Now from (5-11) it follows that

$$H_0(Q) = \text{Range}\, H_Q = \text{Range}\, H_{R\Phi^* + P\Psi^*} \subset \text{Range}\, H_{R\Phi^*}$$
$$+ \text{Range}\, H_{R\Psi^*} \subset H_0(R) + H_0(P)$$
$$\subset H_0(R) \vee H_0(P) = H_0(Q)$$

Hence the equality (5-9) is obtained.

To show the necessity of the strong coprimeness condition $[P, R]_L = I_N$ we assume that P and R are not strongly coprime. The most obvious violation of $[P, R]_L = I_N$ is the existence of a nonzero vector $\eta \in N$ and a point λ in the open unit disc for which

$$P(\lambda)^* \eta = R(\lambda)^* \eta = 0 \qquad (5\text{-}12)$$

But $P(\lambda)^* \eta = 0$ implies that the function $\chi(1 - \bar\lambda\chi)^{-1}$ is in $H_0(P)$ and hence (5-12) implies $H(P) \cap H(R) \neq \{0\}$.

In general (5-12) does not hold and we resort to an approximation argument similar to the one used in the proof of Theorem II 14-11. If $[P, R]_L \neq I_N$ there exists a sequence of points λ_n, $|\lambda_n| < 1$, and a sequence of unit vectors $\eta_n \in N$ for which

$$\lim \|P(\lambda_n)^* \eta_n\| = \lim \|R(\lambda_n)^* \eta_n\| = 0$$

We will show the existence of a sequence $F_n \in H_0(P)$ and a sequence $F'_n \in H_0(R)$ for which $\lim \|F_n\| = \lim \|F'_n\| = 1$ and also $\lim(F_n, F'_n) = 1$. This implies that $H_0(P)$ and $H_0(R)$ have zero angle between them.

The functions $H_n = (1 - |\lambda_n|^2)^{1/2} \chi (1 - \bar{\lambda}_n \chi)^{-1} \eta_n$ are normalized eigenfunctions of the left shift in $H_{0,N}^2$. Let F_n and F_n' be their orthogonal projections on $H_0(P)$ and $H_0(R)$, respectively, and let $G_n = H_n - F_n$ and $G_n' = H_n - F_n'$. It follows from Lemma II 13-7 that

$$G_n(z) = \frac{(1 - |\lambda_n|^2)^{1/2} P(z) P(\lambda_n)^* \eta_n}{1 - \bar{\lambda}_n z}$$

and hence that $\lim \|G_n\| = \lim \|G_n'\| = 0$ and as a consequence that $\lim \|F_n\| = \lim \|F_n'\| = 1$. Now $1 = (H_n, H_n) = (F_n + G_n, F_n' + G_n') = (F_n, F_n') + (F_n, G_n') + (G_n, F_n') + (G_n, G_n')$. The last three terms obviously tend to zero and we have $\lim (F_n, F_n') = 1$ as required.

We recall that the McMillan degree of a proper rational matrix function T is defined as the dimension of the state space of any, and hence by Theorem I 8-4 all, canonical realization of T. Thus δ is a map from the set of proper rational functions into Z_+ which satisfies

$$(a)\ \delta(T_1 + T_2) \le \delta(T_1) + \delta(T_2)$$
$$(b)\ \delta(T_1 T_2) \le \delta(T_1) + \delta(T_2) \tag{5-13}$$

Equalities in (a) and (b) are subject to the coprimeness conditions which guarantee that no pole-zero cancellations occur. An alternative way to define the McMillan degree of a rational function T is to let it be the rank of the associated Hankel matrix. Since the rank of a matrix is the dimension of its range space, then just as we used the determinant function to replace the concept of dimension, we are led to make the following definition.

Let T be a strictly noncyclic function in $H^\infty(B(U, Y))$. We define the *degree* of T, denoted by $\Delta(T)$, by

$$\Delta(T) = d(\overline{\text{Range } H_T}) \tag{5-14}$$

where d is defined by (5-2). Thus if Q is an inner function such that $\{\text{Range } T\}^\perp = QH_{0,Y}^2$ then $\Delta(T) = \det Q$.

Theorem 5-6 Let $T \in H^\infty(B(U, Y))$ be strictly noncyclic then

$$\Delta(\tilde{T}) = \Delta(T) \tag{5-15}$$

PROOF That $\tilde{T}$ is strictly noncyclic is the content of Corollary 3-6 (c). If $T = QH^* = H_1^* Q_1$ are coprime factorizations of T then $\tilde{T} = \tilde{Q}_1 \tilde{H}_1^* = \tilde{H}^* \tilde{Q}$. Thus $\Delta(\tilde{T}) = \det \tilde{Q}_1 = \det Q_1$. Since $S(Q)$ and $S(Q_1)$ are quasisimilar, Q and Q_1 are quasiequivalent and hence $\det Q = \det Q_1$. This proves the theorem.

The degree function Δ defined by (5-14) satisfies the multiplicative analogs of (5-13), that is

$$(a)\ \Delta(T_1 + T_2) | \Delta(T_1) \cdot \Delta(T_2)$$
$$(b)\ \ \ \Delta(T_1 T_2) | \Delta(T_1) \cdot \Delta(T_2) \tag{5-16}$$

To prove this we have to study in detail the ranges of Hankel operators induced by sums and products of strictly noncyclic functions. We begin with the study of products.

Let L, M, and N be three finite dimensional Hilbert spaces. Let A and B be strictly noncyclic functions in $L^\infty(B(N, M))$ and $L^\infty(B(L, N))$, respectively. By Theorem 3-5 the functions A and B have the following factorizations on the unit circle

$$A = PC^* = C_1^* P_1 \tag{5-17}$$

and

$$B = RD^* = D_1^* R_1 \tag{5-18}$$

where P is an inner function in M, P_1 and R inner in N, R_1 inner in L, whereas $C, C_1 \in H^\infty(B(M, N))$ and $D, D_1 \in H^\infty(B(N, L))$. Moreover we assume the factorizations to be coprime, that is, the conditions

$$(P, C)_R = I_M, \qquad (P_1, C_1)_L = I_N \tag{5-19}$$

and

$$(R, D)_R = I_N, \qquad (R_1, D_1)_L = I_L \tag{5-20}$$

are satisfied.

Since A and B are strictly noncyclic both have meromorphic extensions of bounded type to D_e and hence also their product AB has such an extension. So $T = AB$ itself is strictly noncyclic and by the same theorem used before admits factorizations

$$T = QH^* = H_1^* Q_1 \tag{5-21}$$

with

$$(Q, H)_R = I_M, \qquad (Q_1, H_1)_L = I_L \tag{5-22}$$

satisfied.

The analysis of the general case will be based on the two special cases $B = R$ and $B = D^*$.

Lemma 5-7 Let $A \in L^\infty(B(N, M))$ be strictly noncyclic and let R be an inner function in N, then Range $H_A \subset$ Range H_{AR}.

PROOF Let $f \in \bar{H}_N^2$ then $R^* f \in \bar{H}_N^2$ and

$$H_{AR}(R^* f) = P_{H^2{}_M} A R R^* f = P_{H_0{}_M} A f$$

which proves the stated range inclusion.

In this case $T = AR$ and from the coprime factorization (5-21) we obtain Range $H_{AR} = H_0(Q)$.

Lemma 5-8 Let A and AR have the coprime factorizations (5-17) and (5-21) then

$$\det Q = \det P \cdot \det R \qquad (5\text{-}23)$$

if and only if

$$(R, C)_L = I_N \qquad (5\text{-}24)$$

PROOF Assume R and C have a nontrivial greatest left inner factor S. Thus $R = SR_2$ and $C = SC_2$ and $(R_2, C_2)_L = I_N$. Since S is nontrivial, $\det R_2 \mid \det R$ we have $\det R \neq \det R_2$. Now $T = AR = PC^*R = PC_2^*R_2$. By Theorem 3-5 there exists R_3 and C_3 satisfying $(R_3, C_3)_R = I_M$ as well as $C_2^*R_2 = R_3C_3^*$. Since R_2 and R_3 are quasiequivalent the equality $\det R_2 = \det R_3$ holds. Now $T = AR = PC_2^*R_2 = PR_3C_3^*$ it follows that $H_0(Q) = \overline{\text{Range} H_{AR}} \subset H_0(PR_3)$. Thus PR_3 is divisible on the left by Q and hence $\det Q \mid \det P \cdot \det R_3$, and with it (5-23) is impossible.

To prove the converse assume (5-24) holds. As before $T = AR = PC^*R = PR_2C_2^* = QH^*$. From Lemma 5-7, $H_0(P) \subset H_0(Q)$ and hence $Q = PS$ for some inner function S. Thus $PC^*R = QH^* = PSH^*$ and $(S, H)_R = I_M$ and from it the equality $R_2C_2^* = SH^*$. As both factorizations are coprime we have $H_0(R_2) = H_0(S)$ and therefore R_2 and S differ at most by a constant unitary factor on the right. In particular $\det S = \det R_2 = \det R$ and (5-23) is satisfied.

This lemma can be sharpened to yield a result about the range closure of H_{AR}.

Lemma 5-9 Let A and AR admit the coprime factorizations (5-17) and (5-22), respectively. If H_A has closed range then H_{AR} has closed range $H_0(Q)$ with (5-23) satisfied if and only if

$$[R, C]_L = I_M \qquad (5\text{-}25)$$

PROOF For (5-23) to hold the coprimeness condition (5-24) is necessary. Using the notation in the proof of the previous lemma $AR = PC^*R = PR_2C_2^*$ and the last factorization is coprime. For the range closure of H_{AR} $[PR_2, C_2]_R = I_M$ is necessary which implies the necessity of the weaker condition $[R_2, C_2]_R = I_M$ which is equivalent to (5-25).

Conversely assume (5-25) holds then by the previous lemma $\text{Range} H_{AR}$ is dense in $H(Q) = H(PR_2)$. Since we have

$$H(PR_2) = H(P) \oplus PH(R_2) \qquad (5\text{-}26)$$

and as $\text{Range} H_A = H(P) \subset \text{Range} H_R$ it suffices to show that $PH(R_2) \subset \text{Range} H_{AR}$. To this end let $f \in \bar{H}_N^2$ then

$$H_{AR}f = P_{H_{0,M}^2} ARf = P_{H_{0,M}^2} PR_2C_2^*f = PP_{H_{0,M}^2} R_2C_2^*f = PH_{R_2C_2^*}f$$

Now (5-25) implies $[R_2, C_2]_R = I_N$ and with it $\text{Range} H_{R_2C_2^*} = H_0(R_2)$.

Next we assume $B = D^*$ for $D \in H^\infty(B(N, L))$, or equivalently $T = AD^*$.

Lemma 5-10 Let A be strictly noncyclic in $L^\infty(B(N, M))$ and $D \in H^\infty(B(N, L))$. Then AD^* is strictly noncyclic and $\operatorname{Range} H_{AD^*} \subset \operatorname{Range} H_A$.

PROOF Let $f \in \bar{H}_L^2$ then $D^* f \in \bar{H}_L^2$ and

$$H_{AD^*} f = P_{H_{0,M}^2} AD^* f = H_A(D^* f)$$

If $T = AD^* = PC^*D^*$ factors coprimely as before by (5-21) we have $\operatorname{Range} H_{AD^*} = H(Q) \subset H(P)$. The inclusion implies $P = QS$ for some inner function on S.

Lemma 5-11 Let A be strictly noncyclic in $L^\infty(B(N, M))$ and let A and $T = AD^*$ have the coprime factorizations (5-17) and (5-21), respectively. A necessary and sufficient condition for the equality

$$\det P = \det Q \tag{5-27}$$

to be satisfied, up to a constant of absolute value one, is

$$(P_1, D)_R = I_N \tag{5-28}$$

PROOF The proof of necessity follows along the lines of the proof of Lemma 5-8. For the proof of sufficiency we note that

$$AD^* = PC^*D^* = C_1^* P_1 D^* = QH^*$$

Since $P = QS$ it follows that $SC^*D^* = H^*$. Now $(P, C)_R = I_M$ implies $(S, C)_R = I_M$ and hence $SC^* = C_2^* S_2$ with $(S_2, C_2)_L = I_N$. Now the Hankel operator induced by H^* is the zero operator. This implies that

$$P_{H_{0,M}^2} C_2^* P_{H_{0,N}^2} S_2 D^* f = 0 \qquad \text{for all} \qquad f \in \bar{H}_L^2$$

and hence that the operator defined on $H_{0,N}^2$ by $P_{H_{0,M}^2} C_2^* g$ has nontrivial kernel. By an application of Theorem II 14-11 this contradicts the coprimeness relation $(C_2, S_2)_L = I_N$. Necessarily we have therefore that $\operatorname{Range} H_{S_2 D^*}$ is trivial which can occur only if S_2 is constant. Since $Q = PS$ the determinants of P and Q differ at most by a constant of absolute value one.

As in the case of Lemma 5-8 also this lemma can be sharpened to obtain the following.

Lemma 5-12 Let A be strictly noncyclic in $L^\infty(B(N, M))$ having the coprime factorizations (5-17) on the limit circle, and assume $\operatorname{Range} H_A$ is closed. Let $D \in H^\infty(B(N, L))$, then a necessary and sufficient condition for the equality

$$\operatorname{Range} H_{AD^*} = \operatorname{Range} H_A \tag{5-29}$$

to hold is

$$[P_1, D]_R = I_N \tag{5-30}$$

PROOF We begin by proving the necessity of (5-30). We saw already that $(P_1, D)_R = I_N$ is necessary for $\overline{\text{Range}\, H_{AD*}} = \overline{\text{Range}\, H_A}$. In that case $P_1 D^* = D_2^* P_2$ with P_2 inner and $(P_2, D_2)_L = I_L$. Since $AD^* = PC^*D^* = C_1^* P_1 D^* = C_1^* D_2^* P_2$, it follows that for $\text{Range}\, H_{AD*} = H_0(P)$ it is necessary that $[P_2, D_2 C_1]_L = I_L$ holds. Hence the weaker condition $[P_2, D_2]_L = I_L$ is also necessary and this is equivalent to (5-30).

Conversely we assume that (5-30) holds. Thus $\text{Range}\, H_{P_1 D*} = H_0(P_1)$. Clearly also $\text{Range}\, H_{P_1} = H_0(P_1)$. Now for $f \in \bar{H}_L^2$

$$H_{AD*} f = P_{H_{0,M}^2} AD^* f = P_{H_{0,M}^2} C_1^* P_1 D^* f = P_{H_{0,M}^2} C_1^* P_{H_{0,N}^2} P_1 D^* f$$

and so

$$\text{Range}\, H_{AD*} = \{ P_{H_{0,M}^2} C_1^* g \,|\, g \in H_0(P_1) \} = \{ P_{H_{0,M}^2} C_1^* P_{H_0}\,_N P_1 f \,|\, f \in L_N^2 \}$$

$$= \{ P_{H_{0,M}^2} C_1^* P_1 f \,|\, f \in \bar{H}_N^2 \} = \text{Range}\, H_A = H(P)$$

We can combine now the results of the previous lemmas to yield the following theorem.

Theorem 5-13 Let $A \in L^\infty\big(B(N, M)\big)$ and $B \in L^\infty\big(R(L, N)\big)$ be strictly non-cyclic, having the coprime factorizations (5-17) and (5-18), respectively. Let $T = AB$ have the coprime factorizations (5-21).

(a) A necessary and sufficient condition for

$$\det Q = \det P \cdot \det R \tag{5-31}$$

to hold is

$$(C, R)_L = I_N \quad \text{and} \quad (P_1, D_1)_R = I_N \tag{5-32}$$

(b) Assume H_A and H_B have closed range then H_{AB} has closed range and (5-31) is satisfied if and only if

$$[C, R]_L = I_N \quad \text{and} \quad [P_1, D_1]_R = I_N \tag{5-33}$$

PROOF (a) The necessity of conditions (5-32) for (5-31) to hold follows along the lines of the proof of Lemma 5-8, in particular we always have $\det Q \,|\, \det P \cdot \det R$. So we assume (5-32) to hold and consider $PC^* R = C_1^* P_1 R$. As $(R_1, C)_L = I_N$ we have $(P_1 R, C_1)_L = I_N$ and thus the range closure of H_{PC*R} is $H_0(Q')$ for some inner function Q' acting in M. For Q' we have, by Lemma 5-8, $\det Q' = \det P \cdot \det R$. Next we consider $AB = (PC^* R) D^* = (C_1^* P_1 R) D^*$. By Lemma 5-8 equality (5-31) will hold if and only if $(P_1 R, D)_R = I_N$. Since $P_1 R D^* = P_1 D_1^* R_1$ it follows from Lemma 5-8 that $(P_1 R, D)_R = I_N$ is equivalent to $(P_1, D_1)_R = I_N$ and $(R_{1'}, D_1)_L = I_L$ which proves the sufficiency part.

(b) By part (a) already (5-32) is necessary for equality (5-31) to hold. If $C^* R = R_2 C_2^*$ with $(R_2, C_2)_R = I_M$ then $AB = PC^* RD^* = PR_2 C_2^* D^*$. For H_{AB} to have closed range it is necessary therefore that $[PR_2, DC_2]_R = I_M$ andhence the necessity of the weaker condition $[R_2, C_2]_R = I_M$, this last condition being equivalent to $[R, C]_L = I_N$. The necessity of $[P_1, D_1]_R = I_N$ is proved analogously using the representation $AB = C_1^* P_1 D_1^* R_1$.

To prove the converse let us assume the strong coprimeness conditions in (5-33) to hold. Thus $\overline{\text{Range} H_{AB}} = H(Q)$ and (5-31) holds by part (a). It suffices therefore to prove the range closure of H_{AB}. By Lemma 5-9 H_{PC^*R} has closed range. Now $PC^*R = C_1^* P_1 R$ hence, by Lemma 5-12, $[P_1 R, C_1]_L = I_N$ holds. To prove the range closure of H_{AB} it suffices to show that $[P_1 R, D]_R = I_N$. To see this we note that by our assumptions the range of $H_{P_1 D_1^*}$ is closed and since $P_1 RD^* = P_1 D_1^* R_1$ the assumption $[R_1, D_1]_L = I_L$ yields, by another application of Lemma 5-9, the range closure of $H_{P_1 RD^*}$. Hence $[P_1 R, D]_R = I_N$ and H_{AB} has closed range.

We pass now to the analysis of Hankel operators induced by sums of strictly noncyclic functions. So we assume that A and B are two strictly noncyclic functions in $L^\infty(B(N, M))$ having the respective factorizations (5-17) and (5-18) on the unit circle. We assume now the coprimeness relations

$$(P, C)_R = I_N, \qquad (P_1, C_1)_L = I_N \qquad (5\text{-}34)$$

and

$$(R, D)_R = I_M, \qquad (R_1, D_1)_L = I_N \qquad (5\text{-}35)$$

As $A + B$ has a meromorphic extension of bounded type to D_e whenever both A and B have then it is clearly strictly noncyclic. We assume that $A + B$ factors as

$$A + B = SH^* = H_1^* S_1 \qquad (5\text{-}36)$$

and the conditions

$$(S, H)_R = I_M, \qquad (S_1, H_1)_L = I_N \qquad (5\text{-}37)$$

are satisfied.

Theorem 5-14

(c) Let A, B be two strictly noncyclic functions in $L^\infty(B(N, M))$ and assume the factorizations (5-17) and (5-18) hold together with the coprimeness conditions (5-34) and (5-35). Let S be the inner function defined by (5-36) and (5-37). then $\det S | \det P \cdot \det R$ and

$$\det S = \det P \cdot \det R \qquad (5\text{-}38)$$

if and only if

$$(P, R)_L = I_M \quad \text{and} \quad (P_1, R_1)_R = I_N \qquad (5\text{-}39)$$

(b) If (5-34) and (5-35) are replaced by

$$[P, C]_R = I_M \quad \text{and} \quad [P_1, C_1]_L = I_N \qquad (5\text{-}40)$$

and

$$[R, D]_R = I_M \quad \text{and} \quad [R_1, D_1]_L = I_N \tag{5-41}$$

respectively, then $\text{Range}\, H_{A+B} = H_0(S)$ and (5-38) holds if and only if

$$[P, R]_L = I_M \quad \text{and} \quad [P_1, R_1]_R = I_N \tag{5-42}$$

PROOF Since M is finite dimensional and the subspaces PH_M^2 and RH_M^2 are of full range so is their intersection and therefore there exists an inner function Q for which

$$QH_M^2 = PH_M^2 \cap RH_M^2 \tag{5-43}$$

Moreover for some inner functions P' and R' we have

$$Q = PR' = RP' \tag{5-44}$$

From the obvious relations

$$\text{Range}\, H_{A+B} = \text{Range}(H_A + H_B) \subset \text{Range}\, H_A \vee \text{Range}\, H_B$$

$$= H_0(P) \vee H_0(R) = H_0(Q)$$

together with $\overline{\text{Range}\, H_{A+B}} = H_0(S)$ we obtain the inclusion $H_0(S) \subset H_0(P)$ and the consequent factorization $Q = SW$ of Q with W also an inner function. This implies $\det S | \det Q$. Since by Corollary 5-2 $\det Q | \det P \cdot \det R$ (5-37) follows. Now for equality (5-38) to hold it is necessary that $\det Q = \det P \cdot \det R$ which is equivalent, again by Corollary 5-2, to $(P, R)_L = I_M$. Thus $(P, R)_L = I_M$ is a necessary condition for (5-38) to hold but generally not suficient. we obtain another necessary condition, namely the coprimeness relation $(P_1, R_1)_R = I_N$, by considering $\tilde{A} + \tilde{B}$ in place of $A + B$. These two necessary coprimeness conditions of (5-39) turn out to be sufficient. To this end we note that from $B = D_1^* R_1$ it follows that $\text{Ker}\, H_B = R_1^* \bar{H}_N^2$. By restricting H_{A+B} to $\text{Ker}\, B$ we obtain for $f \in \bar{H}_N^2$

$$H_{A+B} R_1^* f = (H_A + H_B) R_1^* f = H_A R_1^* f = P_{H_{0,M}^2} A R_1^* f$$

$$= P_{H_{0,M}^2} C_1^* P_1 R_1^* f = H_{C_1^* P_1 R_1^*} f$$

or $\text{Range}\, H_{C_1^* P_1 R_1^*} = H_A(R_1^* \bar{H}_N^2)$. By Lemma 5-11 the condition $(P_1, R_1)_R = I_N$ implies $H_0(P) = \overline{\text{Range}\, H_{C_1 P_1 R_1^*}}$ and so $H_0(P) \subset \overline{\text{Range}\, H_{A+B}}$ and analogously $H_0(R) \subset \overline{\text{Range}\, H_{A+B}}$. Hence $H_0(P) \vee H_0(R) \subset \overline{\text{Range}\, H_{A+B}}$.

Since the inverse inclusion holds always we must have the equality $H_0(P) \vee H_0(R) = \overline{\text{Range}\, H_{A+B}} = H_0(S)$. The coprimeness condition $(P, R)_L = I_M$ implies now that $H_0(Q) = H_0(P) \vee H_0(R)$ and $\det Q = \det P \det R$ and hence equality (5-38). To prove part (b) we assume H_A and H_B to have closed ranges. For (5-38) to hold conditions (5-39) are necessary by part (a) and imply $\overline{\text{Range}\, H_{A+B}} = H_0(Q)$. From (5-44) together with

$$A + B = PC^* + RD^* = QH^*$$

we obtain

$$H = CR' + DP' \tag{5-45}$$

For H_{A+B} to have closed range it is necessary that $[Q, CR' + DP']_R = I_M$ holds. This implies the necessity of $[R', P']_R = I_M$ which is equivalent to $[P, R]_L = I_M$. The necessity of $[P_1, R_1]_R = I_N$ follows by duality considerations.

Conversely we assume the strong coprimeness conditions (5-42). By our assumptions $\text{Range} H_A = H_0(P)$, $\text{Range} H_B = H_0(R)$ and by Theorem 5-5 $[P, R]_L = I_M$ implies that the angle between $H_0(P)$ and $H_0(R)$ is positive, thus we obtain $H_0(Q) = H_0(P) + H_0(R)$. To complete the proof it suffices to show that $H_0(P)$ and $H_0(R)$ are both included in $\text{Range} H_{A+B}$. As in part (a) $H_{A+B}(\text{Ker} H_B) = H_A(R_1^* \bar{H}_N^2) = \text{Range} H_{AR_1^*}$. We can imply Lemma 5-12 to see that $\text{Range} H_{AR_1^*} = \text{Range} H_A = H_0(P)$. By symmetry also $\text{Range} H_B = H_0(R) \subset \text{Range} H_{A+B}$ and the proof is complete.

We remark that part of the content of Theorem 5-13 and Theorem 5-14 is the proof that the degree function Δ defined by (5-14) indeed satisfies relations (5-16) Moreover equalities in (5-16) are dependent on coprimeness conditions.

The degree theory development so far is closely related to the study of systems connected in series and in parallel and we shall delve into this in more detail.

Let $\Sigma_1 = (A_1, B_1, C_1, D_1)$ and $\Sigma_2 = (A_2, B_2, C_2, D_2)$ be two systems which realize the transfer functions T_1 and T_2, respectively.

The *series connection* of Σ_1 and Σ_2, denoted by $\Sigma_1 \Sigma_2$ is obtained, assuming the obvious compatibility conditions that the output space of Σ_1 coincides with the integral space of Σ_2, by feeding the output of Σ_1 into Σ_2. The dynamic equations are

$$\begin{aligned} x_{n+1}^{(1)} &= A_1 x_n^{(1)} + B_1 u_n \\ y_n^{(1)} &= C_1 x_n^{(1)} + D_1 u_n \end{aligned} \tag{5-46}$$

and

$$\begin{aligned} x_{n+1}^{(2)} &= A_2 x_n^{(2)} + B_2 y_n^{(1)} \\ y_n &= C_2 x_n^{(2)} + D_2 y_n^{(1)} \end{aligned} \tag{5-47}$$

or in matrix form

$$\begin{pmatrix} x_{n+1}^{(1)} \\ x_{n+1}^{(2)} \end{pmatrix} = \begin{pmatrix} A_1 & 0 \\ B_2 C_1 & A_2 \end{pmatrix} \begin{pmatrix} x_n^{(1)} \\ x_n^{(2)} \end{pmatrix} + \begin{pmatrix} B_1 \\ B_2 D_1 \end{pmatrix} u_n$$

$$y_n = (D_2 C_1 \quad C_2) \begin{pmatrix} x_n^{(1)} \\ x_n^{(2)} \end{pmatrix} + D_2 D_1 u_n \tag{5-48}$$

in other words

$$\Sigma_{21} = \left(\begin{pmatrix} A_1 & 0 \\ B_2 C_1 & A_1 \end{pmatrix}, \begin{pmatrix} B_1 \\ B_2 D_1 \end{pmatrix}, (D_2 C_1 \quad C_2), D_2 D_1 \right) \tag{5-49}$$

In the same way given two systems $\Sigma_1 = (A_1, B_1, C_1, D_1)$ and $\Sigma_2 = (A_2, B_2, C_2, D_2)$ having the same input and output spaces we define the *parallel connection* of Σ_1 and Σ_2, denoted by $\Sigma_1 + \Sigma_2$, by feeding the same input to both systems and combining their outputs. The dynamic equations are

$$x_{n+1}^{(1)} = A_1 x_n^{(1)} + B_1 u_n \tag{5-50}$$

$$y_n^{(1)} = C_1 x_n^{(1)} + D_1 u_n$$

$$x_{n+1}^{(2)} = A_2 x_n^{(2)} + B_2 u_n \tag{5-51}$$

$$y_n^{(2)} = C_2 x_n^{(2)} + D_2 u_n$$

and

$$y_n = y_n^{(1)} + y_n^{(2)} \tag{5-52}$$

In matrix form this becomes

$$\begin{pmatrix} x_{n+1}^{(1)} \\ x_{n+1}^{(2)} \end{pmatrix} = \begin{pmatrix} A_1 & 0 \\ 0 & A_2 \end{pmatrix} \begin{pmatrix} x_n^{(1)} \\ x_n^{(2)} \end{pmatrix} + \begin{pmatrix} B_1 \\ B_2 \end{pmatrix} u_n$$

$$\tag{5-53}$$

$$y_n = (C_1 \quad C_2) \begin{pmatrix} x_n^{(1)} \\ x_n^{(2)} \end{pmatrix} + (D_1 + D_2) u_n$$

and hence

$$\Sigma_1 + \Sigma_2 = \left(\begin{pmatrix} A_1 & 0 \\ 0 & A_2 \end{pmatrix}, \begin{pmatrix} B_1 \\ B_2 \end{pmatrix}, (C_1 \quad C_2), (D_1 + D_2) \right) \tag{5-54}$$

Theorem 5-15
(a) The transfer function of the series connection $\Sigma_2 \Sigma_1$ of the systems Σ_1 and Σ_2 is the product $T_2 T_1$ of their respective transfer functions
(b) The transfer function of the parallel connection $\Sigma_1 + \Sigma_2$ of the systems Σ_1 and Σ_2 is the sum $T_2 + T_1$ of their respective transfer functions.

PROOF (a) The transfer function of the series connection of Σ_1 and Σ_2 is given by

$$T(z) = D_2 D_1 + z (D_2 C_1 \quad C_2) \begin{pmatrix} I - zA_1 & 0 \\ -zB_2 C_1 & I - zA_2 \end{pmatrix}^{-1} \begin{pmatrix} B_1 \\ B_2 D_1 \end{pmatrix}$$

Now

$$\begin{pmatrix} I - zA_1 & 0 \\ -zB_2 C_1 & I - zA_2 \end{pmatrix}^{-1}$$

$$= \begin{pmatrix} (I - zA_1)^{-1} & 0 \\ z(I - zA_2)^{-1} B_2 C_1 (I - zA_1)^{-1} & (I - zA_2)^{-1} \end{pmatrix}$$

and the result follows. Part (b) is proved by a similar, even simpler, computation.

We pass now to the study of the series connection of shift realizations.

Theorem 5-16 Let L, M, and N be finite dimensional Hilbert spaces and let $A \in H^\infty(B(M, N))$ and $B \in H^\infty(B(L, M))$ be two strictly noncyclic functions having the factorizations

$$A = PC^* = C_1^* P_1 \tag{5-55}$$

and

$$B = RD^* = D_1^* R_1 \tag{5-56}$$

satisfying the coprimeness conditions

$$(P, C)_R = I_N, \qquad (P_1, C_1)_L = I_M \tag{5-57}$$

and

$$(R, D)_R = I_M, \qquad (R_1, D_1)_L = I_L \tag{5-58}$$

respectively. Then the following statements hold.

(a) The series connection $\Sigma_A \Sigma_B$ of the shift realizations Σ_A and Σ_B of A and B, respectively, is observable if and only if $(R, C)_L = I_M$ holds and exactly observable if and only if $[R, C]_L = I_M$.

(b) The series coupling $\Sigma'_A \Sigma'_B$ of the $*$-shift realizations of A and B is reachable if and only if $(P_1, D_1)_R = I_M$ and exactly reachable if and only if $[P_1, D_1]_R = I_M$.

(c) A sufficient condition for the reachability of $\Sigma_A \Sigma_B$ is $(P_1, D_1)_R = I_M$. If Σ_A and Σ_B are both exactly reachable then $\Sigma_A \Sigma_B$ is exactly reachable if and only if $[P_1, D_1]_R = I_M$.

(d) A sufficient condition for the observability of $\Sigma'_A \Sigma'_B$ is $(R, C)_L = I_M$. If Σ'_A and Σ'_B are both exactly observable then $\Sigma'_A \Sigma'_B$ is exactly observable if and only if $[R, C]_L = I_M$.

PROOF The shift realization of A is given, omitting the constant term for simplicity, by $\Sigma_A = (S_0(P)^*, M_A, \gamma_0(P)^*)$ in the state space $H_0(P)$ where $S_0(P)$ is given by (3-1), $M_A : M \to H_0^2(N)$ is defined by

$$M_A \xi = P_{H_0^2(N)} A \xi \qquad \text{for} \qquad \xi \in M \tag{5-59}$$

and $\gamma_0(P) : N \to H_0^2(M)$ is defined by

$$\gamma_0(P) \eta = P_{H_0(P)} \chi \eta \tag{5-60}$$

which is equivalent to $(\gamma_0(P) \eta)(z) = z(I - P(z) P(0)^*) \eta$. Similar formulas can be derived for Σ_B. Note that $\gamma_0(P)^*$ is given by

$$\gamma_0(P)^* f = f'(0) \qquad \text{for} \qquad f \in H_0(P) \tag{5-61}$$

Also using the transformation $\tau'_p : H_0(P) \to H_0(\tilde{P})$ we have

$$\tau'_p \gamma_0(P) = \Gamma_0(\tilde{P}) \tag{5-62}$$

where $\Gamma_0(P) : N \to H_0(\tilde{P})$ is defined by

$$\Gamma_0(\tilde{P}) \eta = (\tilde{P} - \tilde{P}(0)) \eta \tag{5-63}$$

From the preceding representation of the shift realization of A and B we see that the series connection $\Sigma_A \Sigma_B$ of Σ_A and Σ_B has $H_0(R) \oplus H_0(P)$ as state space, is given by the system

$$\left(\begin{pmatrix} S_0(R)^* & 0 \\ M_A \gamma_0(R)^* & S_0(P)^* \end{pmatrix}, \quad \begin{pmatrix} M_B \\ 0 \end{pmatrix}, \quad (0 \quad \gamma_0(P)^*) \right) \tag{5-64}$$

and has AB as its transfer function.

Assume first that $(R, C)_L = I_M$. By Theorem 3-5, there exists an inner function R' acting in N and $C' \in H^\infty(B(N, M))$ such that $C^*R = R'C'^*$, $(R', C')_R = I_N$ and $\det R = \det R'$. From the factorizations (5-55) and (5-56) we obtain $AB = PC^*RD^* = PR'C'^*D^*$ and this factorization of AB enables us to write down explicitly the shift realization Σ_{AB} of AB. It has $H_0(PR')$ as state space and is given by $(S_0(PR'), M_{AB}, \gamma_0(PR')^*)$. This realization is clearly exactly observable but not necessarily reachable. Reachability is equivalent to $H_0(PR') = \text{Range} H_{AB}$ and this is equivalent to the coprimeness condition $(P_1, D_1)_R = I_M$. To prove our theorem we have to study Σ_{AB} in more detail.

By the vector-valued version of Lemma II 15-12, $H_0(PR')$ has a direct sum representation

$$H_0(PR') = H_0(P) \oplus P H_0(R') \tag{5-65}$$

Hence there exists a unitary map of $H_0(PR')$ onto $H_0(R') \oplus H_0(P)$ given by $f = g + Ph \to (h, g)$ where $g + Ph$ is the unique decomposition of $f \in H_0(PR')$ relative to the direct sum (5-65). From the above representation of $f \in H_0(PR')$ we have, using the fact that $(S_+^* f)(z) = f(z)/z - f'(0)$, and recalling that $h(0) = 0$ for $h \in H_0^2(N)$

$$f(z)/z - f'(0) = g(z)/z - g'(0) + P(z) h(z)/z - P(0) h'(0)$$

$$= g(z)/z - g'(0) + P(z) (h(z)/z - h'(0))$$

$$+ (P(z) - P(0)) h'(0)$$

and hence

$$S_0(PR')^* f = S_0(P)^* g + PS_0(R')^* h + \Gamma_0(P) \gamma_0(R')^* h \tag{5-66}$$

Next, from $f = g + Ph$, it follows, again using the fact that $h(0) = 0$, that $f'(0) = g'(0) + P(0) h'(0)$ which implies that

$$\gamma_0(PR')^* f = \gamma_0(P)^* g + P(0) \gamma_0(R')^* h \tag{5-67}$$

Finally, for $\xi \in L$ let $M_{AB} \xi = AB\xi = g + Ph$ with $g \in H_0(P)$ and $h \in H_0(R)$, then we have

$$h = P_{H_0^2(M)}C^*B\xi = P_{H_0^2(M)}C^*M_B\xi \tag{5-68}$$

and

$$g = M_{AB}\xi - P \cdot P_{H_0^2(M)}C^*M_B\xi \tag{5-69}$$

Thus with respect to the direct sum $H(R') \oplus H(P)$ the shift realization Σ_{AB} of AB is given by

$$\left[\begin{pmatrix} S_0(R')^* & 0 \\ \Gamma_0(P)\gamma_0(R')^* & S_0(P)^* \end{pmatrix}, \begin{pmatrix} P_{H^2(M)}C^*M_B \\ M_{AB} - P \cdot P_{H_0^2(M)}C^*M_B \end{pmatrix}\right. \tag{5-70}$$

$$\left. \begin{pmatrix} P(0)\gamma_0(R')^* & \gamma_0(P)^* \end{pmatrix}\right]$$

As our next step we construct a map $X: H_0(R) \oplus H_0(P) \to H_0(R') \oplus H_0(P)$ which intertwines $\Sigma_A\Sigma_B$ and Σ_{AB}. A comparison of the state generators in the two systems (5-64) and (5-70), which are both lower triangular, indicates that we should look for an intertwining operator X of the

$$X = \begin{pmatrix} W & 0 \\ Z & I \end{pmatrix} \tag{5-71}$$

where $W: H_0(R) \to H_0(R')$ and $Z: H_0(R) \to H_0(P)$ are bounded. For the generator W the natural candidate is the quasi-invertible operator that intertwines the shift and the $*$-shift realizations of the analytic part of $E = C^*R = R'C^*$. These two realizations are given by

$$\left(S_0(R)^*, \Gamma_0(R), M_C^*\right) \tag{5-72}$$

and

$$\left(S_0(R')^*, M_E, \gamma_0(R')^*\right) \tag{5-73}$$

respectively. From the commutativity of the diagram

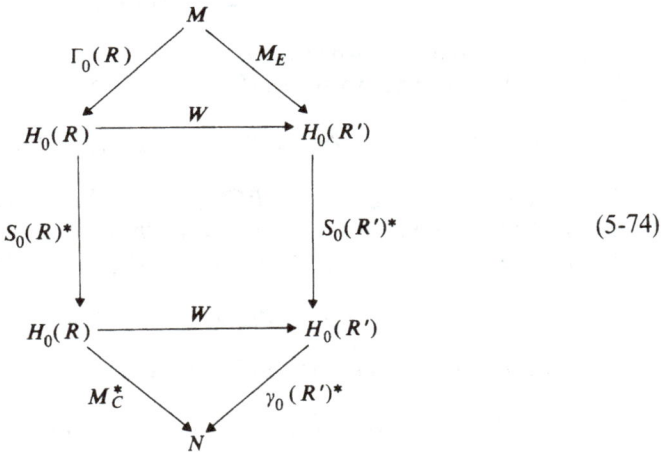

$$\tag{5-74}$$

we have that

$$WT_0(R)\,\xi = WP_{H_\partial^2(M)}R\xi = P_{H_\partial^2(M)}C^*R\xi$$

and since $WS_0(R)^* = S_0(R')^*\,W$ this implies that for each $f \in H_0(R)$

$$Wf = P_{H_0(R')}C^* = P_{H_\partial^2(N)}C^*f \qquad (5\text{-}75)$$

and we take (5-75) to be the definition of W. The coprimeness conditions $(R, C)_L = I_M$ and $(R', C')_R = I_N$ guarantee that W is a quasi-invertible operator.

Now X given by (5-71) intertwines $\Sigma_A\Sigma_B$ and Σ_{AB} if and only if the following relations hold

$$WS_0(R)^* = S_0(R')^*\,W \qquad (5\text{-}76)$$

$$ZS_0(R)^* + M_A\gamma_0(R)^* = \Gamma_0(P)\,\gamma_0(R')^*\,W + S_0(P)^*\,Z \qquad (5\text{-}77)$$

$$P(0)\,\gamma_0(R')^*\,W + \gamma_0(P)^*\,Z = 0 \qquad (5\text{-}78)$$

$$WM_B = P_{H_\partial^2(M)}C^*M_B \qquad (5\text{-}79)$$

and

$$ZM_B = M_{AB} - P\cdot P_{H_\partial^2(M)}C^*M_B \qquad (5\text{-}80)$$

Equalities (5-76) and (5-79) follow directly from the definition of W given by (5-75). We define $Z: H_0(R) \to H_0(P)$ by

$$Zf = P_{H_0(P)}PC^*f \qquad \text{for} \qquad f \in H_0(R) \qquad (5\text{-}81)$$

which immediately implies (5-80).

Now for $f \in H_0(R)$ we have $P(0)\,\gamma_0(R')^*\,Wf = (P\cdot P_{H_0(R')}C^*f)'(0)$ and hence

$$P(0)\,\gamma_0(R')^*\,Wf + \gamma_0(P)^*\,Zf = (P\cdot P_{H_0(R')}C^*f + P_{H_0(P)}PC^*f)'(0)$$
$$= (PC^*f)'(0)$$

as both PC^* and f vanish at zero.

Finally, in order to prove (5-77) we note that for $f \in H_0(R)$

$$\left(ZS_0(R)^* - S_0(P)^*\,Z\right)f = P_{H_0(p)}PC^*P_{H_\partial^2(M)}\chi f P_{H_0(p)}\bar\chi P_{H_0(p)}PC^*f \qquad (5\text{-}82)$$

Take the decomposition of PC^*f, f in $H_0(R)$, relative to $H_0(PR') = H_0(P) \oplus PH_0(R')$. Thus $PC^*f = g + Ph$ with $g \in H_0(P)$ and $h \in H_0(R')$. Clearly we have $g = P_{H_0(p)}PC^*f$ and $h = P_{H_0(R')}C^*f$. This implies the equality

$$P_{H_0(p)}\bar\chi PC^*f = P_{H_0(P)}\bar\chi P_{H_0(P)}PC^*f + P_{H_0(P)}\bar\chi P\cdot P_{H_0(R')}C^*f \qquad (5\text{-}83)$$

Substituting (5-83) back into (5-82) yields

$$\left(ZS_0(R)^* - S_0(P)^*\,Z\right)f = P_{H_0(p)}PC^*P_{H_\partial^2(M)}\bar\chi f - P_{H_0(P)}\bar\chi PC^*f$$
$$+ P_{H_0(p)}\bar\chi P\cdot P_{H_0(R')}C^*f \qquad (5\text{-}84)$$

If we apply equality (5-66) to

$$P_{H_0(p)}\bar{\chi}P \cdot P_{H_0(R')}C^*f = P_{H_0(p)}\bar{\chi}PWf = P_{H_0(p)}P_{H_0^2(N)}\bar{\chi}PWf$$

we obtain

$$P_{H_0(p)}P_{H_0^2(N)}\bar{\chi}PWf = P_{H_0(p)}\{PS(R')^* Wf + \Gamma_0(P)\gamma_0(R')^* Wf\}$$

$$= \Gamma_0(P)\gamma_0(R')^* Wf$$

Also, for the first two terms on the right-hand side of (5-84) we have

$$P_{\boldsymbol{z}_0(p)}PC^*P_{H_0^2(M)}\bar{\chi}f - P_{H_0(p)}\bar{\chi}PC^*f = P_{H_0(p)}\bar{\chi}PC^*(f - \bar{\chi}f'(0)) - P_{H_0(p)}\bar{\chi}PC^*f$$

$$= -P_{H_0(p)}PC^*f'(0) = -PC^*\gamma_0(R)^* f$$

$$= -M_A\gamma_0(R)^* f$$

which proves (5-77).

Since X is quasi-invertible if and only if W is our assumption of the coprimeness relation $(R, C)_L = I_M$ implies the quasi-invertibility of X. Since the shift realization Σ_{AB} of AB is exactly observable it follows, by Theorem 1-7 (c), that Σ is observable. Replacing $(R, C)_L = I_M$ by the stronger coprimeness relation $[R, C]_L = I_M$ guarantees that W, and hence X, is boundedly invertible and so again by Theorem 1-7 (c), the exact observability of Σ_{AB} implies that of the series connection $\Sigma_A\Sigma_B$. Conversely if we assume the observability of $\Sigma_A\Sigma_B$ then X as defined by (5-71), (5-75), and (5-81) is a map that intertwines $\Sigma_A\Sigma_B$ and Σ_{AB}. By Lemma 1-6 (b) X is the only intertwining map which, by Theorem 1-7 (d), is injective. Thus the injectivity of W follows and hence $(R, C)_L = I_M$ has to be satisfied. Similarly if $\Sigma_A\Sigma_B$ is assumed exactly observable then X becomes boundedly invertible which in turn implies $[R, C]_L = I_M$. Thus part (a) of the theorem is proved and part (b) follows directly by duality considerations. Indeed let Σ'_A and Σ'_B be the *-shift realizations of A and B which are unitarily equivalent to $\Sigma_{\tilde{A}}^*$ and $\Sigma_{\tilde{B}}^*$. Thus the series connection $\Sigma'_A\Sigma'_B$ is unitarily equivalent to $(\Sigma_B\Sigma_A)^*$ the adjoint system to the series connection of the shift realizations of $\tilde{B}$ and $\tilde{A}$. The map X intertwining $\Sigma_A\Sigma_B$ and Σ_{AB} has its counterpart now in a map X' that intertwines Σ'_{AB} and $\Sigma'_A\Sigma'_B$. Also note that reachability properties of the system $\Sigma_A\Sigma_B$ are equivalent to observability properties of $\Sigma_{\tilde{B}}\Sigma_{\tilde{A}}$.

Since there exist quasi-invertible maps that intertwine Σ'_A and Σ_A, Σ'_B and Σ_B, respectively, their direct sum, denoted by Ξ, is a map that intertwines $\Sigma'_A\Sigma'_B$ and $\Sigma_A\Sigma_B$. Thus we obtain the series of intertwining maps

$$\Sigma'_{AB} \xrightarrow{X'} \Sigma'_A\Sigma'_B \xrightarrow{\Xi} \Sigma_A\Sigma_B \xrightarrow{X} \Sigma_{AB} \tag{5-85}$$

The transformation X has dense range by construction and is injective if and only if $(R, C)_L = I_M$. X' is injective by construction and has dense range if $(P_1, D_1)_R = I_M$. To obtain bounded invertibility of X and X' the coprimeness conditions have to be replaced by strong coprimeness conditions. The map Ξ is boundedly invertible if and only if the maps intertwining the *-shift and shift realizations of A, and B, respectively, are actually boundedly invertible and this is equivalent to the exact reachability and exact observability of the four systems Σ_A, Σ_B, Σ_A', and Σ_B'. In turn this is equivalent to the strong coprimeness conditions $[P, C]_R = I_N$, $[P_1, C_1] = I_M$, $[R, D]_R = I_M$ and $[R_1, D_1] = I_L$.

We add the obvious remark that in the case of rational functions the map Ξ is always boundedly invertible and hence this theorem provides a complete analysis of the series connection in the finite dimensional case.

Next we analyse the parallel connection of the shift realizations of two strictly noncyclic functions. The analysis is simpler as the parallel coupling of shift systems is also a shift system.

We begin by proving a simple lemma concerning inner functions.

Lemma 5-17 Given a Hilbert space M then an inner function Q acting in $M \oplus M$ is a left inner factor of $\begin{pmatrix} I_M \\ I_M \end{pmatrix}$ if and only if it has, up to a constant right unitary factor, the form

$$Q = \frac{1}{2}\begin{pmatrix} I_M + S & I_M - S \\ I_M - S & I_M + S \end{pmatrix} \tag{5-86}$$

for some inner functions S acting in M.

PROOF Let S be an inner function acting in M then Q defined by (5-86) is also inner and since

$$\begin{pmatrix} I_M \\ I_M \end{pmatrix} = \frac{1}{2}\begin{pmatrix} I_M + S & I_M - S \\ I_M - S & I_M + S \end{pmatrix}\begin{pmatrix} I_M \\ I_M \end{pmatrix} \tag{5-87}$$

it is a left inner factor of $\begin{pmatrix} I_M \\ I_M \end{pmatrix}$.

Conversely we consider the constant unitary operator U in $M \oplus M$ defined by

$$U = \frac{1}{\sqrt{2}}\begin{pmatrix} I_M & I_M \\ I_M & -I_M \end{pmatrix} \tag{5-88}$$

which extends naturally to a unitary operator in $H^2(M \oplus M)$. Obviously we have

$$U\begin{pmatrix} I_M \\ I_M \end{pmatrix} = \sqrt{2}\begin{pmatrix} I_M \\ 0 \end{pmatrix} \tag{5-89}$$

Thus an inner function Q acting in $M \oplus M$ is a left factor of $\begin{pmatrix} I_M \\ I_M \end{pmatrix}$ if and only if UQ is a left factor of $\begin{pmatrix} I_M \\ 0 \end{pmatrix}$. However, left inner factors of $\begin{pmatrix} I_M \\ 0 \end{pmatrix}$ are those associated with full range right invariant subspaces of $H^2(M \oplus M)$ which contain $H^2(M) \oplus \{0\}$. These subspaces are clearly of the form $H^2(M) \oplus SH^2(M)$ for some inner function S acting in M. Thus the corresponding inner functions are of the form $\begin{pmatrix} I_M & 0 \\ 0 & S \end{pmatrix}$. Hence

$$Q = U^* \begin{pmatrix} I_M & 0 \\ 0 & S \end{pmatrix} = \frac{1}{\sqrt{2}} \begin{pmatrix} I_M & S \\ I_M & -S \end{pmatrix}$$

Since Q is unique up to a right constant inner factor by right multiplication by U we obtain the representation (5-86).

Lemma 5-18 Given inner functions P and R acting in M. The following two statements are true.

(*a*) The coprimeness conditions

$$\left[\begin{pmatrix} P & 0 \\ 0 & R \end{pmatrix}, \begin{pmatrix} I_M \\ I_M \end{pmatrix} \right]_L = I_{M \oplus M} \tag{5-90}$$

and

$$(P, R)_L = I_M \tag{5-91}$$

are equivalent.

(*b*) The strong coprimeness conditions

$$\left[\begin{pmatrix} P & 0 \\ 0 & R \end{pmatrix}, \begin{pmatrix} I_M \\ I_M \end{pmatrix} \right]_L = I_{M \oplus M} \tag{5-92}$$

and

$$[P, R]_L = I_M \tag{5-93}$$

are equivalent.

PROOF
(*a*) Let S be a common left inner factor of P and R. Thus $P = SP_1$ and $R = SR_1$. Since

$$\begin{pmatrix} P & 0 \\ 0 & R \end{pmatrix} = \frac{1}{2} \begin{pmatrix} I_M + S & I_M - S \\ I_M - S & I_M + S \end{pmatrix} \cdot \frac{1}{2} \begin{pmatrix} I_M + S & S - I_M \\ S - I_M & I_M + S \end{pmatrix} \begin{pmatrix} P_1 & 0 \\ 0 & R_1 \end{pmatrix}$$

$$= \begin{pmatrix} SP_1 & 0 \\ 0 & SR_1 \end{pmatrix}$$

then together with (5-87) it follows that Q defined by (5-86) is a nontrivial left inner factor of

$$\begin{pmatrix} P & 0 \\ 0 & R \end{pmatrix} \quad \text{and} \quad \begin{pmatrix} I_M \\ I_M \end{pmatrix}$$

Conversely if we assume

$$\begin{pmatrix} P & 0 \\ 0 & R \end{pmatrix} \quad \text{and} \quad \begin{pmatrix} I_M \\ I_M \end{pmatrix}$$

have a common left inner factor then by Lemma 5-17 it is necessarily of the form (5-86). Thus

$$\begin{pmatrix} P & 0 \\ 0 & R \end{pmatrix} = \frac{1}{2}\begin{pmatrix} I_M + S & I_M - S \\ I_M - S & I_M + S \end{pmatrix}\begin{pmatrix} A & B \\ C & D \end{pmatrix}$$

from which we obtain the relations $P = S(A - C)$ and $R = S(L - B)$ which taken together show that S is a common left factor of P and R.

(b) The key to the proof is Theorem III 14-10. If $[P, R]_L = I_M$ holds then there exist P_1 and R_1 in $H^\infty(B(M, M))$ for which $PP_1 + RR_1 = I_N$. This implies in turn that

$$\begin{pmatrix} P & 0 \\ 0 & R \end{pmatrix}\begin{pmatrix} P_1 & -P_1 \\ -R_1 & R_1 \end{pmatrix} + \begin{pmatrix} I_M \\ I_M \end{pmatrix}(RR_1 \quad PP_1) = \begin{pmatrix} I_M & 0 \\ 0 & I_M \end{pmatrix}$$

which shows that the coprimeness relation (5-92) holds.

Conversely assume (5-92) holds then there exist bounded analytic functions $\begin{pmatrix} A & B \\ C & D \end{pmatrix}$ and $(E \quad F)$ for which

$$\begin{pmatrix} P & 0 \\ 0 & R \end{pmatrix}\begin{pmatrix} A & B \\ C & D \end{pmatrix} + \begin{pmatrix} I_M \\ I_M \end{pmatrix}(E \quad F) = \begin{pmatrix} I_M & 0 \\ 0 & I_M \end{pmatrix}$$

This implies in particular that $PA - RC = I_M$ and applying Theorem II 14-10 again we have (5-93).

Theorem 5-19 Let A and B be strictly noncyclic functions in $H^\infty(B(N, M))$ having the coprime factorizations (5-55) and (5-56), respectively. Then

(a) The parallel connection $\Sigma_A + \Sigma_B$ of the shift realizations Σ_A and Σ_B is observable if and only if

$$(P, R)_L = I_M \tag{5-94}$$

and is exactly observable if and only if

$$[P, R]_L = I_M \tag{5-95}$$

(b) The parallel connection $\Sigma'_A + \Sigma'_B$ of the *-shift realizations Σ'_A and Σ'_B is reachable if and only if

$$(P_1, R_1)_R = I_M \tag{5-96}$$

and exactly reachable if and only if

$$[P_1, R_1]_R = I_M \tag{5-97}$$

c) A sufficient condition for the reachability of $\Sigma_A + \Sigma_B$ is (5-96). If Σ_A and Σ_B are exactly reachable then (4-36) is also necessary.

PROOF The parallel connection $\Sigma_A + \Sigma_B$ of Σ_A and Σ_B has state space $H_0(P) \oplus H_0(R)$ and is given by

$$\left(\begin{pmatrix} S_0(P)^* & 0 \\ 0 & S_0(R)^* \end{pmatrix}, \begin{pmatrix} M_A \\ M_B \end{pmatrix}, \left(\gamma_0(P)^* \quad \gamma_0(R)^* \right) \right) \qquad (5\text{-}98)$$

Thus $\Sigma_A + \Sigma_B$ is observable and exactly observable if and only if (5-90) and (5-96) hold, respectively. But these coprimeness conditions are equivalent to (5-91) and (5-93), respectively, by Lemma 5-18, which proves (*a*). Part (*b*) follows by duality considerations.

To prove (*c*) let X_A be the map that intertwines the $*$-shift and shift realizations of A. Similarly we define X_B. X_A and X_B are quasi-invertible maps and hence so is $X_A \oplus X_B$ which is a map from $H_0(P_1) \oplus H_0(R_1)$ into $H_0(P) \oplus H_0(R)$ which intertwines $\Sigma'_A + \Sigma'_B$ and $\Sigma_A + \Sigma_B$. Thus the reachability of $\Sigma'_A + \Sigma'_B$ implies that of $\Sigma_A + \Sigma_B$.

If both systems Σ_A and Σ_B are exactly reachable then X_A and X_B are boundedly invertible and so is $X_A \oplus X_B$. In this case $\Sigma'_A + \Sigma'_B$ is reachable or exactly reachable if and only if $\Sigma_A + \Sigma_B$ has these properties.

6. CONTINUOUS TIME SYSTEMS

The study of infinite dimensional continuous time systems presents some difficulties which are absent from the discrete time case. Probably the greatest one is that of deciding about how large a class of systems one wants to study. Thus while we want to develop a theory of systems whose internal representations are of the form

$$\dot{x}(t) = Ax(t) + Bu(t)$$
$$y(t) = Cx(t) + Du(t) \qquad (6\text{-}1)$$

it is far from obvious what restrictions one wants to impose on the operators involved. As we shall see a strict interpretation of (6-1) limits the input/output relations realizable by such systems and hence in order to obtain a theory that would encompass more general input/output relations one would have to relax the restrictions on the operators, A, B, C, D.

The central theme of this section is the discussion of the realization problem for continuous time systems and to this end we want to use the continuous time analogue of the restricted shift realization, namely, a realization that utilizes the left translation semigroup. This is a natural approach both for its similarity to the discrete time methods and for the universal properties of the left translation semigroup as given by Theorem II 10-18.

Let us study equations (6-1) for a moment. As usual we consider finite input/

finite output systems which means that the spaces U and Y are assumed to be finite dimensional. The state x takes its values in a Hilbert space X. For (6-1) to make sense strongly one assumes A to be the infinitesimal generator of a strongly continuous semigroup, which will be denoted by e^{At}, u is to be continuous function and that the range of B is included in the domain of A. Moreover one assumes that the domain of Z, D_C includes the domain of A, D_A, and that the restriction of C to D_A is continuous. Under these assumptions given an initial state $x_0 \in D_A$ a solution of (6-1) exists and is given by the variation of parameters formula

$$x(t) = e^{At}x_0 + \int_0^t e^{A(t-\tau)}Bu(\tau)\,d\tau \tag{6-2}$$

which yields

$$y(t) = Ce^{At}x_0 + \int_0^t Ce^{A(t-\tau)}Bu(\tau)\,d\tau \tag{6-3}$$

The function $Ce^{At}B$ is the weighting pattern of the system. Reversing our steps we may start with a weighting pattern, which we may not restrict to be a function but allow also distributional values and study input/output relations of the form

$$y(t) = \int_0^t W(t-\tau)u(\tau)\,d\tau \tag{6-4}$$

Under assumptions, which will be made precise later, the Fourier transform of (6-4) can be taken which yields

$$\mathscr{F}y = (\mathscr{F}W) \cdot (\mathscr{F}u) \tag{6-5}$$

The Fourier transform of W will be called the *transfer function* of the system. In the realization problem we seek, given a W or its Fourier transform, a system (A, B, C, D) within a prescribed class whose weighting pattern coincides with W.

By weakening our concept of solution we can enlarge the class of systems under consideration. Thus an X-valued function $x(t)$ will be called a (weak) solution if (6-3) is satisfied. Thus x_0 need not be restricted to D_A and the function u need not be continuous, being locally L^1 is sufficient. Furthermore we do not assume that the range of B is included in D_A. What this amounts to is the interpretation of (6-1) in the weak or distributional sense.

Suppose that we assume both operators $B: U \to X$ and $C: X \to Y$ to be bounded linear operators then, assuming $x_0 = 0$, the weighting pattern $W(t) = Ce^{At}B$ is a continuous function from U to Y.

Requiring a weighting pattern to be continuous is a severe restriction on the theory which will exclude most interesting physical systems. Also the boundedness requirements on B and C exclude the cases where either the controls or the observations are applied at the boundary. Thus we are faced with the need to relax our assumptions on the operators that constitute a system.

We will call a system (A, B, C, D) which has $W(t)$ as its weighting pattern a *regular realization* of W if B and C are bounded linear maps. One way to relax the conditions on B and C is the introduction of *balanced realizations*. A realization (A, B, C, D) is called balanced if $B: U \to X$ is bounded and Range $B \subset D_A$, C is a closed linear operator for which $D_C \supset D_A$ and C restricted to D_A is continuous with respect to the graph topology of D_A, that is the topology induced by the norm $\|x\|_A = \{\|x\|^2 + \|Ax\|^2\}^{1/2}$. An equivalent way of stating it is saying that C is A-*bounded*, that is

$$\|Cx\| \leq K(\|x\| + \|Ax\|) \qquad \text{for all} \qquad x \in D_A \qquad (6\text{-}6)$$

The added generality of studying balanced realizations is only illusory for we have the following result.

Theorem 6-1 A weighting pattern $W(t)$ has a balanced realization if and only if it has a regular one.

PROOF Suppose (A, B, C) is a regular realization of $W(t)$. Define a system (F, G, H) by

$$F = A, \qquad G = (\lambda I - A)^{-1} B, \qquad H = C(\lambda I - A) \qquad (6\text{-}7)$$

where we take λ in $\rho(A)$. Clearly the range of G is in D_A. Moreover since A is continuous on D_A, with respect to the graph topology of D_A, $D_H \supset D_A$ and H is continuous on D_A.

Conversely suppose (F, G, H) is a balanced realization of $W(t)$. Define (A, B, C) by

$$A = F, \qquad B = (\lambda I - F) G, \qquad C = H(\lambda I - F)^{-1} \qquad (6\text{-}8)$$

As $(\lambda I - F)$ is closed it follows that B is an everywhere defined closed transformation. The closed graph theorem guarantees that B is bounded. Similarly $(\lambda I - F)^{-1}$ is a continuous map into D_F and as H is F-bounded the continuity of C follows. This is equivalent to boundedness.

It may be of interest to probe a bit further as to the limits of applicability of regular, or alternately balanced, realizations. For simplicity we take up the case of scalar valued weighting patterns. A function f defined on $(0, \infty)$ is said to be of *exponential order* if there exist positive constants M and ω for which $|f(t)| \leq Me^{\omega t}$. Obviously if f is measurable and of exponential order then $e^{-\alpha t} f(t)$ is in $L^2(0, \infty)$ for large enough α.

Theorem 6-2 A necessary condition that $w(t)$ have a balanced realization is that w be continuous and of exponential order. A sufficient condition is that w be absolutely continuous and $\dot{w}$ be of exponential order (in the sense that $|\dot{w}(t)| \leq Me^{\omega t}$ a.e.).

PROOF If w has a balanced realization then it has also a regular one (A, B, C). Since $w(t) = Ce^{At}B$ it follows that w is continuous. Also since $\|e^{At}\| \leq Me^{\beta t}$ for every strongly continuous semigroup we have $|w(t)| \leq \|C\| \cdot \|B\| Me^{\beta t}$ and w is of exponential order.

To prove the sufficiency part we note that if (A, B, C) is a realization of $w(t)$ then $(A - \lambda I, B, C)$ is a realization of $e^{-\lambda t}w(t)$. Thus without loss of generality we may assume that w and $\dot{w}$ belong to $L^2(0, \infty)$.

Let $V(t)$ be the right translation semigroup in $L^2(0, \infty)$, $V(t)^*$ the left translation semigroup. Let A be the infinitesimal generator of $V(t)^*$ then as was shown in Sec. II-10 we have $D_A = \{f \in L^2(0, \infty) | f$ absolutely continuous and $f' \in L^2(0, \infty)\}$ and $Af = f'$ for $f \in D_A$. Let us define $B: \mathbb{C} \to D_A$ by $B\alpha = \alpha w$. Finally let C be the linear functional defined by $Cf = f(0)$. Certainly C is defined on D_A. Moreover we have, by integration by parts that

$$|f(0)|^2 = -\int_0^\infty f(x) \overline{f'(x)} \, dx - \int_0^\infty f'(x) \overline{f(x)} \, dx$$

$$= -2 \operatorname{Re} \int_0^\infty f(x) \overline{f'(x)} \, dx$$

Applying Schwartz's inequality we have

$$|f(0)|^2 \le 2\|f\| \cdot \|f'\| \le \|f\|^2 + \|f'\|^2 = \|f\|^2 + \|Af\|^2$$

Thus the map C is A-bounded. Now

$$(CV(t)^* \, B\alpha)(x) = \alpha C(V(t)^* \, w)(x) = \alpha Cw(x + t) = \alpha w(t)$$

which shows that (A, B, C) is a balanced realization of w.

The construction of the balanced realization in the previous proof contains the central idea of most approaches to the realization problem. We want a state space model that uses the left translation semigroup. B would embed the weighting pattern in some vectorial $L^2(0, \infty)$ space and C should act as evaluation at zero.

Of course one could just as well approach the realization problem in the frequency domain. To this end let Γ_+ denote the open right half plane $\{\lambda | \operatorname{Re} \lambda > 0\}$ and let $H^2(\Gamma_+; N)$ be the corresponding Hardy space. By the results of Sec. II-12 there exists a unitary map $J: H_N^2 \to H^2(\Gamma_+; N)$ given by

$$(Jf)(w) = \frac{1}{\sqrt{\pi}} \frac{1}{w + 1} f\left(\frac{w - 1}{w + 1}\right) \tag{6-9}$$

Also by the Paley–Wiener theorem the Fourier transform

$$(\mathscr{F}f)(w) = \frac{1}{\sqrt{2\pi}} \int_0^\infty f(t) e^{-wt} \, dt \tag{6-10}$$

is a unitary map of $L^2(0, \infty; N)$ onto $H^2(\Gamma_+; N)$. If $\{V(t)\}$ is the right translation semigroup in $L^2(0, \infty; N)$ then $\hat{V}(t) = \mathscr{F} V(t) \mathscr{F}^{-1}$ is a unitarily equivalent semigroup acting in $H^2(\Gamma_+; N)$ which is given by

$$(\hat{V}(t) F)(w) = e^{-wt} F(w) \tag{6-11}$$

in terms of the boundary values of F on the imaginary axis we have

$$(\hat{V}(t) F)(i\omega) = e^{-i\omega t} F(i\omega) \tag{6-12}$$

The adjoint semigroup is given by

$$\hat{V}(t)^* F = P_{H^2(\Gamma_+; N)}(e^{i\omega t} F) \tag{6-13}$$

With the obvious trivial modifications resulting from the different half plane used, all the results of Sec. II-12 can be applied here. Now the problem of central interest to us is that of realization. As in the case of discrete time systems certain factorizations of the transfer function are intimately connected to some specific realizations. We let $H^2(\Gamma_+; HS(U, Y))$ denote the Hardy space of all analytic functions in the right half plane whose values are Hilbert–Schmidt operators from U into Y. If U and Y are finite dimensional then by a choice of bases in U and Y a function T in $H^2(\Gamma_+; HS(U, Y))$ is just a finite matrix with $H^2(\Gamma_+)$ entries. In such a case the map $M_T: U \to H^2(\Gamma_+; Y)$ defined by

$$M_T \xi = T\xi \tag{6-14}$$

is a bounded map.

Theorem 6-3 Let T be a continuous $B(U, Y)$-valued weighting pattern whose Laplace transform $\hat{T}$ is analytic in the right half plane. If $\hat{T} \in H^1(\Gamma_+; TC(U, Y))$ and on the imaginary axis admits a factorization $T(i\omega) = C(i\omega)^* B(i\omega)$ where $B \in H^2(\Gamma_+; HS(U, V))$ and $C \in H^2(\Gamma_+; HS(Y, V))$ then T has a regular realization.

PROOF We choose $H^2(\Gamma_+; V)$ as our state space and define a system (F, G, H) in the following way. We let F be the infinitesimal generator of the semigroup $\{\hat{V}(t)^*\}$ defined by (6-13) whereas the operators G and H are defined by $G\xi = M_B \xi$ with M_T given by (6-14) and $H^*\eta = M_C \eta$. Thus $H = M_C^*$ is given by

$$Hf = \frac{1}{2\pi} \int_{-\infty}^{\infty} C(i\omega)^* f(i\omega) \, d\omega$$

By our assumptions G and H are bounded operators and $\{\hat{V}(t)^*\}$ is a strongly continuous semigroup. Moreover from the factorization of T it is clear that T is in $H'(\Gamma_+; TC(U, Y))$ so that for each $\xi \in M$

$$H\hat{V}(t)^* G\xi = \frac{1}{2\pi} \int_{-\infty}^{\infty} e^{i\omega t} C(i\omega)^* B(i\omega) \, \xi \, d\omega$$

$$= \frac{1}{2\pi} \int_{-\infty}^{\infty} e^{i\omega t} T(i\omega) \, \xi \, d\omega = T(t) \, \xi$$

and hence we obtained a regular realization of T.

We note that this theorem is stronger than the sufficiency part of Theorem 6-2. Indeed let T and $\dot{T}$ be in $L^2(0, \infty; B(U, Y))$ then, since the Fourier transform of $\dot{T}$ is $i\omega \hat{T}(i\omega)$, it follows that $(1 - i\omega)\hat{T}(i\omega)$ is in $H^2(\Gamma_+; HS(U, Y))$. Since U being finite dimensional, $(1 + i\omega)^{-1} I_U \in H^2(\Gamma_+; HS(U, U))$ it follows that $\hat{T}$ is factorable in the form $\hat{T}(i\omega) = ((1 + i\omega)^{-1} I)^* ((1 - i\omega)\hat{T}(i\omega))$ and hence T is regularly realizable.

In the case of the unit disc the inclusion relations $H^\infty \subset H^2 \subset H^1$ hold. In that case once a realization procedure for H^1 functions is obtained it is automatically valid for H^∞ functions. Passing to a half plane those inclusions fail to hold any more. Thus if we want a realization procedure that will work for bounded analytic transfer functions in the right half plane then the previous procedures have to be modified.

That this is not purely academic follows by considering the ideal delay by α units. The corresponding input/output relations are

$$y(t) = u(t - \alpha) \tag{6-15}$$

and the transfer function is $F(w) = e^{-\alpha w}$ which is in $H^\infty(\Gamma_+)$ but is certainly not in $H^2(\Gamma_+)$.

More generally we will consider input/output relations of the form

$$y(t) = \int_0^t d\mathbf{M}(\tau) u(t - \tau) \tag{6-16}$$

where $\mathbf{M}$ is a finite $B(U, Y)$-valued Borel measure on $[0, \infty)$. If $\mathbf{M}$ is absolutely continuous with respect to the Lebesgue measure then there exists a density matrix $M(\tau)$ for which $d\mathbf{M}(t) = M(t)\,dt$ for which $(M(t)\xi, \eta)$ is in $L^1(0, \infty)$ for all choices of $\xi \in U$ and $\eta \in Y$. In such a case (6-16) reduces to the more familiar input/output relation

$$y(t) = \int_0^t M(\tau) u(t - \tau)\,d\tau = \int_0^t M(t - \tau) u(t)\,d\tau \tag{6-17}$$

In the general case a further reduction is possible. Let $D = \mathbf{M}(\{0\})$ then (6-16) transforms into

$$y(t) = Du(t) + \int_{0^+}^\infty d\mathbf{M}(\tau) u(t - \tau) \tag{6-18}$$

Let us denote by F the Fourier transform of the measure $\mathbf{M}$, that is,

$$F(w) = \int_0^\infty e^{-w\tau}\, d\mathbf{M}(\tau) \tag{6-19}$$

Clearly

$$F(w) = \mathbf{M}(\{0\}) + \int_{0^+}^\infty e^{-w\tau}\, d\mathbf{M}(\tau)$$

Now $F(w)$ is analytic in Γ_+ and bounded there. Moreover by an application of the Lebesque dominated convergence theorem it follows that

$$\lim_{x \to \infty} F(x + iy) - \mathbf{M}(\{0\}) = 0$$

uniformly in y. Thus it is natural to call $\mathbf{M}(\{0\})$ by $F(\infty)$ in the sense that $\lim_{x \to \infty} F(x + iy) = F(\infty)$ uniformly in y. The realization procedure that follows has been developed with this class of transfer functions in mind.

The new concept needed for what follows is that of a rigged structure on a Hilbert space X. By a rigged structure we mean a Hilbert space X and a linear manifold $K \subset X$ which is itself a Hilbert space under another, stronger, norm. If K' is the space of all continuous linear functionals on K then any vector $x \in X$ induces a continuous linear functional l_x on K given by $l_x(y) = (y, x)$ for all $y \in K$. This enables us to view X as embedded in K', the embedding being given by $L: X \to K'$ with $Lx = l_x$. Thus we obtain the inclusions

$$K \subset X \subset K' \tag{6-20}$$

Let M be an operator in H for which K is M^*-invariant. Given any $y \in K$ and $l \in K'$ we have in $l(M^*y)$ a well-defined continuous linear functional on K whose dependence on l is linear. Thus there exists a bounded operator $\tilde{M}$ and K' satisfying

$$(\tilde{M}l)(y) = l(M^*y) \tag{6-21}$$

For linear functionals arising out of vectors in H we have

$$\tilde{M}l_x(y) = l_x(M^*y) = (M^*y, x) = (y, Mx) = l_{Mx}(y)$$

or, as $l_x = Lx$, this can be written as

$$\tilde{M}L = LM \tag{6-22}$$

which shows that $\tilde{M}$ is the continuous extension of M to K' and henceforward we will use the M to denote $\tilde{M}$ as well.

If $B: U \to K'$ and $C: K \to U$ are continuous maps then so are $B^*: K \to U$ and $C^*: U \to K'$ which are given by $(u, B^*k) = (Bu)(k')$ and $(C^*u, k) = (u, Ck)$.

Let now $X, U,$ and Y be Hilbert spaces, and let A be the infinitesimal generator of a strongly continuous contraction semigroup acting in X. We noted before that, as A is closed, D_A becomes a Hilbert space with the inner product induced by the graph norm in D_A, $\|x\|_A^2 = \|x\|^2 + \|Ax\|^2$. If D_A' is the dual of D_A we naturally obtain the rigged structure

$$D_A \subset X \subset D_A' \tag{6-23}$$

and similarly

$$D_{A^*} \subset X \subset D_{A^*}' \tag{6-24}$$

Since both A and A^* generate strongly continuous contractive semigroups $(\lambda I - A)^{-1}$ and $(\lambda I - A^*)^{-1}$ exist as bounded operators for all λ such that $\operatorname{Re}\lambda > 0$. Now $(\lambda I - A)^{-1} X \subset D_A$ and the map $(\lambda I - A)^{-1}$ is continuous relative to the graph topology of D_A. Similarly $(\lambda I - A)^{*-1} X \subset D_{A^*}$. By duality we obtain

$$(\lambda I - A^*)^{-1} D'_A \subset X \qquad \text{and} \qquad (\lambda I - A)^{-1} D'_{A^*} \subset X \qquad (6\text{-}25)$$

Also from the invariance of D_A under the semigroup e^{At} it follows that e^{At} is a continuous map of D'_A into itself.

Generally a *continuous time linear system* will denote a quadruple of operators (A, B, C, D) with A as before $B: U \to D'_{A^*}$, $C: D_C \to Y$ and $D: U \to Y$ where we assume the continuity of B, $D_C \supset D_A$ and the restriction of C to D_A to be continuous. By a *compatible system* we denote a system for which

$$(\lambda_0 I - A)^{-1} BU \subset D_C \qquad (6\text{-}26)$$

for some λ_0 with $\operatorname{Re} \lambda_0 > 0$.

Let us note first that the condition (6-26) is independent of the point λ in $\operatorname{Re} \lambda > 0$.

Lemma 6-4 Let (A, B, C) be a compatible system. Then $(\lambda I - A)^{-1} Bu \subset D_C$ for all $u \in U$ and λ with $\operatorname{Re} \lambda > 0$.

PROOF Since (A, B, C) is assumed compatible then (6-26) holds for some λ_0 with $\operatorname{Re} \lambda_0 > 0$. By the resolvent equation we have

$$(\lambda I - A)^{-1} = (\lambda_0 I - A)^{-1} + (\lambda_0 - \lambda)(\lambda_0 I - A)^{-1} (\lambda I - A)^{-1}$$

Applying this to Bu we have

$$(\lambda I - A)^{-1} Bu = (\lambda_0 I - A)^{-1} Bu + (\lambda_0 - \lambda)(\lambda_0 I - A)^{-1} (\lambda I - A)^{-1} Bu$$
$$(6\text{-}27)$$

Now $Bu \in D'_{A^*}$ and from (6-25) it follows that $(\lambda I - A)^{-1} Bu$ is in X. Now $(\lambda_0 I - A)^{-1} X \subset D_A \subset D_C$ and so the right term in (6-27) is in D_C. But $(\lambda_0 I - A)^{-1} Bu \in D_C$ by the assumption of compatibility and so $(\lambda I - A)^{-1} Bu \in D_C$.

We turn now to showing that all transfer functions of the form (6-19) are realizable by compatible systems. This is done by associating with the continuous time realization problem a discrete time problem which is easier to solve and whose solution suitably transformed yields a compatible continuous time realization.

Assume now F is given by (6-19). The map

$$z = \frac{w - 1}{w + 1} \qquad (6\text{-}28)$$

is a fractional linear transformation that maps the right half plane Γ_+ onto the open unit disc D. The inverse transformation is given by

$$w = \frac{1 + z}{1 - z} \qquad (6\text{-}29)$$

Define now a function Φ in the unit disc by

$$\Phi(z) = F\left(\frac{1 + z}{1 - z}\right) \tag{6-30}$$

Clearly $\Phi \in H^\infty(B(\Theta, Y))$ and moreover the strong nontangential limits of Φ at $z = 1$ exist and are equal to $\Phi(1) = F(\infty)$. By considering Φ to be the transfer function of a discrete time system we can apply the results of Sec. 2 to obtain a Hilbert space realization for Φ. Specifically the shift realization (F, G, H, E) of Φ can be used. Thus we identify the state space M of the realization with $\overline{\text{Range}\, H_\Phi}$, $F = S_0^* \,|\, \overline{\text{Range}\, H_\Phi}$, $G\xi = H_\Phi\xi = (\Phi - \Phi(0))\,\xi$, $Hf = f'(0)$ for $f \in M$ and $E\xi = \Phi(0)\,\xi$.

In terms of these operators we can write

$$\Phi(z) = E + zH(I - zF)^{-1}G \tag{6-31}$$

Since the nontangential limits of Φ at $z = 1$ exist and are equal to $\Phi(1) = F(\infty)$ we obtain the relation

$$\Phi(1) = E + H(I - F)^{-1}G = F(\infty) \tag{6-32}$$

From (6-31) we obtain, using the previously introduced fractional linear transformation that

$$
\begin{aligned}
F(w) = \Phi\left(\frac{w - 1}{w + 1}\right) &= E + \frac{w - 1}{w + 1}H\left(I - \frac{w - 1}{w + 1}F\right)^{-1}G \\
&= E + (w - 1)H\big((w + 1)I - (w - 1)F\big)^{-1}G \\
&= E + (w - 1)H\big(w(I - F) + (I + F)\big)^{-1}G \\
&= E + (w - 1)H(I - F)^{-1}\big(w - (F + I)(F - I)^{-1}\big)^{-1}G
\end{aligned}
$$

Since F is a completely nonunitary contraction $(F - I)^{-1}$ exists as a possibly unbounded, closed operator. Let $A_0 = (F + I)(F - I)^{-1}$ then A_0 is maximal accretive and hence the infinitesimal generator of a strongly continuous contractive semigroup. Moreover the relation

$$(F - I)^{-1} = (I - A_0)/2 \tag{6-33}$$

implies

$$F(w) = E + (w - 1)H(I - A_0)(wI - A_0)^{-1}G \tag{6-34}$$

We define new operators B_0 and C_0 by

$$B_0 = -\frac{1}{2\sqrt{\pi}}(I - A_0)G \tag{6-35}$$

and

$$C_0 = \sqrt{\pi}\,H(I - A_0) \tag{6-36}$$

Thus (6-34) can be rewritten as

$$F(w) = E - C_0(w - 1)(wI - A_0)^{-1}(I - A_0)^{-1} B_0 \tag{6-37}$$

and, by applying the resolvent identity, we obtain the following representation for F

$$F(w) = E - C_0(I - A_0)^{-1} B_0 + C_0(wI - A_0)^{-1} B_0$$
$$= F(\infty) + C_0(wI - A_0)^{-1} B_0 \tag{6-38}$$

In summary, the system $(A_0, B_0, C_0, F(\infty))$ is a realization of the transfer function F, and it will be shown that it is actually a compatible realization.

From (6-35) we have

$$B_0^* = -\frac{1}{2\sqrt{\pi}} G^*(I - A_0^*)$$

and hence, for every $x \in D_{A^*}$

$$\|B_0^* x\| = \left\| -\frac{1}{2\sqrt{\pi}} G^*(I - A_0^*) x \right\| \le \frac{1}{2\sqrt{\pi}} \|G^*\| (\|x\| + \|A_0^* x\|)$$

Hence $B_0^*: D_{A^*} \to U$ is continuous with respect to the graph norm of D_{A^*} and by duality $B_0: U \to D'_{A^*}$ is a continuous map. Similarly, since $C_0 = \sqrt{\pi} H(I - A_0)$ and as $D_{A_0} = \text{Range}(I - A_0)^{-1}$, the boundedness of H implies that $D_{C_0} \supset D_{A_0}$ and that the restriction of C_0 to D_{A_0} is continuous. Also $F(\infty) = E - C_0(I - A_0)^{-1} B_0$ shows that for each $\xi \in U$ we have $(I - A_0)^{-1} B_0 \xi \in D_{C_0}$ and so the compatibility condition is satisfied.

Although we have obtained a compatible realization, the result is unsatisfactory inasmuch as the state space of the realization is a subspace of $H_{0,Y}^2$ of the unit disc whereas we would like the setting of the realization to be either a space of functions analytic in the right half plane or a subspace of some L^2 space on $[0, \infty)$. This is indeed possible and is summarized by the following theorem.

The main realization result is the following.

Theorem 6-5 Let **M** be a complex $B(U, Y)$-valued Borel measure on $[0, \infty)$ and let

$$F(w) = \int_0^\infty e^{-wt} \, d\mathbf{M}(t)$$

(a) The state space system (A, B, C, D) with state space $M = J_0 M_0 = J_0 \overline{\text{Range} H_\Phi}$ where $J_0: H_{0,Y}^2 \to H^2(\Gamma_+; Y)$ is given by

$$(J_0 f)(w) = \frac{1}{\sqrt{\pi}} \cdot \frac{1}{w - 1} f\left(\frac{w - 1}{w + 1}\right) \tag{6-39}$$

and A, B, C, D are given by

$$(Af)(i\omega) = i\omega f(i\omega) \qquad \text{for} \qquad f \in D_A \tag{6-40}$$

$$(B\xi, g) = \frac{1}{2\pi} \int_{-\infty}^{\infty} \left((F(i\omega) - F(1)) \, \xi, g(i\omega) \right) d\omega \tag{6-41}$$

for $\xi \in U$ and $g \in D_{A^*}$

$$Cf = \int_{-\infty}^{\infty} f(i\omega) \, d\omega \qquad \text{for} \qquad f \in D_C \tag{6-42}$$

and

$$D = F(\infty) \tag{6-43}$$

is a compatible realization of F.

(b) The state space system $(\check{A}, \check{B}, \check{C}, \check{D})$ with the state space $M = \mathscr{F}^{-1}M$, $\mathscr{F}$ being Fourier–Plancherel transform, with

$$(\check{A}\varphi)(\rho) = \varphi'(s) \qquad \text{for} \qquad \varphi \in D_A \tag{6-44}$$

$$(\check{B}\xi, \gamma) = \frac{1}{2\pi} \int_{0}^{\infty} \left(\varphi_{\xi}(s), \gamma(s) + \gamma'(s) \right) ds \tag{6-45}$$

for $\xi \in U$ and $\gamma \in D_{\check{A}^*}$

$$(\check{C}\varphi, \eta) = \int_{0}^{\infty} \left(\varphi(s) - \varphi'(s), e^{-s}\eta \right) ds \tag{6-46}$$

for $\varphi \in D_C$ and

$$\check{D} = F(\infty) \tag{6-47}$$

is a compatible realization of F.

PROOF (a) The map J_0 defined by (6-39) is a minor modification of (6-9) and it maps $H^2_{0,Y}$ unitarily onto $H^2(\Gamma_+, Y)$. Under J_0 the right shift in $H^2_{0,Y}$ is mapped onto the multiplication by $(w - 1)/(w + 1)$ operator in $H^2(\Gamma_+, Y)$. If $M = J_0 M_0 = J_0 \overline{\text{Range} H_\Phi}$ then $M^\perp$ is a subspace of $H^2(\Gamma_+; Y)$ invariant under multiplication by all $H^\infty(\Gamma_+)$ functions. If (F, G, H, E) is the shift realization of Φ defined previously then $(J_0 F J_0^{-1}, J_0 G, H J_0^{-1}, E)$ is a unitarily equivalent realization of Φ in M. The operator $A_0 = (F + I)(F - I)^{-1}$ is unitarily equivalent to $A = J_0 A_0 J_0^{-1} = (J_0 F J_0^{-1} + I)(J_0 F J_0^{-1} - I)^{-1}$. Since $F^*\varphi = P_{M_0}\chi\varphi$ it follows that

$$J_0 F^* J_0^{-1} f = P_M \frac{\chi - 1}{\chi + 1} f$$

where χ denotes in both cases the identity function, in D and Γ_+, respectively. Thus $A^* f = -P_M \chi f$ and hence for $f \in D_A$ we obtain, in terms of the boundary values

$$Af = -P_M \bar{\chi} f \tag{6-48}$$

or, somewhat less precisely, $(Af)(i\omega) = P_M(i\omega f(i\omega))$ for all $f \in D_A$. The action of the semigroup generated by A in M is given by $(e^{At}f)(i\omega) = P_M(e^{i\omega t}f(i\omega))$ and this proves (6-40).

To compute B we note that $B = J_0 B_0 = -1/(2\sqrt{\pi}) J_0(I - A_0) G = -1/(2\sqrt{\pi})(1 - A) J_0 G$. Now $(G\xi)(z) = (H_\Phi\xi)(z) = \Phi(z) - \Phi(0)$ which implies that

$$(J_0 G\xi)(w) = \frac{1}{\sqrt{\pi}} \frac{1}{w-1}\left[\Phi\left(\frac{w-1}{w+1}\right) - \Phi(0)\right]\xi = \frac{1}{\sqrt{\pi}}\left[\frac{F(w) - F(1)}{w-1}\right]\xi$$

Hence for each $g \in D_{A^*}$ and $\xi \in U$

$$(B\xi, g) = -\frac{1}{2\sqrt{\pi}}((I - A) J_0 G, g) = -\frac{1}{2\sqrt{\pi}}(J_0 G\xi, (I - A^*) g)$$

$$= -\frac{1}{2\pi}\int_{-\infty}^{\infty}\left(\left(\frac{F(i\omega) - F(1)}{i\omega - 1}\right)\xi, g(i\omega) + i\omega g(i\omega)\right) d\omega$$

$$= \frac{1}{2\pi}\int ((F(i\omega) - F(1))\xi, g(i\omega)) d\omega$$

which proves (6-41).

To calculate the representation of C we note that for $\varphi \in M_0$ we have $H\varphi = \varphi'(0)$. Now the inverse of J_0 is given by

$$(J_0^{-1}f)(z) = 2\sqrt{\pi}\frac{z}{1-z}f\left(\frac{1+z}{1-z}\right) \tag{6-49}$$

so if $\varphi = J_0^{-1}f$ then

$$\varphi'(z) = 2\sqrt{\pi} z\frac{d}{dz}\left(\frac{1}{1-z}f\left(\frac{1+z}{1-z}\right)\right) + \frac{2\sqrt{\pi}}{1-z}f\left(\frac{1+z}{1-z}\right)$$

which implies that $\varphi'(0) = 2\sqrt{\pi} f(1)$. Therefore

$$Cf = C_0 J_0^{-1}f = \sqrt{\pi} H(I - A_0) J_0^{-1}f = \sqrt{\pi} H J_0^{-1}(I - A) f$$

Now evaluation at any point $\lambda \in \Gamma_+$ is a continuous map on $H^2(\Gamma_+; Y)$ and we have for each $f \in H^2(\Gamma_+; Y)$ and $\eta \in Y$ that

$$\frac{1}{2\pi}\int_{-\infty}^{\infty}\left(f(i\omega), \frac{\eta}{i\omega + \bar{\lambda}}\right) d\omega = \frac{1}{2\pi}\int_{-\infty}^{\infty}\frac{(f(i\omega), \eta)}{\lambda - i\omega} d\omega = (f(\lambda), \eta) \tag{6-50}$$

Hence, for $f \in D_A$

$$(Cf, \eta) = (\sqrt{\pi} H J_0^{-1}(I - A), \eta) = \int_{-\infty}^{\infty} (f(i\omega), \eta) d\omega$$

This proves (6-42) while (6-43) follows from (6-38).

To prove part (*b*) we apply the inverse Fourier–Plancherel transform to obtain a time domain realization $(\check{A}, \check{B}, \check{C}, \check{D}) = (\mathscr{F}^{-1}A\mathscr{F}, \mathscr{F}^{-1}B, C\mathscr{F}, D)$ in the state space $\check{M} = \mathscr{F}^{-1}M$. From previous considerations $\check{A}$ is the infinitesimal generator of the left translation semigroup restricted to the left invariant subspace $\check{M}$ of $L^2(0, \infty; Y)$. So $(e^{\check{A}t}\varphi)(s) = \varphi(s + t)$ for all $\varphi \in \check{M}$, and $\varphi \in D_{\check{A}}$ we have $\check{A}\varphi = \varphi'$.

From (6-41) it follows that

$$(B\xi, g) = \frac{1}{2\pi} \int_{-\infty}^{\infty} \big((F(i\omega) - F(1)), \xi, g(i\omega)\big)\, d\omega$$

$$= \frac{1}{2\pi} \int_{-\infty}^{\infty} \left(\left(\frac{F(i\omega) - F(1)}{1 - i\omega}\right)\xi, (1 + i\omega)\, g(i\omega)\right) d\omega$$

and taking the inverse Fourier transform we obtain

$$(\check{B}\xi, \gamma) = \frac{1}{2\pi} \int_{0}^{\infty} \big(\varphi_\xi(s), \gamma(s) + \gamma'(s)\big)\, ds$$

for all $\xi \in U$ and $\gamma \in D'_{\check{A}*}$. Here φ_ξ denotes the function

$$\varphi_\xi = \mathscr{F}^{-1}\left(\frac{F(w) - F(1)}{1 - w}\right) \tag{6-51}$$

Finally for $f \in D_A$ we have, letting $\varphi = \mathscr{F}^{-1}f$

$$(Cf, \eta) = \int_{-\infty}^{\infty} \big(f(i\omega), \eta\big)\, d\omega = \int_{-\infty}^{\infty} \left((1 - i\omega)\, f(i\omega), \frac{\eta}{1 + i\omega}\right) d\omega$$

or

$$(\check{C}\varphi, \eta) = \int_{0}^{\infty} \big(\varphi(s) - \varphi'(s), e^{-s}\eta\big)\, ds$$

Integration by parts yields for differentiable φ, certainly for $\varphi \in D_{\check{A}}$, that
$$\check{C}\varphi = \varphi(0) \tag{6-52}$$

It is of interest to verify directly that we have obtained a realization. To this end we evaluate

$$(D + C(\lambda I - A)^{-1} B)\, \xi = D\xi + C(\lambda I - A)^{-1} B\xi$$

$$= F(\infty) + \frac{1}{2\pi} \int_{-\infty}^{\infty} \frac{(F(i\omega) - F(1))}{\lambda - i\omega}\, d\omega$$

Using the scalar resolvent identity

$$\frac{1}{\lambda - i\omega} = \frac{1}{1 - i\omega} + \frac{(1 - \lambda)}{(1 - i\omega)(\lambda - i\omega)}$$

we can rewrite the last integral as

$$\frac{1}{2\pi}\int_{-\infty}^{-\infty}\frac{(F(i\omega)-F(1))\,\xi}{1-i\omega}\,d\omega + \frac{(1-\lambda)}{2\pi}\int_{-\infty}^{\infty}\frac{(F(i\omega)-F(1))\,\xi}{(1-i\omega)(\lambda-i\omega)}\,d\omega$$

Since $(F(w)-F(1))\,\xi/1-w$ is in $H^2(\Gamma_+,Y)$ it follows from (6-50) that

$$\frac{(1-\lambda)}{2\pi}\int_{-\infty}^{\infty}\frac{(F(i\omega)-F(1))}{(1-i\omega)(\lambda-i\omega)}\,\xi\,d\omega = F(\lambda)-F(1)$$

Thus it remains to evaluate the first integral. To this end we recall that

$$F(w) = \int_0^{\infty} e^{-w\tau}\,d\mathbf{M}(\tau) = F(\infty) + \int_{0^+}^{\infty} e^{-w\tau}\,d\mathbf{M}(\tau)$$

Let us define G by

$$G(w) = F(w) - F(\infty) = \int_{0^+}^{\infty} e^{-w\tau}\,d\mathbf{M}(\tau)$$

Clearly $G(1) = F(1) - F(\infty)$ and so

$$\frac{F(w)-F(1)}{1-w} = \frac{G(w)-G(1)}{1-w}$$

By the Paley–Wiener theorem

$$\frac{F(w)-F(1)}{1-w}\,\xi = \frac{G(w)-G(1)}{1-w}\,\xi$$

is the Fourier–Plancherel transform of an $L^2(0,\infty;Y)$ function, which we denote by φ_ξ. Neither $F(w)\,\xi/(1-w)$ nor $F(1)\,\xi/(1-w)$ is in $H^2(\Gamma_+;Y)$, unless $F(1)\,\xi = 0$, but by restricting ourselves to the half plane $\operatorname{Re}w > 1$ we can identify $F(w)/(1-w)$ as the Laplace transform of the convolution of the function $-e^{-t}$ and the measure $\mathbf{M}$, that is

$$\frac{F(w)\,\xi}{1-w} = \int_0^{\infty} e^{-wt}\int_0^t e^{t-\tau}\,d\mathbf{M}(\tau)\,\xi\,dt$$

Also

$$\frac{F(1)\,\xi}{1-w} = -F(1)\int_0^{\infty} e^{-wt}\cdot e^t\,dt = -\int_0^{\infty} e^{-wt}\int_0^{\infty} e^{t-\tau}\,d\mathbf{M}(\tau)\,\xi\,dt$$

which implies the equality

$$\varphi_\xi(t) = e^t\int_t^{\infty} e^{-\tau}\,d\mathbf{M}(\tau)\,\xi \tag{6-53}$$

We can interpret (6-53) as the variation of parameters formula for the solution of the nonhomogeneous differential equation

$$y'(t) - y(t) = \mathbf{M}\xi \tag{6-54}$$

The differential equation has to be considered in the distributional sense, the solution being actually in $L^2(0, \infty; Y)$.

We can evaluate now the integral

$$\frac{1}{2\pi} \int_{-\infty}^{\infty} \frac{(F(i\omega) - F(1))\xi}{1 - i\omega} d\omega$$

This integral is equal to

$$C\left(\frac{F(w) - F(1)}{1 - w}\right)\xi = \check{C}\varphi_\xi = \int_0^{\infty} (\varphi_\xi(s) - \varphi'_\xi(s)) e^{-s} ds$$

By (6-54) $\varphi_\xi - \varphi'_\xi = \mathbf{M}\xi$ and so

$$\check{C}\varphi_\xi = \int_0^{\infty} e^{-s} d\mathbf{M}(s) \xi = F(1)$$

which shows that we have indeed solved the realization problem. We call the realization provided by Theorem 6-5 the *restricted translation realization*.

As an example of the preceding theorem we obtain a state space realization of the simple delay line. The input/output relation is $y(t) = u(t - a)$ for some $a > 0$. Thus the transfer function is $F(w) = d^{-aw}I$ which is the Fourier transform of the weighting pattern $\mathbf{M} = \delta_a \cdot I$ where δ_a is the Dirac delta function.

To realize $\delta_a \cdot I$ we take our state space to be $L^2(0, a; Y)$. The operator A will be the infinitesimal generator of the left translation semigroup restricted to $L^2(0, a\ Y)$. Thus for $\varphi \in D_A$ we have $A\varphi = \varphi'$. By (6-53) we have in our case

$$\varphi_\xi(s) = e^s \int_s^{\infty} e^{-\tau} d\ \mathbf{M}(\tau) \xi = e^s \int_s^{\infty} e^{-\tau}\delta_a(\tau) \xi$$

$$= \begin{cases} 0 & s > a \\ e^{s-a}\xi & s < a \end{cases}$$

This implies that for $\xi \in U$ and $\gamma \in D_{A^*}$

$$(\check{B}\xi, \gamma) = \int_0^{\infty} (\varphi_\xi(s), \gamma(s) + \gamma'(s)) ds = \int_0^{a} (e^{s-a}\xi, \gamma(s) + \gamma'(s)) ds$$

Now as $\gamma \in D_{\check{A}^*}$ it follows from the results of Sec. II-10 that $\gamma(0) = 0$. Hence integration by parts yields $(\check{B}\xi, \gamma) = (\xi, \gamma(a))$ which is formally equivalent to $\check{B}\xi = \delta_a\xi$.

Now from the variation of parameters formula we obtain

$$x(t, s) = \int_0^t e^{\check{A}(t-\tau)} \check{B} u(\tau) \, d\tau = \int_0^t e^{A(t-\tau)} B u(\tau) \, \delta_a(s) \, d\tau$$

$$= \int_0^t u(\tau) \, \delta_a(s + t - \tau) \, d\tau = u(s + t - a)$$

Assuming u is continuous we immediately obtain $y(t) = \check{C} x(t) = x(\cdot, 0) = u(t - a)$, i.e. we have realized the simple delay.

We turn now to the discussion of Hankel operators. The state space of the shift realization was best characterized as the range of the Hankel operator of the transfer function. We expect the same to be true in the continuous time case if the Hankel operator is suitably defined.

Given $F \in H^\infty(\Gamma_+ ; B(U, Y))$ we define $H_F : H^2(\Gamma_- ; U) \to H^2(\Gamma_+ ; Y)$ by

$$H_F v = P_{H^2(\Gamma_+ ; Y)} F v \qquad (6\text{-}55)$$

Let $\Phi \in H^\infty(B(U, Y))$ and F be related by (6-30) and let $J_0 : H_{0,Y}^2 \to H^2(\Gamma_+ ; Y)$ be defined by (6-39). Then for $u \in \bar{H}_U^2$ we have

$$J_0 H_\Phi u = J_0 P_{H_{0,Y}^2} \Phi u = P_{H^2(\Gamma_+ ; Y)} J_0 \Phi u = P_{H^2(\Gamma_+ ; Y)} F(J_0(u)) = H_F(J_0 u)$$

or

$$J_0 H_\Phi = H_F J_0 \qquad (6\text{-}56)$$

that is, the two Hankel operators H_Φ and H_F are unitarily equivalent. By the Paley–Wiener theorem the Fourier–Plancherel transform maps $L^2(-\infty, 0; Y)$ and $L^2(0, \infty; Y)$ on $H^2(\Gamma_- ; Y)$ and $H^2(\Gamma_+ ; Y)$, respectively. Thus we want to identify $\mathcal{F}^{-1} H_F \mathcal{F}$. Assume again that $F(w) = \int_0^\infty e^{-wt} \, d\mathbf{M}$ for some complex matrix valued measure $\mathbf{M}$. Since the Fourier transform of a convolution is the product of the Fourier transforms we have

$$(\mathcal{F}^{-1}(Fv))(t) = \int_{-\infty}^0 d\mathbf{M}(t - \sigma)(\mathcal{F}^{-1} v)(\sigma) \, d\sigma$$

$$= \int_0^\infty d\mathbf{M}(t + \sigma)(\mathcal{F}^{-1} v)(-\sigma) \, d\sigma$$

or

$$(\mathcal{F}^{-1} H_F \mathcal{F} u)(t) = \int_0^\infty d\mathbf{M}(t + \sigma) u(\sigma) \, d\sigma \qquad (6\text{-}57)$$

With the above identification of all the results concerning reachability, observability, and spectral analysis carry over to the continuous time setting.

To discuss reachability let us consider a pair (A, B) where we assume first that $B : U \to X$ which is more restrictive than the assumption of compatibility. We say that (A, B) is a reachable pair of $\overline{\bigcup_{t \geq 0} K_t} = X$ where K_t is the linear manifold consisting of all vectors of the form $\int_0^t e^{A(t-\tau)} Bu(\tau) \, d\tau$ where the control function u is restricted to some function space, say, $C([0, \infty); U)$.

Lemma 6-6 The pair (A, B) is a reachable pair if and only if

$$\bigcap_{t \geq 0} \operatorname{Ker} B^* e^{A^* t} = \{0\} \qquad (6\text{-}58)$$

PROOF (A, B) is not a reachable pair if and only if there exists a nonzero vector $x \in X$ for which $0 = \left(x, \int_0^t e^{A(t-\tau)} Bu(\tau) \, d\tau\right) = \int_0^t \left(B^* e^{A^*(t-\tau)} x, u(\tau)\right) d\tau$. Choose $u(\tau) = \psi(t) \eta$ where $\eta \in U$ and ψ is any scalar C^∞ function. Thus $x \in \bigcap_{t \geq 0} \operatorname{Ker} B^* e^{A^* t}$.

If we assume that A is the infinitesimal generator of a strongly continuous semigroup of contractions the reachability of (A, B) can be shown to be equivalent to the reachability of an associated discrete time system.

Lemma 6-7 Let A be the infinitesimal generator of a strongly continuous semigroup of completely nonunitary contractions in X and let $T = (A + I)(A - I)^{-1}$ be its Cayley transform. Then (A, B) is reachable if and only if (T, B) is reachable.

PROOF The reachability of the two systems is equivalent to $\bigcap_{t \geq 0} \operatorname{Ker} B^* e^{A^* t} = \{0\}$ and $\bigcap_{t \geq 0} \operatorname{Ker} B^* T^{*n} = \{0\}$, respectively. Thus it suffices to show the equivalence of the later two conditions. To this end we recall that $e^{At} = e_t(T)$ where $e_t(z) = e^{-t(1+z)/(1-z)} = \sum \alpha_n(t) z^n$ and $e_t(T) = \lim_{r \to 1} e_{r,t}(T)$ with $e_{r,t}(z) = e_t(rz)$. Thus

$$B^* e^{A^* t} x = \lim_{r \to 1} \sum_{n=0}^{\infty} \alpha_n(t) r^n B^* T^{*n} x$$

so $x \in \bigcap_{n \geq 0} \operatorname{Ker} B^* T^{*n}$ implies $x \in \bigcap_{t \geq 0} \operatorname{Ker} B^* e^{A^* t}$.

To prove the converse we know that with $\varphi_t(z) = (z - 1 + t)/(z - 1 - t)$ we have $T = \lim_{t \to 0} \varphi_t(e^{At})$ and similarly $T^* = \lim_{t \to 0} \varphi_t(e^{A^* t})$. This shows that T^{*n} has a series expansion in terms of $e^{A^* t}$ which implies the converse inclusion $\bigcap_{t \geq 0} \operatorname{Ker} B^* e^{A^* t} \subset \bigcap_{n \geq 0} \operatorname{Ker} B^* A^{*n}$.

Of course reachability could also be discussed in terms of the reachability operator. We define R on $C_0^\infty((-\infty, 0]; U)$ by

$$Ru = \int_0^\infty e^{At} Bu(-t) \, dt \qquad (6\text{-}59)$$

Since u has compact support R is well defined. It is very simple to check that (A, B) is a reachable pair if and only if the range of R is dense in X.

We pass now to the discussion of compatible systems. In this case R is a priori a map into D'_{A^*}. However, since we restricted R to a space of C^∞-functions, in fact once differentiable would be sufficient for our purposes, it follows by integration by parts that

$$Ru = \int_0^\infty e^{At} Bu(-t)\, dt$$

$$= -(A - \lambda_0)^{-1} Bu(0) + \int_0^\infty e^{At}(A - \lambda_0)^{-1} B(\dot{u}(-t) - \lambda_0 u(-t))\, dt$$

which shows by compatibility that $Ru \in X$. We say now that (A, B) is reachable if R thus defined has range dense in X. The pair (A, B) is an exactly reachable pair if R has an extension to a continuous map of $L^2(-\infty, 0; U)$ onto X.

We can now apply Lemmas 6-6 and 6-7 to compatible systems.

Corollary 6-8 Let (A, B) be a compatible pair. Then
(a) (A, B) is reachable if and only if

$$\bigcap_{t \geq 0} \operatorname{Ker} B^*(\bar{\lambda}_0 I - A^*) e^{A^* t} = \{0\} \tag{6-60}$$

(b) If A is the infinitesimal generator of a completely non-unitary semigroup then (A, B) is a reachable pair if and only if $(T, (I - A)^{-1} B)$ is reachable where T is the Cayley transform of A.

As in the discrete time case we can identify the reachability operator of the restricted translation realization with the corresponding Hankel operator. Indeed if $F \in H^\infty(\Gamma_+; B(U, Y))$ then for $u \in L^2(-\infty, 0; U)$

$$Ru = \int_0^\infty e^{At} Bu(-t)\, dt = \int_0^\infty P_{H^2(\Gamma_+; Y)}\big(e^{i\omega t}(F(i\omega) - F(1))\, u(-t)\, dt$$

$$= P_{H^2(\Gamma_+; Y)} \int_0^\infty e^{i\omega t} F(i\omega)\, u(-t)\, dt - P_{H^2(\Gamma_+; Y)} \int_0^\infty F(1)\, e^{i\omega t} u(-t)\, dt$$

$$= P_{H^2(\Gamma_+; Y)} \int_0^\infty e^{i\omega t} F(i\omega)\, u(-t)\, dt = H_F u$$

We say that $F \in H^\infty(\Gamma_+; B(U, Y))$ is strictly noncyclic if $(\operatorname{Range} H_F)^\perp$ is an invariant subspace of full range. Thus we infer that, for Φ defined by (6-30), Φ is strictly noncyclic if and only if F is. This implies that theorems like Theorem 3-5 and Theorem 3-10 hold fur functions in $H^\infty(\Pi_+; B(U, Y))$.

An analogous discussion holds for observability. The observability operator

O is defined through its adjoint

$$O^*y = \int_0^\infty e^{A^*t}C^*y(t)\,dt$$

For the restricted translation realization we obtain from (6-42) that $(C^*\eta)(i\omega) = \eta$ for all $\eta \in Y$ and hence

$$\int_0^\infty e^{A^*t}C^*y(t)\,dt = \int_0^\infty e^{-i\omega t}y(t)\,dt = (\mathscr{F}y)(i\omega)$$

Thus O^* can be identified with the Fourier transform and so the restricted translation realization is exactly observable.

7. SYMMETRIC SYSTEMS

The realization theory developed in the previous section using the translation semigroup has as its natural domain of applicability the set of strictly noncyclic functions. This by no means exhausts the interesting cases. In particular certain internal symmetry properties of systems are reflected in the corresponding transfer function. Thus we expect a more limited set of transfer functions, or weighting patterns, to be realizable by systems with internal symmetries. In some sense this is an approach contrary in spirit to the shift and translation realizations which were highly nonsymmetric.

The systems to be studied in this section are of the form (A, B, C) with A a, possibly unbounded, self-adjoint operator in the Hilbert space H which generates a strongly continuous semigroup. This is equivalent to the semiboundedness of A from above. That is we assume there exists an $\omega > 0$ such that

$$(Ax, x) \le \omega \|x\|^2 \tag{7-1}$$

for all $x \in D_A$. This implies and is actually equivalent to $\sigma(A) \subset (-\infty, \omega]$. The semigroup e^{At} generated by A will be contractive if and only if $\sigma(A) \subset (-\infty, 0]$. Hence by replacing A by $A - \omega I$, and the semigroup e^{At} by $e^{(A-\omega I)t}$ we may as well assume that A is the infinitesimal generator of a strongly continuous semigroup of contractions. A system (A, B, C) will be called a *self-adjoint system* if the spaces U and Y are equal and if besides the self-adjointness of A we assume $C = B^*$. We say that (A, B, C) is a *stable self-adjoint system* if it is a self-adjoint system and A generates a strongly continuous semigroup of self-adjoint contractions.

Just as shift operators could be utilized to study systems mainly because we had a convenient functional model for them the same is true in the case of systems with self-adjoint generators. The theory of spectral representations for self-adjoint operators provides the tool to study these systems. Let us assume that A has finite multiplicity n. In this case we may assume that A is given in its spectral representation. Thus the Hilbert space H can be identified with $L^2(\mathbf{M})$ for some $n \times n$ matrix measure $\mathbf{M}$. The operator A acts on functions in its domain by

$$(Af)(\lambda) = \lambda f(\lambda) \tag{7-2}$$

and the action of the semigroup is

$$(e^{At}f)(\lambda) = e^{\lambda t}f(\lambda) \tag{7-3}$$

for all $f \in L^2(\mathbf{M})$. If the input space U is finite dimensional then, by a choice of basis, it can be identified with $\mathbb{C}^m$ where $m = \dim U$. Since $B: \mathbb{C}^m \to L^2(\mathbf{M})$ there exists a measurable $n \times m$ matrix valued function $B(\lambda)$ such that

$$(B\xi)(\lambda) = B(\lambda)\,\xi \tag{7-4}$$

for all $\xi \in \mathbb{C}^m$.

We define the reachability operator R by

$$Ru = \int_{-\infty}^{0} e^{-A\tau} Bu(\tau)\, d\tau \tag{7-5}$$

and take its domain of definition to be the space of all bounded measurable $\mathbb{C}^m$-valued functions of compact support. Thus defined R is a map into $L^2(\mathbf{M})$.

We can use the spectral representation of A to obtain a corresponding representation for R. It follows from (7-5) that

$$(Ru)(\lambda) = \int_{-\infty}^{0} e^{-\lambda\tau} B(\lambda)\, u(\tau)\, d\tau = B(\lambda) \int_{\infty}^{0} e^{-\lambda\tau} u(\tau)\, d\tau \tag{7-6}$$

But $\int_{-\infty}^{0} e^{-\lambda\tau} u(\tau)\, d\tau$ is just the Laplace transform $\mu = \mathfrak{L}a$ of a. Thus we can define $\hat{R}$ on the set of all Laplace transforms of permissible inputs by $\hat{R}\hat{u} = Ru$ or equivalently $\hat{R}\mathfrak{L} = \mathfrak{L}R$. So

$$(\hat{R}\hat{u})(\lambda) = B(\lambda)\,\hat{u}(\lambda) \tag{7-7}$$

properly extended by continuity to a function space, which is a B-module, where B is the algebra of bounded Borel measurable functions on $\mathbb{R}$, then R would be a B-homomorphism.

The observability operator or rather its adjoint can be analogously analyzed. Given a state $f \in L^2(\mathbf{M})$ we let

$$(Of)(t) = Ce^{At}f \tag{7-8}$$

where we assume $C = L^2(\mathbf{M}) \to \mathbb{C}^p$, having identified Y with $\mathbb{C}^p$. Let v be any $\mathbb{C}^p$-valued function for which the $L^2(0, \infty; \mathbb{C}^p)$ inner product (Of, v) makes sense. Since

$$(C^*\eta)(\lambda) = C(\lambda)^*\,\eta \tag{7-9}$$

for some measurable $p \times n$ matrix function C. From (7-9) it follows that

$$(Cf, \eta) = (f, C^*\eta) = \int (d\mathbf{M}f(\lambda), C(\lambda)^*\,\eta)$$

$$= \int (C(\lambda)\, d\mathbf{M}f(\lambda), \eta) = \left(\int C(\lambda)\, d\mathbf{M}f(\lambda), \eta \right)$$

or

$$Cf = \int C(\lambda)\,d\mathbf{M}f(\lambda) \qquad \text{for} \qquad f \in L^2(\mathbf{M})$$

and this in turn implies

$$(Of)(t) = Ce^{At}f = \int C(\lambda)\,d\mathbf{M}e^{\lambda t}f(\lambda)$$

or

$$(Of)(t) = \int e^{\lambda t}C(\lambda)\,d\mathbf{M}f(\lambda) \tag{7-10}$$

Thus the representation

$$(O^*v)(\lambda) = C(\lambda)^* \int_0^\infty e^{\lambda t}v(t)\,dt \tag{7-11}$$

holds and in terms of Laplace transforms on $[0, \infty)$ this can be rewritten as

$$(\hat{O}^*\hat{v})(\lambda) = C(\lambda)^* \hat{v}(\lambda) \tag{7-12}$$

where now $\hat{v}$ refers to the Laplace transforms of functions defined on $[0, \infty)$. Thus proper extensions of $\hat{O}^*$ are also B-homomorphisms.

Now the reachability operator R is completely determined by the matrix measure $\mathbf{M}$ and the matrix function B and so it is natural to characterize reachability in those terms. Analogously we want to characterize observability in terms of C and $\mathbf{M}$.

In our context reachability and observability mean that R and O^* have dense range conditions which are obviously equivalent to

$$\bigcap_{t \geq 0} \operatorname{Ker} B^* e^{At} = \{0\} \tag{7-13}$$

and

$$\bigcap_{t \geq 0} \operatorname{Ker} C e^{At} = \{0\} \tag{7-14}$$

respectively.

Let us choose a scalar measure σ such that $\mathbf{M}|\sigma I_n$. Thus $L^2(\mathbf{M})$ can be unitarily embedded in $L^2(\sigma I_n)$ where the embedding operators in $U_{\mathbf{M}}^\sigma$ as defined by (6-15–6-16) of Chap. II. We furthermore define another pair of operators $B': \mathbb{C}^m \to L^2(\sigma I_n)$ and $C'^*: \mathbb{C}^p \to L^2(\sigma I_n)$ by

$$B' = U_{\mathbf{M}}^\sigma B \qquad \text{and} \qquad C'^* = U_{\mathbf{M}}^\sigma C^* \tag{7-15}$$

The introduction of B' and C'^*, which are both multiplication operators, is made in order to remove redundancies in the definition of B and C. In particular B' and C' will be zero on the complement of the support of $\mathbf{M}$. We can apply now Theorem II 6-19 to obtain the following characterization of reachability and observability.

Theorem 7-1 Let (A, B, C) be a system with state space $L^2(\mathbf{M})$ where A, B, and C are defined by (7-2), (7-4), and (7-9), respectively. Let B' and C' be defined by (7-15) and let P be the measurable projection valued function corresponding to $U_{\mathbf{M}}^{\sigma} L^2(\mathbf{M})$. Then the system (A, B, C) is reachable if and only if

$$(B', P^{\perp})_L^{\sigma} = I_n \tag{7-16}$$

and observable if and only if

$$(C', P^{\perp})_R^{\sigma} = I_n \tag{7-17}$$

If we specialize this to the case of a matrix measure of scalar type that is to the case $\mathbf{M} = \mu I_n$ then obviously σ can be identified with μ. Since P is identically equal to I_n we have $P^{\perp} = 0$ and so $(B, P^{\perp})_L^{\mu} = I_n$ if and only if there exists no μ-nontrivial projection valued function R such that $B = RB$. This is obviously the case if and only if B has full row rank μ-a.e. Summarizing this we have obtained.

Corollary 7-2 Let $\mathbf{M} = \mu I_n$ and let the system (A, B, C) be as in Theorem 7-1. Then (A, B, C) is reachable if and only if B has full row rank μ-a.e. and observable if and only if C has full column rank μ-a.e. That is

$$\text{rank } B(\lambda) = n \qquad \mu\text{-a.e.} \tag{7-18}$$

and

$$\text{rank } C(\lambda) = n \qquad \mu\text{-a.e.} \tag{7-19}$$

respectively.

From this corollary we can immediately deduce the general reachability conditions. To this end we utilize the canonical spectral representation of A. A is unitarily equivalent to the multiplication by λ operators in $L^2(\mathbf{N})$ where $\mathbf{N}$ is the diagonal matrix measure

$$\mathbf{N} = \begin{pmatrix} v_1 + \cdots + v_n & & & \\ & v_2 + \cdots + v_n & & \\ & & \ddots & \\ & & & v_n \end{pmatrix} \tag{7-20}$$

Let $\sigma = v_1 + \cdots + v_p$ and let n_i be the Radon–Nikodym derivative of v_i with respect to σ and let $E_i = \{\lambda \,|\, n_i(\lambda) \neq 0\}$ then σ-a.e. the measurable projection valued function P that satisfies $U_{\mathbf{N}}^{\sigma} L^2(\mathbf{N}) = PL^2(\sigma I_n)$ is given by

$$P(\lambda) = \begin{pmatrix} \chi_{E_1}(\lambda) + \cdots + \chi_{E_n}(\lambda) & & & \\ & \chi_{E_2}(\lambda) + \cdots + \chi_{E_n}(\lambda) & & \\ & & \ddots & \\ & & & \chi_{E_n}(\lambda) \end{pmatrix} \tag{7-21}$$

With this notation we can state the general reachability and observability conditions as follows.

Theorem 7-3 Let (A, B, C) be the system as in Theorem 7-1. Then with respect to the canonical spectral representation the system (A, B, C) is reachable if and only if

$$\text{Rank } B(\lambda) = k \qquad v_k \text{ -a.e.} \tag{7-22}$$

for all $k = 1, \ldots, n$. Analogously the system is observable if and only if

$$\text{Rank } C(\lambda) = k \qquad v_k \text{ -a.e.} \tag{7-23}$$

for all $k = 1, \ldots, n$.

As is clear from the single example at the end of Sec. 1 two systems with self-adjoint generators that realize the same transfer function need not be similar. Therefore for isomorphism results we have either to strengthen the reachability (or alternatively observability) conditions or to tighten the relation between the operators B and C. We begin by studying the first possibility modeling our approach on the exact reachability conditions introduced previously.

Let R be the reachability operator of the system (A, B, C) with the state space $L^2(\mathbf{M})$ Let σ be a positive measure on $(-\infty, 0]$. We say the system (A, B, C) is *σ-exactly reachable* if R can be extended by continuity to a bounded operator from $L^2(\sigma I_n)$ onto $L^2(\mathbf{M})$. *σ-exact observability* is defined analogously. Note that the σ-exact reachability of the system (A, B, C) shows that R is a B-homomorphism.

Lemma 7-4 Let (A, B, C) be as in Theorem 7-1. Then if (A, B, C) is σ-exactly reachable and $\sigma \ll \rho$ then (A, B, C) is also ρ-exactly reachable.

PROOF By Lemma II 6-14 there exists a B-homomorphism $\hat{\bar{R}}: L^2(\sigma I_n) \to L^2(\sigma I_n)$ for which $\hat{R} = (U_\mathbf{M}^\sigma)^* \hat{\bar{R}}$. If $\sigma \ll \rho$, $\hat{\bar{R}}$ can be lifted to a B-homomorphism $\hat{\bar{R}}: L^2(\rho I_n) \to L^2(\rho I_n)$ for which

$$\hat{\bar{R}} U_\sigma^\rho = U_\sigma^\rho \hat{\bar{R}} \tag{7-24}$$

Thus $(U_\mathbf{M}^\sigma)^* \hat{\bar{R}}$ provides the necessary bounded extension of $\hat{R}$ to a B-homomorphism of $L^2(\rho I_n)$ onto $L^2(\mathbf{M})$.

To get a characterization of σ-exact reachability in terms of the matrix measure $\mathbf{M}$, the measure σ and the function B we first show that $L^2(\mathbf{M})$ can be considered as a subspace of $L^2(\sigma I_n)$. Actually we prove a bit more and the result may be of independent interest. This is the converse to Lemma II 6-2.

Theorem 7-5 Let $\mathbf{M}$ and $\mathbf{N}$ be two $n \times n$ positive matrix measures and let $X: L^2(\mathbf{M}) \to L^2(\mathbf{N})$ be a B-homomorphism. Then
(a) If X is one to one then $\mathbf{M} | \mathbf{N}$.
(b) If X has dense range then $\mathbf{N} | \mathbf{M}$.

PROOF (a) By Theorem II 5-9, there exists an isometry, which can be checked to be B-homomorphism, $U: L^2(\mathbf{M}) \to L^2(\mathbf{N})$. Let E be any Borel set with compact closure, χ_E its characteristic function and $\xi \in \mathbf{C}^n$. Then $\chi_E \xi$ belongs to any $L^2(\mathbf{M})$ space. Since U is a B-homomorphism we have

$$(U(\chi_E \xi))(\lambda) = \chi_E(\lambda)(U\xi)(\lambda) = \chi_E(\lambda) J(\lambda) \xi$$

for some measurable matrix function J. Since U is isometric we have

$$\int (d\mathbf{N}\chi_E(\lambda) J(\lambda) \xi, J(\lambda) \xi) = \int (d\mathbf{M}\chi_E(\lambda) \xi, \chi_E(\lambda))$$

or

$$\int_E (J(\lambda)^* \, d\mathbf{N} J(\lambda) \xi, \xi) = \int_E (d\mathbf{M}\xi, \xi) \tag{7-25}$$

Since (7-25) holds for arbitrary vectors $\xi \in \mathbf{C}^n$ and every Borel set E then

$$d\mathbf{M} = J^* d\mathbf{N} J \tag{7-26}$$

or $\mathbf{M} | \mathbf{N}$. Part (b) follows by duality.

Corollary 7-6 Let (A, B, C) be a system as in Theorem 7-1. If (A, B, C) is σ-exactly reachable then $\mathbf{M} | \sigma I_n$.

An easy application of Theorem II 6-19 yields the characterization of σ-exact reachability and observability.

Theorem 7-7 Let (A, B, C) be a system in $L^2(\mathbf{M})$ where A, B, and C are defined by (7-2), (7-4), and (7-9), respectively, and let B' and C' be defined by (7-15). Then
(a) (A, B, C) is σ-exactly reachable if and only if $\mathbf{M} | \sigma I_n$ and

$$[B', P^\perp]_L^\sigma = I_n \tag{7-27}$$

where P is the orthogonal projection valued function for which $PL^2(\sigma I_n) = U_{\mathbf{M}}^\sigma L^2(\mathbf{M})$.
(b) (A, B, C) is σ-exactly observable if and only if $\mathbf{M} | \sigma I_n$ and

$$[C', P^\perp]_R^\sigma = I_n \tag{7-28}$$

where P is as in part (a).

The concept of σ-exact reachability allows us to state and prove a theorem analogous to the isomorphism result of Theorem 1-9.

Theorem 7-8 Let (A, B, C) and (A_1, B_1, C_1) be two systems with finite multiplicity self-adjoint generators and assume both systems realize the same weighting pattern. If both systems are observable and σ-exactly reachable then they are isomorphic.

PROOF Let R, R_1 and O, O_1 be the respective reachability and observability operators of the two systems. Since both systems realize the same weighting pattern we have $OR = O_1 R_1$ and by the assumption of observability $\operatorname{Ker} O = \operatorname{Ker} O_1 = \{0\}$. This implies that $\operatorname{Ker} R = \operatorname{Ker} R_1$ and hence also $\operatorname{Ker} \hat{R} = \operatorname{Ker} \hat{R}_1$.

Define a map $X: L^2(\mathbf{M}) \to L^2(\mathbf{M}_1)$ by

$$X\hat{R} = \hat{R}_1 \tag{7-29}$$

$\hat{R}^*$ the pseudoinverse of $\hat{R}$ is defined and bounded on all of $L^2(\mathbf{M})$ and as $\hat{R}$ is onto $L^2(\mathbf{M})$ it follows that $X = \hat{R}_1 \hat{R}^*$ and X is a B-homomorphism of $L^2(\mathbf{M})$ onto $L^2(\mathbf{M}_1)$.

From the definition of X we have $XB(\lambda) = B_1(\lambda)$ and $XA = A_1 X$ follows from the fact that X is a B-homomorphism and hence is given by a multiplication operator that intertwines scalar multiplication operators. Finally, since $OR = O_1 R_1$, using (7-29) we have $O_1 R_1 = O_1 XR = OR$. As R is surjective we deduce that $O_1 X = O$ which in turn implies $C_1 X = C$.

We expect to obtain stronger statements by limiting the class of systems under consideration. Hereby we focus our attention on stable self-adjoint systems. Our first object is the characterization of the class of weighting patterns that are realizable by such systems. To this end we introduce complete monotonicity. A scalar function w defined on $(0, \infty)$ is called *completely monotonic* if it is C^∞ and satisfies $(-1)^n w^{(n)}(t) \geq 0$ for all $n \geq 0$ and $t \geq 0$. This extends easily to a Hilbert space H operator valued function W. We say W is completely monotonic if the scalar function $(W(t) x, x)$ is completely monotonic for every $x \in H$. Scalar completely monotonic functions have analytic extensions to the right half plane and so W has a weakly analytic extension. Since weak and uniform analyticity are equivalent [29] it follows that a completely monotonic operator valued function is actually infinitely differentiable in the uniform operator topology.

Theorem 7-9 A U-valued weighting pattern W defined on $[0, \infty)$ is realizable by a stable self-adjoint system if and only if it is completely monotonic.

PROOF Suppose the stable self-adjoint system (A, B, B^*) realizes W in the state space H. Let E be the spectral measure of A. Then, by te spectral theorem

$$W(t) = B^* e^{At} B = B^* \int_{-\infty}^{0} e^{\lambda t} E(d\lambda) B = \int_{-\infty}^{0} e^{\lambda t} B^* E(d\lambda) B$$

Hence

$$(-1)^n W^{(n)}(t) = \int_{-\infty}^{0} (-\lambda)^n e^{\lambda t} B^* E(d\lambda) B$$

or

$$(-1)^n (W^{(n)}(t) x, x) = \int_{-\infty}^{0} (-\lambda)^n e^{\lambda t} \left\| E(d\lambda) Bx \right\|^2 \geq 0$$

Conversely assume W is completely monotonic. We apply an integral representation theorem of Bernstein [120] to the effect that the class of completely monotonic functions is in a one-to-one correspondence with the set of Laplace transforms of finite nonnegative measures supported on $(-\infty, 0]$. Thus every completely monotonic function is uniquely representable in the form $\int_{-\infty}^{0} e^{\lambda t} d\mu$. Since W is assumed completely monotonic it follows that for every $\xi \in U$ there exists a measure μ_ξ for which

$$(W(t) \, \xi, \, \xi) = \int_{-\infty}^{0} e^{\lambda t} \, d\mu_\xi \tag{7-30}$$

By polarization we have for each ξ and η in U the existence of a unique finite complex Borel measure $\mu_{\xi, \eta}$ such that

$$(W(t) \, \xi, \, \eta) = \int_{-\infty}^{0} e^{\lambda t} \, d\mu_{\xi, \eta} \tag{7-31}$$

By essentially the same method used in the proof of the spectral theorem there exists a complex U-operator valued measure $\mathbf{M}$ for which

$$(\mathbf{M}(\sigma) \, \xi, \, \eta) = \mu_{\xi, \eta}(\sigma) \tag{7-32}$$

for every Borel set σ. Since $\mu_{\xi, \xi}$ is nonnegative for every $\xi \in U$ then the measure $\mathbf{M}$ is also nonnegative. From (7-31) it follows that

$$W(t) = \int_{-\infty}^{0} e^{\lambda t} \, d\mathbf{M} \tag{7-33}$$

To get a realization out of the above representation we assume, without loss of generality, as this can be achieved by rescaling, that $\mathbf{M}((-\infty, 0]) \leq I$. Theorem II 10-21 guarantees the existence of a dilation space $K = U$ and a spectral measure E in K such that for every Borel set σ

$$\mathbf{M}(\sigma) = PE(\sigma) | U \tag{7-34}$$

where P is the orthogonal projection of K onto H. In particular this implies the factorization

$$\mathbf{M}(\sigma) = B^* E(\sigma) \, B \tag{7-35}$$

where B is the injection of U into K and $B^*: K \to U$ satisfies $B^* x = Px$ for every $x \in K$. Using the spectral measure E in K we define a self-adjoint operator A by

$$A = \int \lambda E(d\lambda) \tag{7-36}$$

The system (A, B, B^*) is a self-adjoint system and it realizes W as

$$B^* e^{At} B = B^* \int e^{\lambda t} E(d\lambda) \, B = \int e^{\lambda t} B^* E(d\lambda) \, B = \int e^{\lambda t} \, d\mathbf{M} = W(t)$$

We did not have to assume the finite dimensionality of U. However, if U is finite dimensional it can be identified, through a choice of an ortho-normal basis, with $\mathbb{C}^n$. Thus $\mathbf{M}$ is in this case a nonnegative matrix measure and the dilation space can be identified with $L^2(\mathbf{M})$. The operator A is taken to be the multiplication by λ operator in $L^2(\mathbf{M})$ whereas $B: U \to L^2(\mathbf{M})$ is the embedding which takes a vector $\xi \in U$ into the constant function ξ, that is, $(B\xi)(\lambda) = \xi$. The spectral measure E of the self-adjoint operator A is given by

$$(E(\sigma)f)(\lambda) = \chi(\lambda)f(\lambda)$$

where χ_σ is the characteristic function of σ.

For every Borel set σ we have

$$(B^*E(\sigma)B\xi, \xi) = (E(\sigma)B\xi, B\xi) = \int_\sigma (d\mathbf{M}\xi, \xi) = (\mathbf{M}(\sigma)\xi, \xi)$$

and so the factorization (7-35) follows. As a consequence of Theorem 7-3 this realization is canonical.

For self-adjoint systems the conditions for observability and reachability coincide. Previously we characterized reachability in terms of $\mathbf{M}$ and B. Since the spectral measure E determines A uniquely we can obtain an equivalent characterization in these terms.

Theorem 7-10 Let (A, B, B^*) be a self-adjoint system with state space χ and let E be the spectral measure of A. Then the system is canonical if and only if

$$\bigcap_{\sigma \in \Sigma} \operatorname{Ker} B^*E(\sigma) = \{0\} \tag{7-37}$$

the intersection taken over the set Σ of all Borel subsets of $(-\infty, 0]$.

PROOF It suffices to show that $\bigcap_{\sigma \in \Sigma} \operatorname{Ker} B^*E(\sigma) = \bigcap_{t \geq 0} \operatorname{Ker} B^*e^{At}$. Assume $x \in \bigcap_{\sigma \in \Sigma} \operatorname{Ker} B^*E(\sigma)$ then

$$B^*e^{At}\chi = \int_{-\infty}^0 e^{\lambda t}B^*E(d\lambda)\chi = 0$$

and hence the inclusion $\bigcap_{\sigma \in \Sigma} \operatorname{Ker} B^*E(\sigma) \subset \bigcap_{t \geq 0} \operatorname{Ker} B^*e^{At}$. To prove the converse inclusion let $x \in \bigcap_{t \geq 0} \operatorname{Ker} B^*e^{At}$ then $\int_{-\infty}^0 e^{t}B^*E(d\lambda)x = 0$ for all $t \geq 0$ and the uniqueness part of Bernstein's theorem implies $x \in \bigcap_{\sigma \in \Sigma} \operatorname{Ker} B^*E(\sigma)$. This proves the theorem.

Theorem 7-11 Let (A, B, B^*) be a canonical self-adjoint realization of a transfer function Γ then the realization is spectrally minimal.

PROOF Since A is self-adjoint A has the representation $A = \int \lambda E(d\lambda)$ with respect to a uniquely determined spectral measure E. The spectral measure $E((c, b))$ of an open interval (a, b) can be obtained from the resolvent function

$R(\lambda; A)$ by

$$E((a, b)) = \lim_{\delta \to 0} \lim_{\varepsilon \to 0} \frac{1}{2\pi i} \int_{a+\delta}^{b-\delta} [R(\lambda - i\varepsilon; A) - R(\lambda + i\varepsilon; A)] \cdot d\lambda \qquad (7\text{-}38)$$

Hence for every vector $\xi \in U$

$$\| E((a, b)) B\xi \|^2 = (B^* E((a, b)) B\xi, \xi)$$

$$= \lim_{\delta \to 0} \lim_{\varepsilon \to 0} \int_{a+\delta}^{b-\delta} [(\Gamma(\lambda - i\varepsilon) \xi, \xi) - (\Gamma(\lambda + i\varepsilon) \xi, \xi)] \, d\lambda$$

$$(7\text{-}39)$$

Suppose (a, b) is an open interval which is included in the domain of analyticity of Γ. Then, by Cauchy's theorem, the previous equality $E((a, b)) B\xi = 0$. Since the semigroup e^{At} commutes with the spectral measure E it follows that

$$E((a, b)) e^{At} B\xi = e^{At} E((a, b)) B\xi = 0$$

Reachability of the system is equivalent to the set of vectors of the form $e^{At} B\xi$ spanning the state space H. Hence we can conclude $E((a, b)) = 0$ as $(a, b) \subset \sigma(A)$. Thus $\sigma(A) \subset \sigma(\Gamma)$ and as the reverse inclusion holds always the proof is complete.

A direct consequence of the spectral minimality property for canonical self-adjoint systems is the fact that the spectra of the generators in two different canonical self-adjoint realizations of a given transfer function coincide. Actually an isomorphism result can be proved and as a step in that direction we prove the following.

Lemma 7-12 Let (A, B, B^*) and (A_1, B_1, B_1^*) be two canonical self-adjoint realizations with Γ and Γ_1 their respective transfer functions and E and E_1 the spectral measures of A and A_1, respectively. Then the transfer functions Γ and Γ_1 coincide if and only if for every Borel set σ on the real line we have

$$B^* E(\sigma) B = B_1^* E_1(\sigma) B_1 \qquad (7\text{-}40)$$

PROOF If (7-39) holds then

$$\Gamma(z) = B^* (zI - A)^{-1} B = \int (z - \lambda)^{-1} B^* E(d\lambda) B$$

$$= \int (z - \lambda)^{-1} B_1^* E_1(d\lambda) B_1 = B_1^* (zI - A_1)^{-1} B_1 = \Gamma_1(z)$$

The converse follows from (7-37) for open intervals and hence by standard measure theoretic techniques for all Borel sets.

We can state and prove now the state space isomorphism for self-adjoint systems.

Theorem 7-13 Let (A, B, B^*) and (A_1, B_1, B_1^*) be two canonical self-adjoint systems in the Hilbert spaces H and H_1, respectively. A necessary and sufficient condition that the two systems realize the same transfer function is that they are unitarily equivalent.

PROOF The sufficiency part is trivial. To prove the converse we note that every self-adjoint operator has a unitarily equivalent spectral representation. Thus there is no loss of generality in assuming that both operators are given in their canonical spectral representation. So by Theorem II 6-8 we can identify the state space of the two systems with $\oplus_{j=1}^n L^2(v_j I_j)$ and $\ominus_{j=1}^n L^2(v_j^{(1)} I_j)$ where the v_j are mutually singular. The supports of v_j are the sets of multiplicity j. By Lemma 7-12 equality (7-40) holds for all Borel sets σ which implies, integrating over the supports of v_j and $v_j^{(1)}$, that v_j and $v^{(1)}$ are equivalent measures. It follows that $dv_j = h_j^2 dv_j^{(1)}$ for some measurable function h_j. Moreover for λ in E_j the support of v_j we have

$$B(\lambda)^* B(\lambda) = h_j(\lambda)^2 B_1(\lambda)^* B_1(\lambda) \qquad (7\text{-}41)$$

Define a map $U_j(\lambda)$ in E_j by

$$U_j(\lambda) B(\lambda) = h_j(\lambda) B_1(\lambda) \qquad (7\text{-}42)$$

By reachability B and B_1 have full rank a.e. with respect to v_j (or $v_j^{(1)}$) and hence $U(\lambda)$ is invertible a.e. Moreover, clearly U_j is unitary. Let $U: \oplus L^2(v_j I_j) \to \oplus L^2(v_j^{(1)} I_j)$ be defined by $U = \oplus_{j=1}^n U_j$ then U is a unitary $\mathscr{B}$-homomorphism, thus intertwines A and A_1. Condition (7-41) is equivalent to $UB = B_1$ and so U provides the unitary equivalence.

NOTES AND REFERENCES

Almost simultaneously with the development of finite dimensional system theory attention has been directed to the study of infinite dimensional problems. Of the early work in this direction one should mention Balakrishnan's [3, 4]. That modern operator theory is relevant to system theory has been recognized concurrently by several researchers, namely, Dewilde [22], Helton [69–71], Baras and Brockett [5, 6], and the author [42–50, 54]. Essentially it is this body of work that motivated the book and constitutes the content of this chapter. It was Helton [69] who introduced the concept of exact reachability and proved the version of the state space isomorphism theorem appearing in Theorem 1-9.

The realization procedure using shift operators has been constructed by the author in the scalar case [42] and by Helton in the general case using Hankel operators [69]. The results concerning restricted shift systems follow the author's [43–45]. Theorem 3-5 characterizing strictly noncyclic functions is motivated by and based on the fundamental paper of Douglas, Shapiro, and Shields [28], as well as Kriete [79]. That Theorem II 14-11 can be applied to study Hankel operator ranges has been observed first by Clark in the scalar case [69] and worked out in detail by the author in [43, 44].

Spectral minimality and its importance in system theory have been stressed by Baras and Brockett [6]. The proof of the spectral minimality of the shift realization of strictly noncyclic functions is due to the author.

Section 5 is based on several of the author's papers, namely [47–49] which deal with ranges of Hankel operators induced by sums and products and with series and parallel connection of systems. Some work on degree theory for infinite dimensional systems has been done also by Dewilde [22] who also recognized the role of the determinant of an inner function as a proper substitute for a degree function.

The study of the infinite dimensional realization problem in continuous time began with Balakrishnan. The first rigorous work seems to be that of Baras and Brockett [6]. The idea of using rigged Hilbert spaces is due to Helton [71] who introduced also compatible systems. We follow Hedberg's [66] approach to the continuous time realization problem. This is done by associating with it a discrete time realization problem and appropriately transforming the solution. Such an association appears also in the study of controllability in [41, 43]. The restriction to transfer functions that are Fourier transforms of measures is done for technical reasons. There is no doubt, however, that the theory can be developed in much greater generality.

Section 7 is based largely on [16, 54]; Theorem 7-3 is due to Fattorini [35]. Infinite dimensional systems with other symmetry constraints are discussed in [17, 36, 84].

REFERENCES

1. Ahern, P. R. and D. N. Clark: On functions orthogonal to invariant subspaces, *Acta Math.*, **124**, 191–204, 1970.
2. Akhiezer, I. N. and I. M. Glazman: "Theory of Linear Operators in Hilbert space," F. Ungar, New York, 1961.
3. Balakrishnan, A. V.: Linear systems with infinite dimensional state spaces, Symposium on System Theory, Polytechnic Institute of Brooklyn, 1965.
4. Balakrishnan, A. V.: System theory and stochastic optimization, Proc. Nato Advanced Study Institute of Network and Signal Theory, Bournemouth, England, 1972.
5. Baras, J. S.: Algebraic structure of infinite dimensional linear systems in Hilbert Space, Proc. of CNR-CISM Symposium—Advanced School on Algebraic System Theory, Udine, Springer, 1975.
6. Baras, J. S. and R. W. Brockett: H^2-functions and infinite-dimensional realization theory, *SIAM J. Control*, **13**, 221–241, 1975.
7. Baras, J. S., Brockett, R. W., and P. A. Fuhrmann: State space models for infinite-dimensional systems, *IEEE Trans. Automatic Control*, **19**, 693–700, 1974.
8. Baras, J. S. and P. Dewilde: Invariant subspace methods in linear multivariable distributed systems and lumped distributed network synthesis, *Proc. IEEE* **64**, 160–178, 1976.
9. Barnett, S.: Matrices, polynominals and linear time invariant systems, *IEEE Trans. Automatic Control*, **18**, 1–10, 1973.
10. Beals, R.: "Topics in Operator Theory," Univ. Chicago Press, Chicago and London, 1971.

11. Beurling, A.: On two problems concerning linear transformations in Hilbert Space, *Acta Math.*, **81**, 239–255, 1949.
12. Birkhoff, G. and G. C. Rota: "Ordinary Differential Equations," Blaisdell, Waltham, Mass. 1962.
13. Bochner, S. and K. Chandrasekharan: "Fourier Transforms," Princeton Univ. Press, Princeton N.J., 1949.
14. De Branges, L. and J. Rovnyak: The existence of invariant subspaces, *Bull. Amer. Math. Soc.*, **70**, 718–721, 1964; and **71**, 396, 1965.
15. Brockett, R. W.: "Finite Dimensional Linear Systems," Wiley, New York, 1970.
16. Brockett, R. W. and P. A. Fuhrmann: Normal symmetric dynamical systems, *SIAM J. Control*, **14**, 107–119, 1976.
17. Brodskii, M. S.: "Triangular and Jordan representations of linear operators," *Amer. Math. Soc.*, Providence R.I., 1971.
18. Brown, A.: A version of multiplicity theory, in "Topics in Operators Theory" (ed. C. Pearcy) *Amer. Math. Soc.*, Providence R.L., 1974.
19. Callier, F. M. and C. D. Nahum: Necessary and sufficient conditions for the complete controllability and observability of systems in series using the coprime decomposition of a rational matrix, *IEEE Trans. Circuits and Systems,* **22**, 90–95, 1975.
20. Carleson, L.: Interpolation by bounded analytic functions and the corona problem, *Ann. of Math.*, **76**, 547–559, 1962.
21. Cooper, J. L. B.: One parameter semi-groups of isometric operators in Hilbert space, *Ann. of Math.*, **48**, 827–842, 1947.
22. Dewilde, P.: Input–Output description of roomy systems, *SIAM J. Control,* **14**, 712–736, 1976.
23. Douglas, R. G.: On majorization, factorization and range inclusion of operators on Hilbert Space, *Proc. Amer. Math. Soc.,* **17**, 413–415, 1966.
24. Douglas, R. G.: "Banach Algebra Techniques in Operators Theory," Acedemic Press, New York, 1972.
25. Douglas, R. G.: Canonical models, in "Topics in Operators Theory" (Ed. C. Pearcy), *Amer. Math. Soc.,* Providence R.I., 1974.
26. Douglas, R. G. and J. W. Helton: Inner dilations of analytic matrix functions and Darlington Synthesis, *Acta Sci. Math.,* **34**, 301–310, 1973.
27. Douglas, R. G., Muhly, P. S. and C. Pearcy: Lifting commuting operators, *Mich. Math. J.,* **15**, 385–395, 1968.
28. Douglas, R. G., Shapiro, H. S. and A. L. Shields: Cyclic vectors and invariant subspaces for the backward shift, *Ann. Inst. Fourier, Grenoble* **20, 1**, 37–76, 1971.
29. Dunford, N. and J. T. Schwartz: "Linear Operators," Vols. **1. 2.** Interscience, New York, 1957, 1963.
30. Duren, P. L.: "Theory of H^p-Spaces," Academic Press, New York, 1970.
31. Dym, H. and H. P. McKean: "Fourier Series and Integrals," Academic Press, New York, 1972.
32. Eckberg, A.: A characterization of linear systems via polynomial matrices and module theory, *MIT Report ESL-R-528*, 1974.
33. Eilenberg, S.: "Automata, Languages and Machines," vol. A. Academic Press, New York, 1974.
34. Embry, M. R.: Factorization of Operators on Banach Space, *Proc. Amer. Math. Soc.,* **38**, 587–590, 1973.
35. Fattorini, H. O.: On complete controllability of linear systems, *J. Diff. Eq.,* 391–402, 1967.
36. Feintuch, A.: "Realization theory for symmetric systems," *J. Math. Anal. Appl.* **71**, 131–146, 1979.
37. Fillmore, P. A.: "Notes on Operator Theory," Van Nostrand, New York, 1968.
38. Forney, G. D.: Minimal bases of rational vector spaces with applications to multivariable linear systems, *SIAM J. Control*, **13**, 493–520, 1975.
39. Fuhrmann, P. A.: On the corona theorem and its applications to spectral problems in Hilbert space, *Trans. Amer. Math. Soc.,* **132**, 55–66, 1968.
40. Fuhrmann, P. A.: A functional calculus in Hilbert space based on operator valued analytic functions, *Israel J. Math.,* **6**, 267–278, 1968.
41. Fuhrmann, P. A.: On weak and strong reachability and controllability of infinite-dimensional linear systems, *J. Opt. Th. Appl.,* **9**, 77–87, 1972.

42. Fuhrmann, P. A.: On realizations of linear systems and applications to some questions of stability, *Math. Syst. Th.*, **8**, 132–141, 1974.
43. Fuhrmann, P. A.: Exact Controllability and observability and realization theory, *J. Math. Anal. Appl.*, **53**, 377–392, 1976.
44. Fuhrmann, P. A.: Realization theory in Hilbert space for a class of transfer functions, *J. Funct. Anal.*, **18**, 338–349, 1975.
45. Fuhrmann, P.A.: On Hankel operator ranges, meromorphic pseudo-continuation and factorization of operator valued analytic functions, *J. Lond. Math. Soc.,* **(2)** 13, 323–327, 1975.
46. Fuhrmann, P. A.: On controllability and observability of systems connected in parallel, *IEEE Trans. Circuits and Systems*, Cas-22, 57, 1975.
47. Fuhrmann, P. A.: On canonical realizations of sums and products of nonrational transfer functions, *Proc. 8th Princeton Conf. on Information Sciences and Systems,* 213–217, 1974.
48. Fuhrmann, P. A.: On generalized Hankel operators induced by sums and products, *Israel J. Math.*, **21**, 279–295, 1975.
49. Fuhrmann, P. A.: On series and parallel coupling of infinite dimensional linear systems, *SIAM J. Control*, **14**, 339–358, 1976.
50. Fuhrmann, P. A.: Some results on controllability, *Ricerche di Automatica*, **5**, 1–5, 1974.
51. Fuhrmann P. A.: Algebraic system theory; an analyst's point of view, *J. Franklin Inst.*, **301**, 521–540, 1976.
52. Fuhrmann, P. A.: On strict system equivalence and similarity, *Int. J. Control*, **25**, 5–10, 1977.
53. Fuhrmann, P. A.: Simulation of linear systems and factorization of matrix polynomials, *Int. J. Control*, **28**, 689–705, 1978.
54. Fuhrmann, P. A.: Operator measures, self-adjoint operators and dynamical systems, *SIAM J. Math. Anal.* **11**, 1980.
55. Gantmacher, F. R.: The Theory of Matrices," Chelsea, New York, 1959.
56. Gohberg, I. C. and I. A. Feldman: "Convolution Equations and Projection Methods for Their Solution", *Amer. Math. Soc.*, Providence R.I., 1974.
57. Gohberg, I., Lancaster, P. and L. Rodman: Spectral analysis of matrix polynomials, I. Canonical forms and divisors, *Linear Algebra Appl.*, **20**, 1–44, 1977.
58. Gohberg, I., Lancaster, P. and L. Rodman: Spectral analysis of matrix polynomials, II. The resolvent form and spectral divisor, *Linear Algebra Appl.*, **21**, 65–88, 1978.
59. Gohberg, I., Lancaster, P. and L. Rodman: Representations and divisibility of operator polynomials, *Canadian J. of Math.*, **30**, 1045–1069, 1978.
60. Gohberg, I. C. and L. E. Lerer: Resultants of matrix polynomials.
61. Halmos, P. R.: Normal dilations and extensions of operators, *Summa Brasil.*, **2**, 125–134, 1950.
62. Halmos, P. R.: "Introduction to Hilbert Space and the Theory of Spectral Multiplicity," Chelsea, New York, 1951.
63. Halmos, P. R.: Shifts on Hilbert Spaces, *J. Reine Angew. Math.*, **208**, 102–112, 1961.
64. Halmos, P. R.: "A Hilbert Space Problem Book," Van Nostrand, New York, 1967.
65. Halperin, I.: The unitary dilation of a contraction operator, *Duke math. J.*, **28**, 563–571, 1961.
66. Hedberg, D. J.: "Operator Models of Infinite Dimensional Systems," *Ph.D. Thesis*, Dept. of System Science UCLA, 1977.
67. Helson, H.: "Lectures on Invariant Subspaces," Academic Press, New York, 1964.
68. Helson, H. and D. Lowdenslager: Prediction theory and Fourier Series in Several Variables, *Acta Math.*, **99**, 165–202, 1958; II, ibid. **106**, 175–213, 1961.
69. Helton, J. W.: Discrete time systems, operator models and scattering theory, *J. Funct. Anal.*, **16**, 15–38, 1974.
70. Helton, J. W.: A spectral factorization approach to the distributed stable regulator problem: the algebraic Ricatti equation, *SIAM J. Control*, **14**, 639–661, 1976.
71. Helton, J. W.: Systems with infinite-dimensional state space: The Hilbert space approach, *Proc. IEEE*, **64**, 145–160, 1976.
72. Hille, E. and R. S. Phillips: Functional Analysis and Semigroups, *Amer. Math. Soc.*, Providence, 1957.
73. Hoffman, K.: "Banach Spaces of Analytic Functions," Prentice Hall, Englewood Cliffs, N.J., 1962.

74. Jacobson, N.: Lectures in Abstract Algebra, Vol. 2 "Linear Algebra," Van Nostrand, Princeton, 1953.
75. Jacobson, N.: "Basic Algebra I," W. H. Freeman, San Francisco, 1974.
76. Kalman, R. E.: "Lectures on Controllability and observability," *CIME Summer Course* 1968; Cremonese, Roma 1969.
77. Kalman, R. E., Falb, P. L. and M. A. Arbib: "Topics in Mathematical System Theory," McGraw-Hill, New York, 1969.
78. Katznelson, Y.: "An Introduction to Harmonic Analysis," Wiley, New York, 1968.
79. Kriete, T. L.: A generalized Paley–Wiener theorem, *J. Math. Anal. Appl.*, **36**, 529–555, 1971.
80. Lang, S.: "Algebra," Addison Wesley, Reading, Mass., 1965.
81. Lax, P. D.: Translation Invariant Subspaces, *Acta Math.*, **101**, 163–178, 1959.
82. Lax, P. D. and R. S. Phillips: "Scattering Theory," Academic Press, New York, 1967.
83. Livsic, M. S.: On the spectral resolution of linear non-selfadjoint operators, *Math. Sb.*, **34** (76), 145–199, 1954.
84. Livsic, M. S.: "Operators, Oscillations, Waves. Open Systems," *Amer. Math. Soc. Translations*, **34** 1973.
85. Lorch, E. R.: "Spectral Theory," Oxford University Press, New York, 1962.
86. MacDuffee, C. C.: "The Theory of Matrices," Chelsea, New York, 1946.
87. MacDuffee, C. C.: Some applications of matrices in the theory of equations, *Amer. Math. Monthly*, **57**, 154–161, 1950.
88. MacLane, S. and G. Birkhoff: "Algebra," Macmillan, New York, 1967.
89. Moeller, J. W.: On the spectra of some translation invariant subspaces, *J. Math. Anal. Appl.*, **4**, 275–296, 1962.
90. Moore, B. III: Canonical forms in linear systems, *Proc. Eleventh Allerton Conference*, University of Illinois, 36–44, 1973.
91. Moore, B. III and E. A. Nordgren: On quasi-equivalence and quasi-similarity, *Acta Sci. Math.*, **34**, 311–316, 1973.
92. Naimark, M. A.: Positive definite operator functions on a commutative group, *Izvestija Akad. Nauk SSSR*, **7**, 237–244, 1943.
93. Naimark, M. A.: "On a representation of additive operator set functions," *Doklady Akad. Nauk SSSR*, **41**, 359–361, 1943.
94. Nelson, E.: "Topics in Dynamics I: Flows," Princeton Univ. Press, Princeton N.J., 1968.
95. Nikolskii, N. K. and B. S. Pavlov: Eigenvector bases of completely nonunitary contractions and the characteristic function, *Math. USSR-Izvestija*, **4**, 91–134, 1970.
96. Nirenberg, L.: "Functional Analysis," Lecture Notes of the CIMS, New York, 1961.
97. Nordgren, E. A.: On quasi-equivalence of matrices over H^∞, *Acta Sci. Math.*, **34**, 301–310, 1973.
98. Plessner, A. I.: "Spectral Theory of Linear Operators I, II," F. Ungar, New York, 1969.
99. Riesz, F. and B. Sz.-Nagy: "Functional Analysis," F. Ungar, New York, 1955.
100. Rosenbrock, H. H.: "State Space and Multivariable Theory," Wiley, New York, 1970.
101. Rota, G. C.: On models for linear operators, *Comm. Pure and Appl. Math.*, **13**, 469–472, 1960.
102. Rowe, A.: The generalized resultant matrix, *J. Inst. Math. Appl.*, **9**, 390–396, 1972.
103. Sarason, D.: A remark on the Volterra operator, *J. Math. Anal. Appl.*, **12**, 244–246, 1965.
104. Sarason, D.: Generalized interpolation in H^∞, *Trans. Amer. Math. Soc.*, **127**, 179–203, 1967.
105. Schaffer, J. J.: On unitary dilations of contractions, *Proc. Amer. Math. Soc.*, **6**, 322, 1955.
106. Schatten, R.: "Norm Ideals of Completely Continuous Operators," Springer, Berlin, 1960.
107. Shapiro, H. S. and A. L. Shields: On some interpolation problems for analytic functions, *Amer. J. Math.*, **83**, 513–532, 1961.
108. Sherman, M. J.: Operators and inner functions, *Pacific J. Math.*, **22**, 159–170, 1967.
109. Sontag, E. D.: Linear systems over commutative rings: a survey, *Ricerche di Automatica*, **7**, 1–34, 1976.
110. Sontag, E. D.: On linear systems and noncommutative rings, *Math. Sys. Th.*, **9**, 327–344, 1976.
111. Stone, M. H.: Linear transformations in Hilbert space and their applications to analysis, *Amer. Math. Soc.*, Colloquium Pub. **15**, 1932.
112. Sz.-Nagy, B.: Sur les contractions de l'espace de Hilbert, *Acta Sci. Math.*, **15**, 87–92, 1953.

113. Sz.-Nagy, B.: Isometric flows in Hilbert space, *Proc. Cambridge Phil. Soc.*, **60**, 45–49, 1964.
114. Sz.-Nagy, B.: Diagonalization of matrices over H^∞, *Acta Sci. Math.*, **38**, 223–238, 1976.
115. Sz.-Nagy, B. and C. Foias: "Harmonic Analysis of Operators on Hilbert Space," North Holland, Amsterdam, 1970.
116. Sz.-Nagy, B. and C. Foias: Operateurs sans multiplicite, *Acta Sci. Math.*, **30**, 1–18, 1969.
117. Sz.-Nagy, B. and C. Foias: Modele de Jordan pour une classe d'operateurs de l'espace de Hilbert, *Acta Sci. Math.*, **31**, 91–117, 1970.
118. Sz.-Nagy, B. and C. Foias: On the structure of intertwining operators, *Acta Sci. Math.*, **35**, 225–254, 1973.
119. Van der Waerden, B. L.: "Modern Algebra," F. Ungar, New York, 1949.
120. Widder, D. V.: "The Laplace Transform," Princeton Univ. Press, Princeton, N.J., 1946.
121. Wiener, N. and P. Masani: The prediction theory of multivariate stochastic processes, I, *Acta Math.*, **98**, 111–150, 1957; II, ibid. **99**, 93–139, 1958.
122. Wold, H.: "A study in the Analysis of Stationary Time Series," Almqvist and Wiksell. Stockholm, 1938.
123. Wolovich, W. A.: "Linear Multivariable Systems," Springer Verlag, New York, 1974.
124. Wonham, M. W.: "Linear Multivariable Control," Springer Verlag, Berlin, 1974.
125. Yosida, K.: "Functional Analysis," Springer Verlag, Berlin, 1968.
126. Zadeh, L. and C. A. Desoer: "Linear System Theory," McGraw-Hill, 1963.

INDEX